Patent No. 2281
F. A. Giles' Design for a Top Plate of a Watch
Patented March 13, 1866

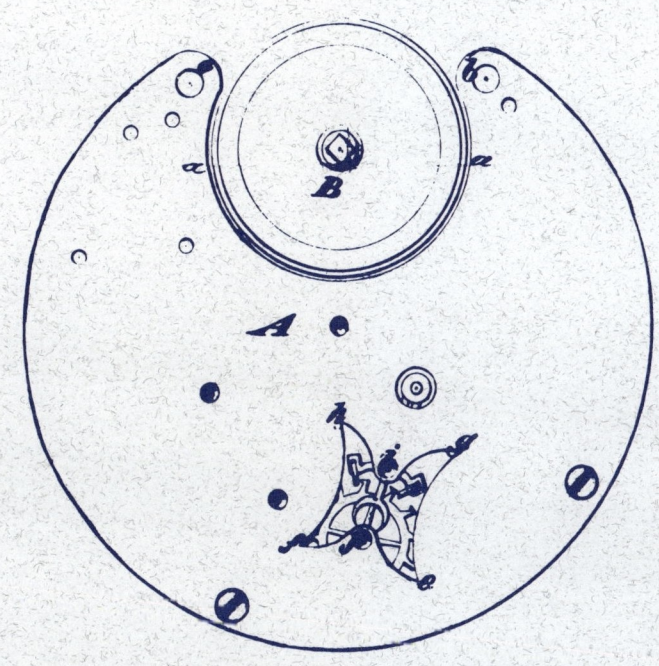

Marion

A History of The United States Watch Company

Frederick Asa Giles
Founder and President
September 8, 1834 - June 18, 1879

Marion

A History of The United States Watch Company

William Muir & Bernard Kraus

Tempus Vitam Regit
(Time Rules Life)

National Association of Watch & Clock Collectors, Inc.
Columbia, PA

NAWCC Special Publication Number 1
A Limited Printing of 2000 Copies
First Edition

©William Muir, 1980
©William Muir and Bernard Kraus, 1985
©National Association of Watch & Clock Collectors, Inc., 1985

No part of this publication may be reproduced, stored in retrieval systems, or transmitted in any form or by any means, except for quotation of brief passages in criticism, without express written permission of NAWCC, Inc.

Library of Congress Card Number 85-61588
ISBN 0-9614984-0-4

Printed in the United States by
Mifflin Press, Inc.
Columbia, PA

This book is gratefully dedicated
to the great-grandson of Frederick Asa Giles,
G. Robert Burns,
and his gracious wife, Carolyn.

Contents

Editor's Foreword 9
Acknowledegements 11
List of Charts and Tables 13
Introduction 15
Chronology 17

HISTORICAL
   1. The Early Years 23
   2. The Money Machine 29
   3. The Watch Factory Scheme 35
   4. Marion 43
   5. First Fruits 49
   6. William Alexander 59
   7. Fayette Stratton 63
   8. The Apex 69
   9. The People of Marion 81
  10. Three-Quarter and Quarter Plate 85
  11. The Panic of '73 95
  12. Giles, Wright & Co. 101
  13. Ellis Elias 105
  14. Royal Gold and Empire City Schemes 109
  15. Riding the Wreck Down 115
  16. Aftermath 121

PRODUCTION
  17. Products and Patents 127
  18. Prices of Movements and Cases 131
  19. USWC Grade Descriptions
      "Frederic Atherton & Co." 137
      "Fayette Stratton" 138
      "George Channing" 139
      "Edwin Rollo" 141
      "United States Watch Co." 142

19. USWC Grade Descriptions (cont'd)
    "Marion Watch Co." 147
    "A. H. Wallis" 148
    "Wm. Alexander" 150
    "S. M. Beard" 151
    "Henry Randel" 152
    "John W. Lewis" 154
    "Asa Fuller" 155
    "R. F. Pratt" 156
    "Chas. G. Knapp" 156
    "Young America" 157
    "J. W. Deacon" 157
    "G. A. Read" 158
    "North Star" 159
    "I. H. Wright" 159
    "A. J. Wood" 160
    Extras, Dials, Parts 161

20. Honors and Testimonials 165
21. Special Order Watch Production 169
22. The Centennial Commemoratives 173
23. Related Watch Companies
    Empire City Watch Company 179
    Royal Gold American Watch 182
    Howard Brothers of Fredonia 183
24. Production Estimates 185
25. USWC Miscellaneous Data 193

Epilogue 199
Illustration Credits 201
Bibliography 203
Index 206

* * *

Editor's Foreword

It was the late 1960's when I saw a Marion watch for the first time. I do not remember the grade, but do remember the impressive butterfly cut out in the top plate and how very little information was available on the company. Marion, New Jersey, was no longer on the map and it was sometime before I found out it had been absorbed by Jersey City.

My interest was piqued when I read Henry Abbott's description of the company and their "United States Watch Co." grade as "the finest movement made by the company, if not the finest then made in this country." Later I read Charles Crossman's description of this particular grade as "the highest price movement made in the United States." At that time I had no idea how long it would be before I ever saw a "United States" grade, or that in 1985 I would be asked to be editor of a book on the company.

I certainly had no idea that, by the late 1960's, the two authors of this book, William Muir in New Jersey and Bernard Kraus in Maryland, had already started independent research efforts on the Marion company. And, they did not know each other or that there was anyone else doing research on the company.

This book by these two co-authors is the first comprehensive history published on the United States Watch Company (USWC) of Marion, New Jersey. It is also the first work produced by the National Association of Watch and Clock Collectors, Inc. (NAWCC) with "special publication" status.

It has been said that a book needs three things to be successful: (1) a good story, (2) a good story, and (3) a good story. This work reads like a best-selling novel, bringing to life the Maiden Lane jewelry district in New York City and its New Jersey neighbors, Jersey City and Newark, during the era of expansion that followed the Civil War. It is the story of the Giles and Wright families of New England, and many other colorful personalities that affected their lives. It is the story of a highly successful wholesale jewelry concern on Maiden Lane and how it built one of the most elaborate watch factories ever constructed in the United States. It is the story of how and why this factory produced a diversity of products from the finest and most expensive to the poorest and least expensive. It is the story of economic growth and panic . . . pioneers and politicians . . . inventors and innovators . . . entrepreneurs and workaholics . . . rogues and con artists . . . frauds and fakes . . . schemes and strategies . . . reformers and crusaders. It is the story of a watch company that had the potential to be a significant competitive threat to the American (Waltham) Watch Company. It is the story of how and why a man's dream and pursuit of excellence and elegance met with disaster.

In dealing with history, it's great to have a good story, but more important is documentation. To achieve an excellent rating, the documentation needs to be weighted heavily to primary sources. The authors merit excellent ratings for both story and documentation. By dealing directly with the descendants of Frederick Giles and sources such as original correspondence, factory price schedules, advertisements and parts lists, the authors were able to clarify confusion that has surrounded Marion and its products for many years. With this type of research as the base and after many years of effort, the authors present in this work an accurate chronology of all important USWC events, correct product descriptions, and the very first reliable production estimates for all its various grades and many of its special products.

Since USWC production records have not surfaced, production estimates are based on the statistical principle of a "stratified random sample." In this case, the total "population" is the total production of USWC with "strata" easily determined by the various grades. Each grade has "sub-strata" in the form of production runs. So the problem was to gather a random sample of surviving examples with verifiable serial numbers and descriptions in sufficient quantity to establish production runs for the various grades. Simple and easy stops here, because of statistical challenges to develop production estimates. Complicating factors include (1) bias and error considerations, (2) all watches designated for a production run were not produced, (3) all watches produced were not sold, (4) all watches produced and sold have obviously not survived, and, finally, (5) the overall sample size is relatively small.

It was late in 1976 when the NAWCC Research Committee first got involved with this project. Sometime in 1977, as a member of that Committee, I was appointed as project coordinator to assist both authors as needed in the completion of their work. Both manuscripts were received by 1980, but still lacked product illustrations of all the various USWC grades. After reviewing both manuscripts, Dr. Douglas H. Shaffer, then Editor of the NAWCC "Bulletin," had the good wisdom to see the potential the combined works had to become an NAWCC "special publication." While Doug Shaffer made the first editorial revisions, the Research Committee assisted and also solicited the needed product illustrations. We obtained the last photographs, then completed the final editorial revisions and layout in August, 1985.

This book actually is the result of help from many contributors recognized in the Acknowledgements. The authors set the stage for their work in the Introduction where they begin to clarify the frustrations that have plagued USWC historians since the time of Messrs. Crossman and Abbott. The clarification process continues with a Chronology which sets forth many important USWC events in chronological sequence. Details of the fascinating USWC story unfold in the two major sections, Historical and Production. An interesting 1896 perspective of the USWC follows in the Epilogue. Finally, sections are included for Illustration Credits and a comprehensive Bibliography and Index.

A comprehensive book such as this is the product of many people's efforts. I want to extend my gratitude to many of the behind-the-scene, "unsung heroes" . . . especially the hard work of the NAWCC Editorial Department staff, to the Officers and members of NAWCC, and all others who extended help and encouragement to complete the project. It is hoped that this effort will inspire others to complete, or start, their own research work . . . the field of horological history has many stones yet to be overturned. The late J. E. Coleman, a pioneer NAWCC horological historian, said it some years ago . . . "answers, if there are any, lie in the cooperative effort."

Eugene T. Fuller
FNAWCC, Research Committee
Sugar Land, Texas

Acknowledgements

When research for this work began a little over 16 years ago it was hoped that it could be finished in a short period of time. Unfortunately, it soon became apparent that documentation was difficult to find and that what did appear tended to raise questions about the accepted versions of the story. Ultimately as the mass of data grew an accurate chronology began to emerge, and as it did information became relatively easy to acquire. The largest items were the court records from no less than two different bankruptcy actions, yet despite these important finds two vital elements were lacking. Neither the records of any of the Giles' firms nor any relevant correspondence could be located. Then in 1977 Eugene Fuller and Henry C. Wing assisted this author in arranging a visit with Frederick Giles' great-grandson, G. Robert Burns, and his gracious wife, Carolyn. The Burnses, who had become extremely interested in the history of their family, not only shared with me their collection of family correspondence, but also provided many of the photographs that grace this work.

I am deeply indebted to many other people who have contributed their time and efforts to this work. First, Robert Burns' mother and Frederick Giles' granddaughter, the late Christabel G. Burns, who contributed a number of important insights into the family as well as a number of rare clippings and other items. Second, the late Freeman H. McMillan, who had much to do with the genesis of this work and who spent many hours gathering data and reading early drafts. Third, the staffs of the New Jersey Room of the Jersey City Public Library, the New York Historical Society, the 42nd Street Branch and the Annex of the New York Public Library, the New York City Archives, the National Archives, and the William Paterson College Library, who are, of course, that group of professionals who do so much to aid any author. Among the many individuals I must cite are Dick Ziebell, James Henderson, former engineer for the City of Jersey City, Edwin Battison, the late Edwin B. Burt, the late Carl Prytula, and Barry Ted Moskowitz, attorney at law, who helped shed light on some of the 19th Century legal technicalities. Finally, my wife, Ann, who listened to theories, became an expert on the Elias brothers, read endless city directories and periodicals and then corrected and typed the final manuscript. When the final layout for the entire book was completed, Ann prepared the comprehensive index that accompanies this work.

William Muir
Wayne, New Jersey

Without the help and understanding from so many people all over the United States, this work simply could not have been completed. One significant problem that had to be overcome was the need for good photos of the many products produced by the company over its tenure in the market. No one individual, or organization, had examples of all grades, especially those models that were originally produced in very small quantity. We did receive many photos that did not lend themselves to good reproduction, but still they provided good data that was used in production estimates. It should be noted that the photo accumulation task continued even after I had completed my segment of the project by NAWCC Research Committee member and editor of this book, Eugene T. Fuller. After many contacts, letters and phone calls to places as far away as Hawaii, we finally obtained an excellent cross section of photographs on every USWC major grade, their special products, plus many unique examples never before photographed.

Lack of product photos was just one of many problems that needed to be overcome. In order to develop statistically reliable production estimates, we needed a broad, random sample of surviving examples with verifiable serial numbers and thoroughly accurate product descriptions. Once again, we needed the help of many people. The late Col. George E. Townsend was an early contributor of a very large block of serial numbers and excellent descriptions. Chris Bailey and the American Clock & Watch Museum, Dr. Robert L. Ravel, Adin Mathews, and the NAWCC Research Committee also provided significant serial numbers and related data. And there were many others who took the time to send in data on their USWC examples.

A very special thank you to Henry C. Wing, Jr., who provided some very key data and photos, and to Gene Fuller who was always there with the helpful assist and the necessary, gentle prod when it was needed. Ward Francillon and Dorian Clair made the necessary arrangements and photographed some of the important Marion items in the Dr. Barclay Stephens collection. I also wish to especially thank John Wilson, Roy Ehrhardt, plus Dick Hoban, and Frederick L. Orr for significant data and many of the excellent product photos that help give life to the Marion story. Carlene Stephens and the Smithsonian Institution also provided good information and some key illustrations. To the many others who contributed data, their knowledge and photographs, their time and effort, I will always be grateful.

Bernard G. Kraus
Silver Spring, Maryland

Other contributors who the authors and editor jointly wish to acknowledge are:

Vincent Angell
R. C. Appleton
Gene L. Bagwell
Donald S. Bass
Dana Blackwell
Richard Bovard
Kyle Britt
Dr. Warner D. Bundens, Jr.
Christie's, New York
   Jonathon Snellenburg
Melvin Conrad
Francis R. Crouch
Dren M. Duffy
Gerald Edleman
Bernard J. Edwards, Sr.
Jim Eldridge
Tom Engle
John Fossette
Charles Foster
Henry B. Fried
Del Gantz
Dr. James W. Gibbs
Bill Guido
Orville Hagans

Michael C. Harrold
Dr. William C. Heilman, Jr.
Donald Hoke
T. B. Jackson
Samuel W. Jennings
Lloyd Koenig
Ira Leonard
Robert Levy
Dr. Ernest Lewis
Tim Manly
Robert J. Matz
Don Miller
NAWCC Editorial Dept.
   Dr. Douglas H. Shaffer
   Terence M. Casey
   Amy J. Smith
NAWCC Museum
   Stacy B. C. Wood, Jr.
   Donald J. Summar
Dennis Nichinson
Warren H. Niebling
Chas. S. Porter
Maylene Rabeneck
Rockford Time Museum
   Seth G. Atwood
   Karon Anderson

Irv Roth
Miles F. Sandler
Leonard B. Sax
I. Schnell
William L. Scolnik
Glen A. Smith
Ozzie Sons
Robert G. Spence
Kenneth A. Sposato
St. Paul Dispatch and
Pioneer Press
   William C. Schneider
Dr. Snowden Taylor
Dick Titus
Manfred Trauring
Richard J. Wagner
Jack E. Wallace
Ralph Warner
Alvin Weeks
Charles C. Weige
Paul Wing
Robert M. Wingate
Art Zimmerla

List of Charts and Tables

Chart I.	Giles Family Enterprises, 1857-1893	14
Table I.	Sales & Profits, Giles, Wales & Co. vs. Waltham Watch Co., 1862-1865	37
Chart II.	Sketch, Marion and Surrounding Area	44
Table II.	USWC Movement Styles and Sizes	127
Table III.	USWC Grades and Movements, 1st and 2nd Generation	130
Table IV.	Summary of USWC Models and Prices, 1870-1873	134
Table V.	Some Original Owners of USWC Watches, 1867-1872	166
Table VI.	Some Jeweler's Contract USWC Products	170
Table VII.	USWC 1876 Centennial Product Contrast	176
Table VIII.	Empire City and Equivalent USWC Grades	179
Table IX.	Relative USWC Feature Scarcity	186
Table X.	Model/Grade Groups Ranked by Relative Production Scarcity	187
Table XI.	Model/Grade Groups with Serial Number Allocation	188
Table XII.	Employees and Production Capacities	193

* * *

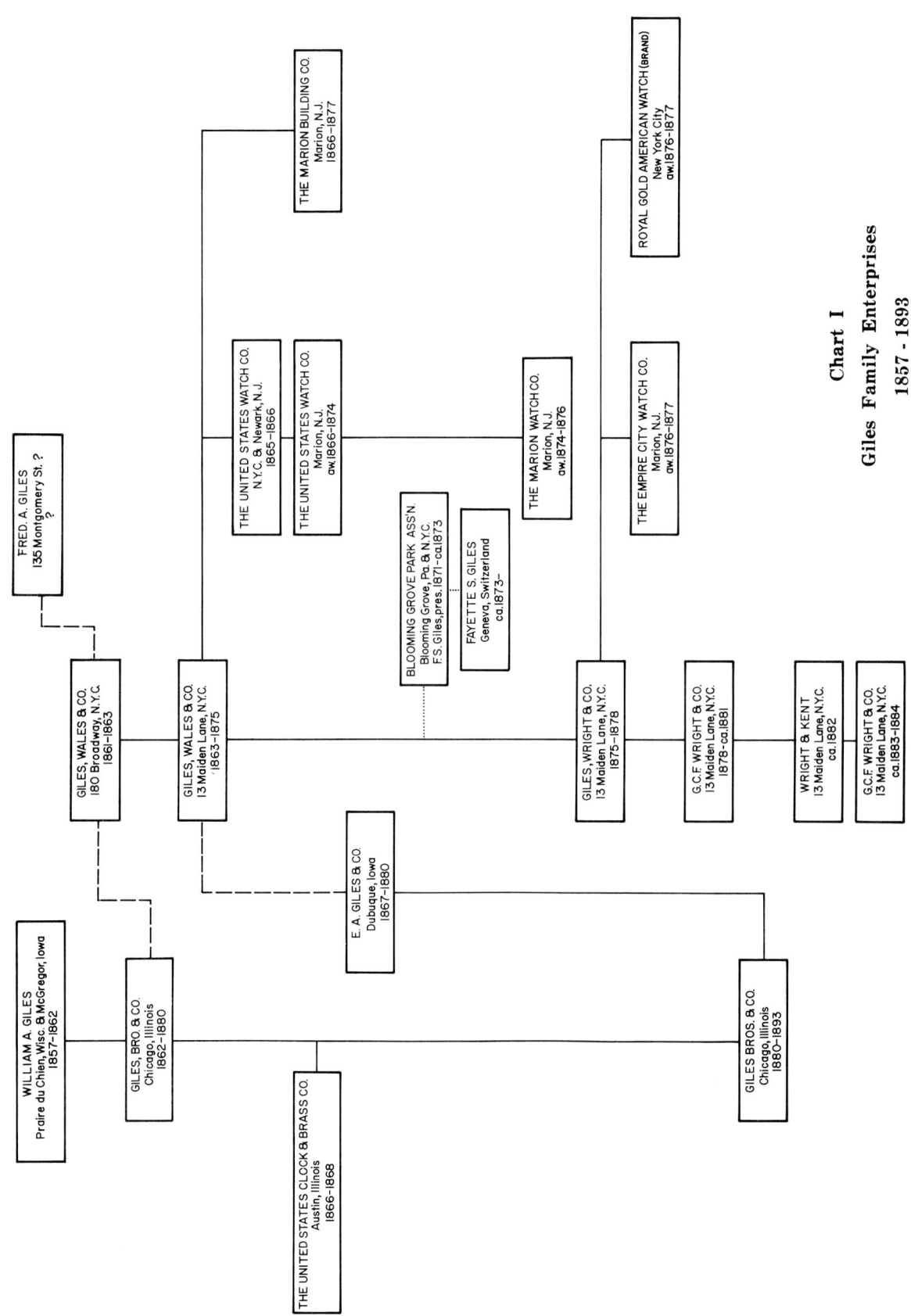

Chart I
Giles Family Enterprises
1857 - 1893

Introduction

Disaster was not an infrequent visitor to the early American watch manufacturing firms. In that world of machinery and bankruptcy the more orderly processes of industry and commerce were often the exception rather than the rule. Watch factory madness was a disease somewhat akin to "Gold Fever." Men believed that a small expenditure of cash and a little Yankee ingenuity would make them rich overnight. The result was all too often tragedy, for watch factories were not bonanzas, but heart-breaking affairs that gobbled up fortunes in a matter of months, or burned, or sometimes even destroyed the dreamers that built them. There were, of course, a few men who with luck or sufficient resources managed to succeed; and there were a few for whom neither luck nor wealth prevailed over the odds against success. While Frederick A. Giles, the founder of the United States Watch Company, was one of the latter, his failure was in a sense unique since it was on a scale that was almost without equal. Yet, for reasons that have only recently become apparent, the details of his tragedy have never been fully or accurately reported.

We owe to Charles S. Crossman, a salesman and practical watchmaker, most of the information or misinformation that we have about all of the early watch manufacturing ventures in the United States.[1] In order to understand both the strengths and faults of his pioneer work, "A Complete History of Watch and Clock Making in America," it is necessary to understand his methods of information gathering.[2] He relied almost exclusively upon a technique which is now referred to as "oral history," or to put it in Charles Crossman's own words, "an original record of verified facts, from an unbiased standpoint."[3] The technique as he pursued it was to interview many of the long term managers and technicians within the watch manufacturing industry. Unfortunately, after he acquired his "original record" he all too often failed to verify that record against other sources. This was brought to his attention even before he had an opportunity to lay down his pen. Crossman soon discovered that Edward Howard and Aaron Dennison were not on the best of terms;* neither were Ambrose Webster and D. D. Lock.[4] Worse still, the agents of the Keystone Standard Watch Company demanded a retraction.[5]

One of the problems that later scholars have had in dealing with Crossman's work is that he only occasionally bothers to cite his sources. In the instance of his history of the United States Watch Company the picture is somewhat mixed. While he does, in a casual way, identify the pictorial and written materials that he used,[6] Crossman does not tell us the names of the individuals that he consulted for the bulk of his text. However, if one reads the complete set of articles covering the history of American watchmaking, at least part of his possible sources for the United States Watch article become evident. Article number twenty-six contains, among other things, a biography of James M. Bottum, inventor of the foot-powered watchmaker's lathe as well as the first man to produce tempered hairsprings in this country. Crossman very clearly states that the origin of his information about Bottum was William Atherton Wales, the former secretary-treasurer of the United States Watch Company and Frederick Giles' former business partner.[7] Further, in article twenty-nine are found the biographies of two former employees of the United States factory, Frank Leman[8] and John Logan.[9] In each instance Crossman infers that he consulted the individuals about their biographies. While, of course, it cannot be proved that Crossman got his information for the United States Watch Company article from any of these men, it seems reasonable that he would have utilized such obvious sources.

It has always been assumed that Crossman arrived in New York too late to witness any of the events that surrounded the demise of the United States Watch Company. Recently it has become evident that such was not the case. In his "Life Sketches" Crossman states that he arrived in New York approximately two years before October 1, 1880. This, following his text, would place him at the firm of Aiken, Lambert & Company, 23 Maiden Lane, sometime in the latter part of 1878.[10] If one considers that the final phase of the bankruptcy action against Frederick Giles and his partner George C. F. Wright did not take place until April, 1879,[11] and further, that the successor firms to Giles, Wales & Co. continued to operate at 13 Maiden Lane until early 1884,[12] it then becomes obvious that Crossman at least ought to have had access to gossip about the United States Watch Company affair. Those who doubt this point should consider that the entire jewelry industry on Maiden Lane could be found at that time in the space of two blocks and that it was a relatively close knit community. Thus it should be apparent that it would have been difficult for Crossman to have avoided the subject.

The Crossman version of the United States Watch Company's history is a curious document. This, however, only becomes evident when the facts of the matter are totally laid out. It is only when the last paragraph is contrasted with the truth that one begins to ask questions about the whole. How could an historian know of the existence of Giles, Wright & Co. or the

---

*A. L. Dennison to Charles S. Crossman, October 9, 1886, in the New York Public Library. Edward Howard's objection to Dennison's being referred to as "the father of the American watchmaking industry" caused him to write a letter of reply to Dennison's obituary. Letter of January 22, 1895, Edward Howard to the *Boston Globe* quoted in *Jewelers' Circular*, January, 1895, pp. 66-67.

name Elias and not realize that his dating is absurd? The Crossman version is not only truncated, but it attempts to bar probing questions by burying the reader in a wealth of superficial facts. Nowhere does it seriously attempt to examine or even explain what actually happened at Marion.

There was, of course, that other pioneer historian of the American watch manufacturing scene, Henry G. Abbott. The name Henry G. Abbott was a pseudonym adopted by George H. A. Hazlitt, the thirty-year-old editor of the Chicago based periodical, *The American Jeweler*.[13] More important, however, is the conjecture that Abbott borrowed much of his information from Crossman. But Abbott's *Watch Factories of America* was published several months before the watch manufacturing segment of Crossman's work had run its course in print.[14] Thus it seems most likely that Abbott used many of the same sources as Crossman, rather than plagiarizing his work. In the instance under discussion, the history of the United States Watch Company, Abbott's article contains several accurate details that do not appear in the Crossman work. Further, his source seems to have been Giles Brothers of Chicago, the former western agents of the factory. Despite Abbott's use of different sources we are again led to the false conclusion that the United States Watch Company went into bankruptcy in 1872. In December, 1903, *The American Jeweler* began to run a completely revised edition of *Watch Factories of America*.[15] Unfortunately, tragedy struck and Abbott died before he had an opportunity to rewrite his history of the United States Watch Company. Thus, we are left wondering what might have happened if Abbott had lived a few more months.

In addition to historical inaccuracies, Crossman and Abbott are the source of several product description errors that still plague historians and collectors. Most notable of these is their incorrect description of the "United States Watch Co." grade as a 16 size, three-quarter plate, gold train movement. Frederick Giles contributed as much as anyone to this problem with confusing descriptions in his price lists and introduction of a quarter plate and bridge movement to a market familiar with full plate and three-quarter plate designs. Even a knowledgeable watch person could look at a Giles quarter plate and bridge style and incorrectly refer to it as a three-quarter, or split three-quarter movement — they still do. The confusion is compounded by the fact that the Giles 16 size, like Nashua and early Waltham equivalents, actually measures a good 17 size. In defense of Messrs. Crossman and Abbott, one must remember they were attempting to provide an overview of the American watch industry, not an in-depth study. This general goal was accomplished in admirable fashion, and without their pioneer efforts the modern student would have no starting point.

Our third historian, Charles T. Higginbotham, also qualifies as a pioneer, but more in the field of watch manufacturing than its history. During his early years he had been employed by the backers of the United States factory, Giles, Wales & Co. Thus his comments about the factory, which appear in his "Incidents in the American Watchmaking Industry," ought to be of special interest.[16] Unfortunately, Higginbotham used the Crossman-Abbott chronology.[17] There is, however, a saving grace to his work. In his story of the United States Watch Company and throughout his series are sprinkled many delightful anecdotes.[18] Some of those outside the United States segment appear to be thinly veiled tales about the company. Although these tidbits are, of course, gossip, some of them have stood the test of documentation and thus must be treated with respect.

The most significant objectives of this current work are to establish an accurate, documented chronology of events and major historical aspects surrounding the USWC, and to provide correct descriptions for all their major products and reasonably reliable production estimates. To logically accomplish these goals, the work is presented in two major sections, (1) Historical, and (2) Production.

INTRODUCTION REFERENCES

1. Charles S. Crossman, *Life Sketch of Charles S. Crossman*, quoted in Dr. William Barclay Stephens, "Charles S. Crossman 1856-1930," NAWCC BULLETIN, 1953, 5:420.
2. Charls S. Crossman, "A Complete History of Watch & Clock Making in America," *Jewelers' Circular and Horological Review* (in 50 installments) July 1886-January 1891; also reprinted as, *The Complete History of Watch Making in America* (installments 1-30), Adams Brown Company, n.d., in a reformatted version.
3. Crossman, Adams Brown Reprint, p. 1.
4. Ambrose Webster to Charles S. Crossman, January 12, 1888, in the New York Public Library.
5. Atkinson Brothers, "Mr. Crossman Corrected." *Jewelers' Circular*, January, 1888, pp. 420-1. Crossman's error in this instance was quite serious in that he had stated that the Keystone Standard Watch Company sold directly to the consumer. In the business climate of the 1880's such a charge could have cost both the Keystone firm and their wholesale distributors, Atkinson Brothers, a great deal of business with their retail watchmaker customers.
6. Crossman, Adams Brown Reprint, pp. 83-90.
7. Ibid., p. 173.
8. Ibid., p. 203.
9. Ibid., p. 210.
10. Crossman, *Life Sketch*, p. 420.
11. *Final Discharge from Bankruptcy Frederick A. Giles and George C. F. Wright*, In the Matter of Frederick A. Giles and George C. F. Wright, U.S. District Court, District of New Jersey, Case No. 1197, April 1, 1879, National Archives Record Group No. 21.
12. "Trade Gossip," *Jewelers' Circular*, March, 1884, XV:62.
13. William Muir, "Will the Real Henry G. Abbott Please Stand Up?" NAWCC BULLETIN, December 1969, 14:10.
14. Henry G. Abbott, *Watch Factories of America, Past and Present*, George K. Hazlitt & Co., 1888.
15. Henry G. Abbott, "Watch Factories of America, Past and Present" (Revised). *American Jeweler*, December, 1903-February, 1906.
16. Charles T. Higginbotham, "Incidents in the American Watchmaking Industry," *National Jeweler and Optician*, January-August, 1912.
17. Higginbotham, January, 1912, p. 27.
18. Higginbotham, March, 1912, pp. 181-2.

Chronology

1834	September 8. Frederick Asa Giles, founder of the USWC, is born in Angelica, New York (page 23).
1836	William Alexander Giles, brother of Frederick, is born in New Salem, Massachusetts.
1841	Fayette Stratton, brother of Frederick, is born. The birthdate of another brother, Charles K., is unknown.
1843	Edwin A., the youngest brother, is born.
1843	Prescott, Frederick's father, dies at the age of 44.
1844	Anna, Frederick's youngest sister, is born. Sue, an older sister, had been born sometime between 1829 and 1834.
1844	Lemira, Frederick's mother, dies and shortly after his parents' deaths, he is apprenticed to a jeweler in Amherst, Massachusetts, by his grandparents.
1861	Frederick and William A. Wales form Giles, Wales & Co. and begin business at 180 Broadway in New York City (page 25).
1862	William and Charles Giles form Giles, Bros. & Co. at 142 Lake Street in Chicago (page 59).
1862	June 24. Frederick marries Julia M. Wright at Montague, Massachusetts (pages 26-27).
1863	Giles, Wales & Co. reorganizes into a new and expanded partnership and move to 13 Maiden Lane. Fayette and Frederick's brother-in-law, George Channing Fuller Wright, are added as junior partners (page 30).
1865	February 22. Charter for USWC is drawn in New York City with trustees, F. A. Giles, W. A. Wales, George C. F. Wright, Frank H. Page, and Daniel M. Wells (page 35).
1865	April 25 and August 15. Frederick issued patents on stem winding and setting watches (pages 39-41, 127-128).
1865	May. Work underway on USWC tooling in Newark, NJ, at Dickinson & Rowden's machine shop under supervision of James H. Gerry (page 39).
1865	May 9. Frederick issued design patent for top plate and balance cock of a watch (pages 39, 128).
1865	Possible "model of 1865" could have been based on Frederick's patents of April 25 and May 9, 1865 (pages 39-40).
1865	August. Land acquisition completed and preliminary work started on the USWC Factory at Marion, New Jersey (page 46).
1866	February 5. Two Senate Bills are placed before the New Jersey Senate, one to establish USWC works in Bergen, New Jersey, and the other to incorporate the Marion Building Company with directors F. A. Giles, W. A. Wales, George C. F. Wright, George F. Pratt, Daniel M. Wells, Benjamen G. Clarke, and Alexander H. Wallis (page 44).
1866	March 13. Frederick issued design patent on butterfly opening in the top plate (pages 39, 49-50, 127-128).
1866	Marion House built. Two years later, in 1868, the St. James Hotel will be added to the complex (page 47).
1866	Factory completed, the Gerrys depart, and William B. Learned is appointed as superintendent (pages 48-49).

1866	William Alexander Giles invests in United States Clock and Brass Co. in Chicago, which will only last for two years (page 61).
1866	Fayette Stratton Giles goes to Switzerland to manage the Giles, Wales & Co. agency in Chaux-de-Fonds (page 63).
1867	Edwin A. Giles establishes E. A. Giles & Co. in Dubuque, Iowa (page 63).
1867	Fayette marries Bertha Faigaux in Switzerland (page 64).
1867	July. America's first mass-produced stem winding watch, the 19-jewel "Frederic Atherton & Co." grades make their appearance (pages 49-50, 137).
1868	Fayette is instrumental in getting F. Wilmot of St. Imier, Switzerland, to go to USWC for one year to instruct in the art of gilt and nickel damaskeening (page 65).
1868	December 22. Improved stem winding mechanism developed to correct problems in the 1865 version (page 53).
1869	February. Gilt version of 19-jewel "United States" grade introduced as the USWC prestige item market entry (pages 55-56, 142-143).
1869	June 14. William B. Learned demoted, and Henry J. Lowe takes over as the superintendent (pages 65-66).
1869	December. America's most expensive watch, the first nickel, 19-jewel "United States" grades are produced (pages 77-78, 142-143).
1870	First "Knapp" and "Pratt" grades produced for USWC by the firm of Bourquin in Bienne, Switzerland (pages 66-67).
1870	Apex year for both Giles, Wales & Co. and the USWC (page 69).
1870	October 1. President Grant visits the plant and is presented a gold "stem-winder worth about five hundred dollars" complete in a rosewood gift box (page 71).
1870	USWC participates in three different expositions and wins first-place awards for watches and dials (pages 71, 165).
1870	USWC establishes a fine reputation for its dial making operations, especially for their hand painted, fancy dials (pages 74-77, 161-163).
1870	July. USWC employees number 315 with a reported production capacity of 150 movements per day (pages 69, 193).
1871	Optimism fades and the USWC begins to have problems, especially in achieving production (page 81).
1871	**Fayette leaves Giles, Wales & Co. and the USWC to pursue other interests** (page 81).
1871	September. Fifteen employees steal watch parts valued at one thousand dollars (pages 82-83).
1871	USWC begins a plan to remodel its product line and add ¾ and ¼ plate models (pages 83, 85, 130).
1871	Late 1871 or early 1872. USWC establishes three significant firsts in the area of railroad watches (page 168).
1872	March. New Jersey Supreme Court upholds lower court's decision in favor of William B. Learned over the USWC (page 83).
1872	USWC sells some surplus movements to Ellis Elias for his "Great Geneva Watch Company" sale (page 83).
1872	Competitive developments from the American Watch Co., New York Watch Co., National Watch Co., E. Howard & Co., and the Swiss result in dealer complaints about USWC full plate, slow train watches (page 85).

1872	September. USWC rushes to publish a new price schedule showing a complex line of remodeled full plate, ¾ plate, and ¼ plate and bridge grades even though it will be early 1873 before these new models are ready for sale in reasonable quantity (pages 94, 134-136).
1873	May. USWC introduces a line of 15-jewel "United States" models at prices probably designed to be more competitive with the Waltham "American Watch Co." grades (pages 88-89, 142-147).
1873	September. The beginning of 1873-1875 financial panic, a severe blow to the United States economy but especially painful to the USWC (page 95).
1873	December 30. Frederick Giles is issued a patent on his design for a reversible mainspring barrel (pages 86, 128).
1874	Heavy price cutting hits the American watch manufacturing industry which is especially damaging to the higher priced products of the USWC (page 95).
1874	Some key employees, like George E. Hart and John Logan, begin to leave the USWC (page 96).
1874	July 20. The USWC undergoes a financial reorganization and the Marion Watch Company is incorporated in New Jersey. Prices are then lowered on many grades (page 96).
1874	Giles, Wales & Co. attempt to carry the brunt of financial obligations, but by December, creditors begin to file suits. On December 21, an involuntary bankruptcy petition is filed against F. A. Giles, W. A. Wales, and George C. F. Wright (page 97).
1875	February 1. William A. Wales withdraws from the partnership and leaves the USWC before the bankruptcy proceedings are resolved. Giles, Wales & Co. then becomes Giles, Wright & Co. (page 101).
1875	By mid year, the USWC finds itself in a tragic circle of declining funds, wearing machinery, declining standards of workmanship on all grades, and disappearing key employees (pages 100-102).
1875	June. Sylvanus Sawyer, a major USWC stockholder, purchases the newest and best 16 size ¼ plate and bridge tooling from Frederick, and with Henry J. Lowe, goes to Fitchburg, Massachusetts, to establish the Union Watch Co. (page 102).
1875	December 31. The debts of Giles and Wright had risen to over $470,000. New bankruptcy proceedings are instituted (pages 102-103).
1876	January 5. Giles and Wright, with William Wyse and Andrew Wood, incorporate the Empire City Watch Company. Giles and Wright get financial assistance from Ellis Elias of New York City who needs watches in another of his "operations" (page 105).
1876	January 18. Giles and Wright are declared bankrupt and their assets are placed in the hands of William Muirheid, U.S. commissioner for the district (page 103).
1876	The USWC "string saving" stock of reject and/or imperfect movements becomes the nucleus for rework and sale by Elias as Royal Gold American Watches (pages 109-110, 182-183).
1876	The USWC stock of satisfactory and/or unsold movements provide the source for Empire City Watch Company products (pages 110-113, 178-181).
1876	Giles and Wright decide to pass on the Centennial Exhibition in Philadelphia as the Empire City and Royal Gold efforts require all their attention and meager resources (pages 173-177).
1876	Around mid year, Frederick sells the ¾ plate machinery to George Hart, his former mechanical superintendent. Hart, in turn, sells the equipment, once modified, to the Auburndale Watch Company (pages 114-115).
1877	By early in the year, confronted with verbal attacks from the reformer, Anthony Comstock and increasing demands from Giles, Ellis Elias decides to terminate his arrangement with the USWC (page 113).
1877	By mid year, the affairs of the Marion Building Company, USWC, Marion Watch Company and Empire City Watch Company are a complex myriad of financial entanglements surrounded by complete confusion and ambiguity (pages 116-117).

1877   While the creditors attempt to sort things out and with the factory now closed, Frederick uses the St. James Hotel as headquarters for disposing of some completed Empire City stock (page 117).

1877   June 2. Captain "Clubber" Williams arrests Elias at his General Average Store on charges of conducting a lottery. George C. F. Wright "abandons ship" in the aftermath of negative publicity on the Elias affair and then returns to New York City (page 117).

1877   Cumulative total production for the ten-year period from 1867 would appear to be some 50,000 . . . considerably less than the 300,000 originally thought produced by the USWC (page 187).

1878   March. James Alexander begins efforts to dispose of the USWC factory building, remaining machinery and movement stock on behalf of the creditors (page 121).

1878   By mid year, a very weary and sick Frederick leaves the St. James Hotel and returns to Montague (page 118).

1878   August. The Howards of Fredonia, NY, and Hart, Sloan & Co. of Newark, NJ, buy most of the remaining machinery and movement stock (page 121).

1879   June 18. Frederick Giles dies in Montague of tuberculosis at the age of 44 (page 119).

1879   June 22. A. H. Wallis, long-time supporter and stockholder of USWC, dies at the age of 61 (page 122).

1880   The two firms, E. A. Giles & Co. (Edwin) of Dubuque, Iowa, and Giles, Bros. & Co. (William and Charles) of Chicago, Illinois, consolidate in Chicago due to the strain caused by the collapse of Frederick's empire. Charles will later develop his anti-magnetic watch protector (page 123).

1881   June. Ellis Elias dies in New York City at the St. Cloud Hotel at the age of 42 (page 122).

1884   February 8. George C. F. Wright, now operating the old Giles, Wales firm in his own name at 13 Maiden Lane, dies at the age of 42 (pages 122-123).

1891   December 30. I. H. Wright, George's brother and loyal supporter of Frederick to the end, dies in Montague (page 123).

1896   February 22. William A. Wales dies in New York (page 123).

1896   August. The "Newark Call" newspaper writes an "epitaph" for the USWC that indicates by then the old factory buiding had seen nearly a dozen attempts to establish a going business but all had failed. This contributed to the notion that the factory was not only a "white elephant," but also "hoodooed" (pages 199-200).

1897   December 5. Fayette Stratton Giles dies in North Carolina at the age of 56 (page 123).

1913   December 17. William, who seems to have outlasted everyone and in the end pays back the debt owed Frederick by educating his daughter and granddaughter, dies at the age of 77 in Phoenix, Arizona (page 124).

1925   After a variety of attempts to establish a going business ranging from electric lights to electric traction, textiles to smoking pipes and school furniture, the once elegant iron and glass USWC factory was dismantled . . . "leaving only a vague imprint on the ground" (page 124).

* * *

HISTORICAL

FRED. A. GILES,

IMPORTER AND JOBBER OF

WATCHES, DIAMONDS, JEWELRY, &c.,

NO. 135 MONTGOMERY STREET,

(SECOND FLOOR,)

Would respectfully call the attention of the trade to his large, fine and well selected stock of the above goods, consisting of fine English, Swiss, French and American WATCHES; fine DIAMOND GOODS; Chain, Swiss and American JEWELRY, &c., &c.

An examination of Goods and Prices is respectfully solicited. my5-1&2p

2. *This ad clipping, source unknown, raises the unanswered questions of when and where. Most likely, at some time before the formation of Giles, Wales & Co., Frederick was in business for himself. It is also likely that the where was New York City.*

Chapter 1
The Early Years

In the early years of railroading in these United States, the Paterson and Hudson River Railroad was seeking a cheap and convenient route from Paterson, New Jersey, to Powles Hook (now Exchange Place in Jersey City) and the Hudson River ferries to New York City. Across the railroad's intended path lay the formidable rock mass of Bergen Hill. Early plans called for the use of stationary engines to haul the cars to the top of the western slope, however by 1834 the Paterson people were able to arrange a less complex solution. It was agreed to realign the Paterson and Hudson River roadbed so that it would meet at the base of Bergen Hill with the New Jersey Railroad, which faced similar problems in laying track from Newark to Powles Hook. It was further agreed that from the point of junction the railroads would dig "Shanley's" Cut through the hill and thus both would have, at reasonable cost, their outlet to New York City. The rail junction and its associated station were first given the name Bergen, after the city in which they were then located. Had this state of affairs continued, the land surrounding Bergen Junction, despite the fact that the western half was salt marsh, would have become a valuable piece of property. These prospects were short lived, however, when in the early 1860's the Paterson and Hudson, now a segment of the New York and Erie Railroad, completed its tunnel through Bergen Hill about a half mile north of Shanley's Cut.[1] From that point on, only the land north of the New Jersey Railroad retained value, while the area south of the railroad, soon to be called Marion, became a relatively unimportant strip of land which dips down the western slope of Bergen Hill into the mosquito-infested marshes of the Hackensack River. An ideal locale for Francis Marion, "The Swamp Fox" of revolutionary fame, but as a site for a watch factory it will always seem a bit improbable.

It is difficult to make generalizations about the United States Watch Company and its related firms. However, two families were always at the center of the fortunes of these businesses. These were the Giles family of New Salem, Massachusetts, and the Wright family of nearby Montague. How, or even when, the Giles family first came to settle in New Salem is unclear. The local cemetery has a fair number of 18th Century Giles gravestones, but despite a lot of trying, genealogists have not been able to establish their relationships. The earliest identifiable generation of the branch of the family that concerns us is that of John and Hepzabeth Giles.[2] On October 25, 1794, their son Daniel, a farmer, married a Miss Hannah Learned.[3] This union was blessed with at least five children: Anne and Fanny, twins, born in 1795, Prescott, born in 1799, Hannah, born in 1803, and Sally, born in 1809.[4] Prescott, who is important to our narrative, married a Miss Lemira D. Stratton of Athol, Massachusetts, in 1829.[5] During the first five years of their marriage only one child was born, a daughter, Sue.* Sometime between 1832 and 1834 Prescott and his wife left their farm in New Salem and journeyed to Angelica, New York. There on September 8, 1834, their first son, Frederick Asa, was born.[6] The Gileses soon returned to New Salem where Prescott resumed his farming. In the years that followed their return Prescott and Lemira had five more children: William Alexander, born in 1836,[7] Charles K., whose date of birth is unknown, Fayette Stratton, born in 1841,[8] Edwin A., born in 1843,[9] and Anna, born in 1844.[10]

Tragedy and misfortune seem to have taken a special interest in the Giles family. No matter how successful they became, the malevolent fates were prepared to destroy all. When it began was never recorded, but its effects are all too evident. Old Daniel's farm was cursed with rocks and infertile soil and consequently heavy debts. Prescott refused to accept his father's fate and struggled to achieve better. When he had at last gained the status of successful farmer, the 1837 depression took it all. Prescott never quite recovered from this shock and he died at age forty-four in 1843. The fates, however, had just begun this round; in 1844 Lemira followed her husband, shortly after giving birth to their last child, Annie. As is normal in such instances the aged grandparents took over the care of the children. That is with one exception — Frederick, who was apprenticed out. Even this sad new arrangement did not last and in 1847 Daniel followed his son and daughter-in-law. Thus, the family which occupied the barren old farm was finally reduced to Hannah, the seventy-one-year-old matriarch, and the remaining six Giles children. Of these only Sue and William were old enough to be of immediate assistance.[11]

When he was at his apex Frederick was fond of recounting the struggles of his youth for visiting newspaper reporters. It might be expected that these tales would slightly embroider the facts, but if anything he generally understated his accomplishments. The simple truth was that the first thirty-seven years of Frederick's life were to fit neatly into the Horatio Alger mold. Of course the press occasionally confused things by

---

*This was deduced from *Harper's Weekly*, June 25, 1870, p. 414, and Sue Giles to Julia M. Giles, September 24, 1882. Burns Collection.

making a hash of the details. Still, nothing could obscure the fact that he rose from an impoverished orphan to a highly successful importer.

Shortly after the deaths of Frederick's parents he was apprenticed to a jeweler in Amherst, Massachusetts. This jeweler, whose name is unrecorded, assisted him in obtaining an education by allowing him to attend one of the local high schools. At fourteen Frederick was entrusted with the management of the store. If this was an attempt by his employer to persuade him to remain after his apprenticeship, then it was soon to prove a complete failure. Within two years the itch for broader experience caused Frederick to quit Amherst for Boston. His new locale, however, was apparently no more satisfying than the old. In early 1852 he left Boston for New York where he soon found work with the jewelry firm of Seth P. Squire at 182 Bowery. His stay with Mr. Squire was brief and by the late summer he was working as chief clerk for William T. and Thomas V. Gendar, jewelers at 228 Greenwich Street.[12] The Gendars seem to have developed quite a liking for their energetic young clerk and he was soon involved in their land dealings.* Further, it was through Thomas Gendar, a resident of Hoboken, that Frederick first became aware of the possibilities of that other world across the Hudson River — New Jersey.[13] The Gendars may have also fueled their young clerk's dreams in another direction. They were, or at least so they claimed, involved in the manufacturing of thermometers, hydrometers, watches and jewelry.[14] What the term "manufacturing" connotes in this instance is open to question; however whether the Gendars were jobbers or factors or whatever, the effect would have been to give Frederick direct experience in the manner in which watches and other goods were assembled at this period.

Despite the fact he was now working far from New Salem, Frederick in one sense never left home. A part of his income was always sent back to support the household and to educate the younger children. Both William and Fayette were able to attend New Salem Academy due to his largesse. Although his dedication to business precluded many trips there, central Massachusets was always home. Even when it came to romance Frederick's heart turned homeward. He had begun corresponding with a Miss Julia M. Wright of Montague, Massachusetts.† How the Wright family of Montague and the Giles family of New Salem first came into contact is uncertain. Whether they met through Frederick's acquaintance with Julia or whether some other factor entered in has not been discovered in family correspondence. There was, however, by the latter half of the 1850's a close relationship growing between the

---

*There are at least two land warranties, involving property in Virginia and New Hampshire, from this period that were signed by Frederick. These documents are in the Burns Collection.

†The earliest extant letter is January 6, 1856, Frederick Giles to Julia M. Wright (Freddie to Julia), Burns Collection. This letter mentions another four weeks earlier, from Julia. While it is obvious from the January 6th letter that the relationship is becoming more than merely friendly, the genesis of the correspondence is not explained.

3. Julia M. (Wright) Giles

families. The Wrights were one of those ancient New England families who traced their origins back to the Mayflower. At the time when Frederick and Julia began writing to each other the household consisted of Dr. George Wright, a physician, his wife, Julia, as well as Julia M. and her three brothers: Charles P., Isaac Henry and George Channing Fuller. It would be several years however, before the affairs of the two families became so intertwined that in many ways they became one.

For all his ties with New Salem and his family, Frederick's early years were nomadic. But he was a traveler driven by a goal — wealth and success. Not that this dream or even his methods for obtaining it were unique. His generation would later be noted for its belief that hard work and determination were the only prerequisites for obtaining riches. Even his friendship with the Gendars could not hold him for long. Early in 1855 he left them and briefly joined Young, Stebbins and Co. as a traveling salesman.[15] Then later the same year he took a similar position with the wholesale jewelry importing firm of Platt and Brother, 20 Maiden Lane, New York City.[16] The Platts had a business very much to the twenty-one-year-old Giles' taste. It was a large prosperous operation which sent salesmen all over the country as well as maintaining buyers in Europe. The ambitious, hard working young man must have impressed the Platts as well; on his first two selling trips for them he set an incredible sales record.[17]

24

4. George Channing Fuller Wright

When Prescott and Lemira traveled to Angelica, New York, in the 1830's they would have had to cover at least part of the approximately four hundred mile distance from New Salem by wagon. With the exception of the segment from Albany to Schenectady, New York, traversed by the Mohawk and Hudson, railroads did not as yet exist.[18] Canals, however, did and much of the distance west of Albany could be traveled in relative comfort by canal packet. Twenty-one years later, when Frederick began his travels for the Platts, the picture had changed greatly. The railroads had progressed from their infancy to robust youth and now crisscrossed much of the eastern United States. There was an exception to this progress; north central Massachusetts was still isolated from a direct route west by the Hoosac Range of the Berkshires. Even twenty years later, when at great human cost the Hoosac Tunnel was dug, New Salem remained isolated.[19] The only way there, was to travel to Greenfield or Athol or Montague by rail and then find local conveyance to New Salem. But Frederick did not make this pilgrimage often; his travels were now to Cleveland, Chicago, Galena, Milwaukee, Cincinnati, Louisville and New Orleans.[20] In fact, on one banner sales trip for the Platts he traveled all the way to California.[21]

Just as the railroads had created a centripetal force in the United States, states rights and slavery had created a centrifugal one. In economic terms it became a race of capital intensive industry against labor intensive. Had it been possible for affairs to run their natural course, the outcome would have been both peaceable and foreordained. Alas, sectional rivalries had become sectional hatreds and in such a climate war becomes the master, peace the fool.

Business was already shaken from the panic of 1857, and with war's approach in the early months of 1861 commerce and industry slowed noticeably. The jewelry industry on Maiden Lane, always sensitive to economic conditions, came to a virtual standstill. Trade fell off to such an extent that many firms were compelled to lay off their employees.[22] It was certainly an unlikely time to begin a wholesale jewelry importing business. Yet this is what Frederick Giles did. Despite the business slump the new firm, Giles, Wales & Co., was unable to locate on Maiden Lane, but rather had to settle for a store around the corner at 180 Broadway, next door to Frederick's residence, the Howard Hotel.[23] Most, if not all of the financing, seven thousand dollars, was provided by Frederick, while William Atherton Wales, the other partner, brought his twenty years' experience in the jewelry trade.* Wales, who was born circa 1830, first apprenticed to Jacob M. Crooker, a jeweler in Waterville, Maine, in 1845. Regardless of the fact that Crooker later took him into partnership, Wales left Maine in 1850 for New York City.[24] Interestingly enough, upon arriving in New York Wales again be-

---

*See Table I.

5. William Atherton Wales

came an apprentice; this time his master was James M. Bottum. During the period of Wales' apprenticeship Bottum was manufacturing his pioneer foot-powered watchmaker's lathe and had begun experimenting with the process of tempering hairsprings.[25] Immediately before the formation of Giles, Wales & Co. Wales had been manager of Roger, Smith and Company's New York office.[26]

The concerns attendant to the start of the business were not the only thing pressing on Frederick's mind in late 1861. His youngest brother, Edwin, had enlisted as a private in the 27th Massachusetts Infantry Regiment. Early in November, as it headed south to join General Burnside at Annapolis, the Regiment paused for a few hours in Jersey City. Frederick, William Wales and some friends went across the river to greet them. There was a wait of three hours until at last a little after noon the vessel carrying the troops arrived. The rest of the day was occupied by a collation for the troops and by a delightful incident where Frederick gave his eighteen-year-old brother the numerous gifts that he had brought, which Edwin immediately distributed among his friends. Finally, just before the Regiment departed Frederick and Edwin had a long talk. That evening Frederick reported to Julia in a letter that "He (Edwin) seemed in excellent spirits, but much affected when I bade him goodbye and I felt the parting deeply too, but I knew that it were worse than useless, to indulge in vain regrets."[27] Frederick's concern was not unrealistic. Three months later the 27th was engaged in the Battle of Roanoke Island where their brigade took one hundred thirty-two casualties. A month later on the 11th of March, 1862, they fought in the Battle of New Berne, North Carolina, where the brigade suffered another one hundred eighty-two casualties. Despite the losses these battles were among the few bright spots in a time of disaster for the Union Army.[28]

Frederick's meeting with Edwin points up a side of the elder brother's personality that must be understood because it governs so many of the events in this story. It has already been noted that with the death of his parents Frederick was forced to assume a part of the financial responsibility for the family. As the years passed this role and that of surrogate parent tended to grow.[29] Further, rather than fight against or chafe under these responsibilities he embraced them wholeheartedly. Circumstances, however, dictated that he must be separated from the family for long periods, thus in many respects he became an outsider. His siblings might owe him much, but their loyalty was not always returned in the same coin. Yet the fault was not completely on their side for Frederick was not entirely sensitive to his brothers' pride and need for independence. Thus it is not altogether surprising that when he assisted William and Charles to establish their own jewelry business, William in particular refused to make any public acknowledgement of the fact.[30] For his part Frederick tried to ignore these slights and continued on as if nothing had happened. When Dr. Wright died in October of 1859 he extended his largesse to the Wright family as well.[31] Evidence of this can be seen in the fact that when Giles, Wales & Co. was founded

6. Frederick A. Giles, about the time Giles, Wales & Co. was organized.

George Channing Fuller Wright, Julia's younger brother, was taken in as a clerk.[32] Later, Fayette, Edwin and Henry Wright, Julia's elder brother, would be provided for as well.

Frederick and Julia had by this time, 1862, been corresponding for seven years and it had long been apparent that they would eventually marry. There had been, however, several causes for delay. At the outset, Julia had been too young for a woman of her social class to marry and Frederick had not been in the economic position to contemplate it. Then the death of Julia's father doubtlessly delayed matters further. Ultimately, Frederick's young brother, William, married first and was by 1861 a father.[33] But by this time Frederick and Julia, although much in love, would not be hurried. Certainly a part of the problem now centered around the fact that Julia and her mother had become extremely close and neither wished to be separated.[34] The other aspect of this phase of the relationship was Frederick's almost total absorption in his business. His working day began early, paused briefly for a nap in the afternoon, and then continued well after midnight. In fact, his letters to Julia were often written at night and yet they generally mention various tasks that he still has to accomplish for the day.* Ultimately these impediments, which might well have stymied other couples, proved to be the very conditions that made a successful and happy marriage possible. Where other marriages might have foundered on the separations caused by divergent responsibilities, this one was to thrive.

There were four Wright children who survived to adulthood and Charles was the second eldest of these. Among other things, by the early 1860's he had achieved the distinction of being the town clerk of Montague. In June of 1862, however, only one aspect of his duties interested him — marriage licenses. On the 24th Frederick and Julia were to be married and on the next day, Wednesday, he planned to marry his love, Martha.[35] While his sister's plans were of long standing, his own had only recently been arrived at. As late as May he was unsure of when Martha would set the date. This was particularly irksome since he felt that Julia's departure would leave their mother alone and without assistance. Despite this concern he was still able to have a little fun needling Frederick in a half comic letter giving his sister's marriage plans his blessing. Charles, with typical Wright sense of humor was not unmindful that the custom of the shivaree ought to be kept. One wonders how the couple took his suggestion of meeting them on their "tour."[36] In the end Martha's decision turned the joke around and a bemused Charles signed Frederick and Julia's marriage certificate on June 23rd.[37] Later he would sign another important paper conveying the two and one-half acre Wright homestead to Frederick.[38]

---

*The time varies, the latest one being midnight, i.e.: Frederick Giles to Julia Giles, January 5, 1868, 12 p.m., and Frederick Giles to Julia M. Wright, January 6, 1856, 8:30 p.m., Burns Collection.

CHAPTER 1 REFERENCES

1. Walter Arndt Lucas, *From the Hills to the Hudson* (New York, 1944), Chap. 4 & 9.
2. The Essex Institute, *Vital Records of New Salem, to the End of the Year 1849* (Salem, 1927), p. 21.
3. Ibid., p. 75.
4. Ibid., pp. 21 & 75
5. Ibid., p. 75.
6. This information is verified by several sources including *Record of Death — Frederick A. Giles*, Commonwealth of Massachusetts, June 18, 1879.
7. *Biographical Sketches of the Leading Men of Chicago* (Chicago, 1868), p. 511.
8. Thomas W. Herringshaw, *National Library of American Biography* (Chicago, 1909), Vol. 2.
9. *Vital Records of New Salem*, p. 21.
10. Ibid., and Anna Giles to Julia M. Wright, October 31, 1859, Burns Collection.
11. There are several versions of this story, the most accurate being William A. Giles' story in *Biographical Sketches of the Leading Men of Chicago*. Also see: *The Gazette*, June 18, 1870; *Jersey City Journal*, May 27, 1871; *Harper's Weekly*, June 25, 1870.
12. *Jersey City Journal*, May 27, 1871; note that details such as addresses have been drawn from the relevant city directories.
13. Rode, *New York City Directories 1852-3* and *1853-4*, and Trow, *New York City Directory 1854-5*.
14. Doggett, *New York City Directories 1848-9* and *1849-50*. These directories list the Gendars as manufacturers of thermometers, hydrometers, watches and jewelry.
15. *Jersey City Journal*, May 27, 1871.
16. Ibid., also *Harper's Weekly*, June 25, 1870, p. 414.
17. "Fred. A. Giles — Obituary," unidentified clipping, probably Greenfield, Massachusetts, newspaper, Burns Collection. Also Frederick A. Giles to Julia M. Wright, January 6, 1856, Burns Collection.
18. Hank Wieland Bowman, *Pioneer Railroads* (Greenwich, Conn., 1954), pp. 42-45. In 1834 this railroad possessed John B. Jervis' incredible engine "Experiment." This locomotive was the first mile-a-minute machine ever built.
19. John O'Connell, *Railroad Album* (Chicago, 1954), p. 125.
20. Frederick A. Giles to Julia M. Wright, January 6, 1856, Burns Collection.
21. "Fred. A. Giles — Obituary," unidentified clipping, probably Greenfield, Massachusetts, newspaper, Burns Collection.
22. Albert Ulmann, *Maiden Lane* (New York, 1931), pp. 84-88.
23. Frederick Giles to Julia M. Wright, November 28, 1861, Burns Collection and Trow, *New York City Directory*, 1862-3. There is a newspaper advertisement in the Burns Collection for "Fred. A. Giles, No. 135 Montgomery Street — Importer and Jobber of Watches, Diamonds, Jewelry, etc." Unfortunately it is a clipping and there is no way of knowing the date or even the city.
24. *Jersey City Journal*, May 27, 1871.
25. Crossman, Adams Brown Reprint, p. 173.
26. *Harper's Weekly*, June 25, 1870, p. 414.
27. Frederick Giles to Julia M. Wright, November 2, 1861, Burns Collection.
28. Robert Underwood Johnson and Clarence Clough Buel, eds., *Battles and Leaders of the Civil War* (New York, 1956), I:640-670.
29. Anna Giles to Julia Wright, October 31, 1859; Sue Giles Moore to Julia Giles, September 24, 1882, Burns Collection. One suspects that Sue, the elder sister, had on several occasions reminded Frederick's brothers of their obligations.
30. *Biographical Sketches of Leading Men of Chicago*.
31. Wright Family Bible.
32. *Jersey City Journal*, May 27, 1871.
33. Frederick Giles to Julia M. Wright, September 29, 1861, Burns Collection.
34. Charles P. Wright to Frederick Giles, May 25, 1862, Burns Collection.
35. Wright Family Bible.
36. Charles P. Wright to Frederick Giles, May 25, 1862, Burns Collection.
37. *Certificate of Marriage*, Frederick A. Giles and Julia M. Wright, date of issue June 23, 1862. Burns Collection. The marriage was performed on June 24, 1862, by John B. Green.
38. *Mortgage*, conveying homestead of George Wright, deceased, Charles P. Wright to Frederick A. Giles, Burns Collection.

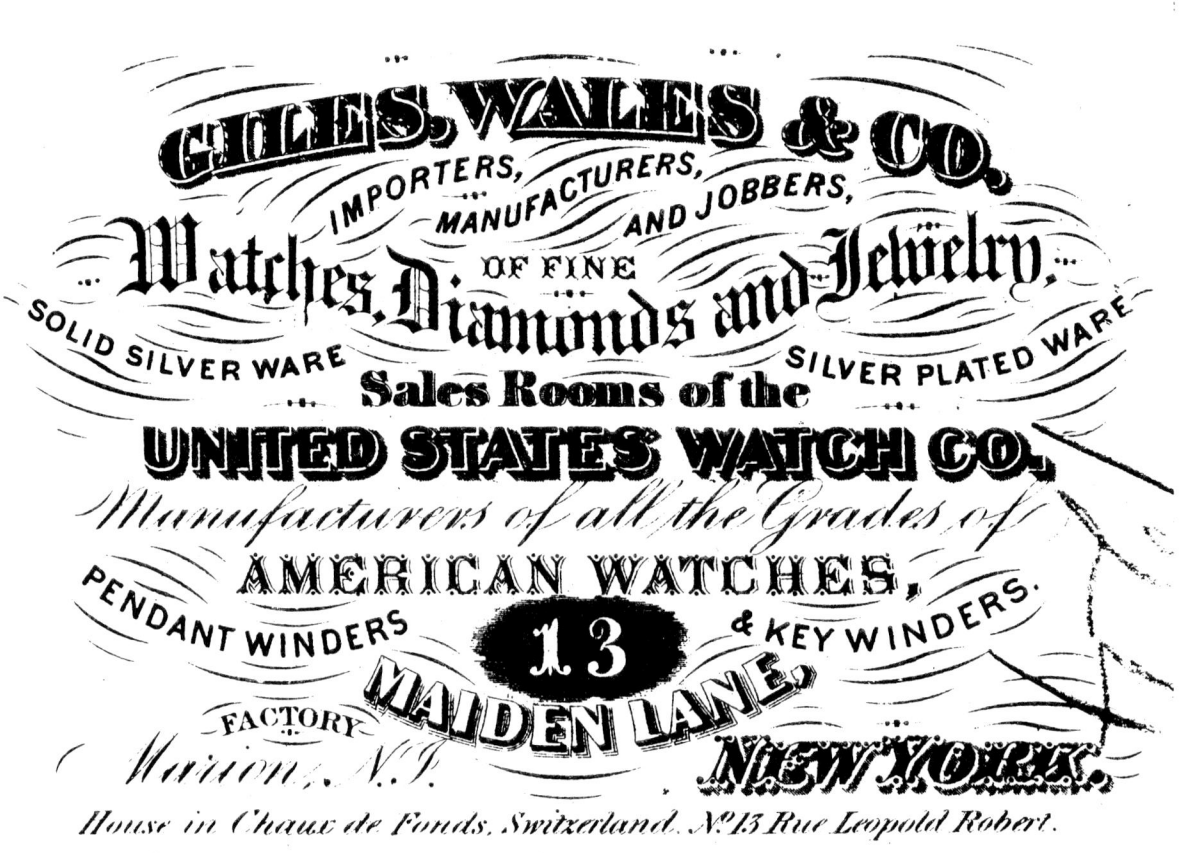

7. Giles, Wales & Co. Trade Card, circa 1869.

Chapter 2

The Money Machine

The economic stalemate of late 1861 and early 1862 was not the kind of situation that could last for long — not in the hot-house of war. But while it did, the situation disoriented even the shrewdest of businessmen. At Waltham, Royal E. Robbins, Treasurer and Chief Executive of the American Watch Company, fired Aaron L. Dennison, the firm's founder, when the latter suggested manufacturing a soldier's watch.[1] American businessmen were inexperienced with the economics of war.

On Maiden Lane, the revival came first as a trickle. Here and there a new breed of buyer began to appear — the sutler. Sutlers were a Civil War version of that later military fixture, the Post Exchange. Once the army had begun to gather, it began to develop appetites. These the sutlers filled with pies, herrings, cider, jewelry, and the myriad of odds and ends that men away from home demand.[2] This, however, was but one of the factors that effected economic change. In Washington, Salmon P. Chase, the Secretary of the Treasury, had loosed powerful forces when he decided that the war should be financed by loans rather than taxes. While the banks subscribed to the loans, both the government and the banks had great difficulty in selling the loan certificates to private investors. The banks' reserves, which were at that time in specie, were much reduced as a result. Thus when the Union's early military reverses led fearful depositors to make heavy withdrawals of specie, the situation became impossible. On December 20, 1861, the banks were forced to suspend further payments in gold or coin. In January, 1862, the Federal government followed suit. One month later legal tender money unredeemable in specie, the "Greenback," was introduced.[3]

Robbins and Appleton may have been reduced by the economic difficulties of 1861 to serving soup to the troops on Staten Island, Maiden Lane may have been prostrated, but at 180 Broadway, Giles and Wales ignored the economic trends.[4] The young partners did not have the financial resources to allow themselves to be paralyzed by events. If business would not come to them, then they must pursue it. The following from one of Frederick's letters will explain their tactics: "Today is Thanksgiving, but I worked till three o'clock, although the store was closed, but I made the acquaintance last night of a first rate man from Pittsburgh, Pa. and to make a sure thing of selling him, I took him to the store *today* and sold him by gas light, about $1,000, and have made a good customer of him."[5] There must have been numerous instances when Frederick and Billy Wales used their youth and energy to advantage, for by the end of 1862 they were able to enter the tidy sum of one hundred seventeen thousand one hundred four dollars and seventy-two cents in gross sales into their cash books.*

After their "tour" Frederick and Julia had settled down in the Howard Hotel, but events would limit this period of togetherness. By early 1863 business conditions had improved, although they were still not what could be called normal.[6] Not that the war news entirely justified even this modest swing of the economic pendulum. Only in the west could the Union point to much success; the Mississippi was slowly but inexorably falling to their arms. In the east, the Army of the Potomac was blundering its way from defeat to defeat. The Civil War was, however, as much an economic conflict as a military one, and there at least some semblance of order had been reached. Washington had found a way to sell its bonds — they simply turned things over to an agent, Jay Cooke. It is amazing sometimes what a well run merchandising campaign can do.[7] Interestingly enough, after the government had suffered its earlier failure to sell the loan, it had restored, at least in part, to taxation to finance the war. Even by present day standards the breadth of these taxes is surprising. They increased customs duties, they levied taxes on insurance, bankers, gold watches, brokers, manufacturing, teachers and even incomes.† While these taxes were successful in that they provided needed revenue, one aspect placed an extra burden on importers, such as Giles, Wales & Co. — the new customs duties had to be payed in specie.[8]

Giles, Wales & Co.'s early success during the secession recession placed them in an extremely advantageous position when conditions improved. Instead of recovering from losses as many of their competitors were forced to do they were able by 1863 to convert their profits into additional working capital. However, rather than merely pouring these profits into the firm, it was decided to reorganize into a new and expanded partnership. Unlike the old firm where Giles had provided the capital, each of the four partners in the new arrangement was able to contribute funds. Frederick was still the major source of financing, providing eighteen thousand dollars. William Wales, using his earn-

---

*See Table I, page 37.
†*Industry Comes of Age*, pp. 22-24. Kirkland states on p. 23 in reference to the income tax that "For a time toward the close of the war, the statutory exemption was $600, the sum required to provide a family 'with the bare necessities of life,' and the rate reached 10 per cent upon incomes between $600 and $5,000, 12.5 per cent on the excess above $5,000, and 15 per cent on the excess above $10,000.

8. North side of Maiden Lane, from corner of Broadway, No. 1 through No. 19, circa 1872.

9. North side of Maiden Lane, No. 21 through No. 39, circa 1872.

ings from the first year and a half of business, put up nine thousand five hundred dollars. Of the new partners, Fayette Stratton Giles, Frederick's younger brother, contributed one thousand dollars; while Julia's brother, George Channing Fuller Wright, who had been a clerk in the old firm, provided nine hundred dollars. As before, the arrangement was on a share and share alike basis, that is with the exception of return on capital. Each partner was to receive a return of ten per cent per annum on the amount he had invested. The reorganization did not stop, however, with the refurbished partnership, but also included a new store. It was not to be a very distant move. In fact it was only around the corner, from one side of the Howard to just a few doors beyond on the other. Thus began the twenty-one-year occupancy of 13 Maiden Lane by Giles, Wales & Co. and their successors.[9]

By the mid 19th Century, New York's jewelry industry had begun to spread northward from Maiden Lane to Fulton and John Streets. However, its most important part remained the two narrow blocks of Maiden Lane that led east from Broadway to William Street.

Here could be found the headquarters and sales offices of the wholesale and manufacturing jewelry industry in the United States. Those unfamiliar with the business practices which result from the high cost of land and floorspace in New York City might assume that these blocks were quite long or that the buildings used by the industry were quite tall, but neither was the case. Businessmen in that city had long ago reduced the maximum utilization of square footage to an art form. Even today visitors are often taken back by the tiny floor space that many firms occupy. Although Maiden Lane is no longer the jewelry center and most of the buildings of a century ago are gone, there are areas of lower Manhattan such as Mulberry Street where similar ancient four- to six-story brick commercial structures still stand.* In the old days the ground floor level

---

*Most of the buildings on Maiden Lane were brick with cast iron facades. Much information about these buildings may be found in *The Origins of Cast Iron Architecture in America*, New York, 1970. (This is largely a reprint of the Badger Iron Works 1865 catalog.)

10. *South side of Maiden Lane, from corner of Broadway, No. 1½ through No. 18, circa 1872.*

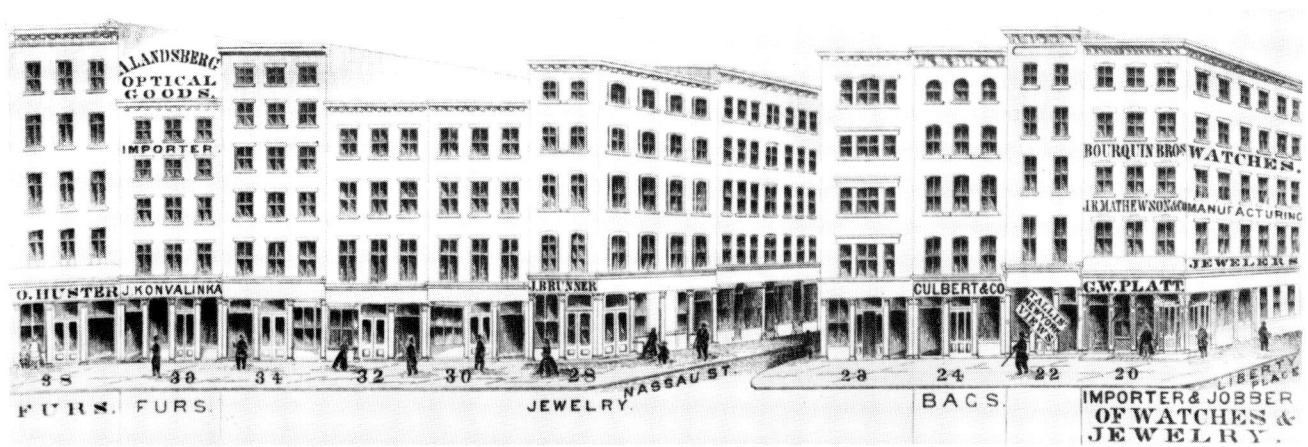

11. *South side of Maiden Lane, No. 20 through No. 38, circa 1872.*

of such buildings would be occupied by those firms that could afford the luxury of show windows. Above were those who could or must avoid such expense. The four-story establishment at number 13, for instance, was occupied by Giles, Wales & Co. on the ground floor and the jeweler's material house of Sussfeld Lorsch and Company on the upper three stories. Actually, number 13 was almost unique in that it had only two occupants. Across the street at number 16 no less than nine firms filled the premises, including Joseph Fahys; Falkman, Pollack & Co.; Albert J. Pratt, and Samuel D. Burbank. Down the block another four-story marvel of space utilization, number 21, housed the operations of Cross & Beguelin; Louis S. Fellows & Schell; Marius J. Paillard; Henry Harrison; Church, Metcalf & Co.; and Steere & Crooker. And so it went from building to building.[10]

Perhaps if a certain degree of claustrophobia has not begun to clutch the more sensitive reader, then a bit of commentary about the nature of the Maiden Lane jewelry trade is in order. Jewelry is really much too limited a term to describe the nature of the businesses that were centered there. Although importers of watches and jewelry made up the largest contingent, there were also manufacturing jewelers, jobbers, clock dealers, watch case makers, dealers in American watches, hair jewelry and music boxes. Further, if the out-of-town buyer was unsatisfied with this list, he could extend it by visiting the adjoining streets where, besides more of the aforementioned, were chronometer makers, refiners, silversmiths and even a case spring maker.[11] Despite the fact that many dealers listed themselves as manufacturers, most of the large scale operations had, by the 1860's, been moved elsewhere, in particular to Newark. Still, most firms found it expedient to do at least some of their manufacturing on Maiden Lane. It might be assumed, for instance, that dealers and importers of watches had little to do outside of a stray custom case or so, but nothing could be further from the truth. Although English watches generally arrived cased, Swiss and American ones did not. Further, lower grade Swiss watches were often imported unfinished, and as tariffs rose these watches were imported as parts.[12]

Where and how Giles, Wales & Co. obtained their

original stock of merchandise in 1861 is not entirely clear. Although they referred to themselves as importers, their later practices suggest that the firm purchased its fancy goods and jewelry locally. Watches, however, were another matter — these were imported from both England and Switzerland. And unlike jewelry they were never purchased for resale, but taken on consignment.[13] In truth, even if Giles, Wales had wanted to do otherwise their thin capital made it impossible.* Of course they could have purchased on credit, but at that point in their history it was against the firm's policy.[14] Aside from the financial details there remains still another question — who made the arrangements with the various manufacturers? Did Frederick go to Europe to conduct the initial negotiations or were they handled by an agent?

The question of whether Frederick went to Europe in 1860 in order to make the necessary contacts revolves around a single reference. One of his obituaries states that, "In 1860, he took passage in the Great Eastern and went to Switzerland, where he established an agency that made their firm the most extensive importers of the highest grade of Swiss watches."[15] Since it is a certainty that Frederick was in Europe three years later and there is no other supporting data for an 1860 trip, one is led at first to the conclusion that the report was merely in error about the date. After all, rounding off numbers is a pretty common occurrence. This notion will not work, however, for the great iron ship was not available for Frederick's 1863 visit to Switzerland. In August of 1862 the Great Eastern struck a rock off New York and did not resume service until after Giles had arrived in Chaux-de-Fonds.[16] Thus we are left with no alternative but to accept the possibility of an 1860 trip.

There is another reason for believing that Frederick started his business by traveling aboard the Great Eastern — it fits his personality. Never willing to settle for less when more would do, he was ever a disciple of the cult of bigness. Isambard Kingdom Brunel's creation, the absolute leviathan of its age, met and set any standard for size. This twenty-two thousand five hundred-ton ship completely dwarfed anything that plied or would ply the seas for the next fifty years. As might be expected, the monster ship was more of a curiosity than a commercial success. In essence it should have served Frederick as a caution about size and its abuse, but he was a young man dazzled by inexperience.

While conjecture might surround Frederick's possible trip to Switzerland in 1860, there can be no doubt about his visit there in 1863. This is one of those periods from which several pieces of his correspondence have survived. Shortly after the reorganization of Giles, Wales & Co. in March he was on his way to Europe. Julia did not accompany him, but returned to Montague where Martha, Charles' wife, was expecting a child.[17] There were, of course, other reasons for her not traveling to Europe; the voyage, always dangerous, now had the additional danger of Confederate commerce raiders.* Anyway Frederick, absorbed in business, had little time for other concerns.

The summer of 1863 was a season of difficulty. In the United States the war had reached a thunderous pitch with the events of Vicksburg, Gettysburg and the Draft Riots happening almost simultaneously. At Locle, Frederick had heard vague hints of the gathering storm, but since the news was three weeks behind he could only express his misgivings at the distant thunder in his letters. His business affairs in Switzerland had also produced a storm — of lesser proportions, but of no less concern to him. On Wednesday, July 1, he reported in a letter to his partners that, "I went yesterday to St. Imier to see Kiesner. He says now he will consign his goods direct to us, but the N. E. Co. will undoubtedly use all of their influence to prevent it, so it is for the interest of their own stock to prevent our obtaining goods to compete with them. They are quite bitter against Mr. Favre, for having taken the liberty to move his office without waiting for their consent. They are to have a meeting tomorrow at which I suppose they will come to some conclusion as to the course they will pursue. Kiesner says he has 180 doz. watches nearly ready which he will send us. Still I don't count on it as certain, for I am confident that if he should see any of the Co. they would use every means in their power to prevent his consigning goods to us. Whether they would succeed or not I don't know. We have this much in our favor, Kiesner does not like them and is well disposed toward us as he is convinced we have the ability to sell his goods in large quantities. David Perret also says he will give us all the goods we want. Have been to Chaux de Fonds today. Have been laying a good many ropes and shall commence to pull them soon."[18]

At least one of Frederick's clients decided that Giles, Wales & Co. was not the place to consign his watches. Frederick reported that, "In the matter of Roulets Watches which I wrote you about in a former letter, you need give yourself no extra trouble to sell them in preference to other goods as he will *not consign* any more under any circumstances."[19] This loss was minor, however, since that firm's products were really not up to the standard that Giles was trying to maintain.

Sometimes ignorance is not only a blessing, but it can also produce ironies that make the best of men seem ridiculous in hindsight. The week of July 12, 1863, was to witness one of the more violent civil upheavals in American history, the New York Draft Riots. In Switzerland on that same date began a ten-day "Federal Shooting" celebration. Frederick, since he was ignorant of the former, chafed about the latter which had totally interrupted his program of wooing the Swiss manufacturers. On Saturday, the 18th, he was forced

---

*This fact is quite evident from the firm's bankruptcy records. Incidentally, Trows *New York City Directory 1862-3* in the instance of George W. Platt describes fancy goods as, "Cutlery, combs, beads, bags, brushes, china, toys, perfumes, tea and coffee sets."

*Frederick's letters show considerable concern about Confederate raiders, and perhaps this worry was not totally unfounded — a year later, June 19, 1864, the U.S.S. Kearsarge sank the commerce raider C.S.S. Alabama off Cherbourg.

ABOVE. 12. Dial view, "Giles, Wales & Co., Chaux-de-Fonds" two-train jump-quarter chronograph and minute repeater, in heavy 18k hunting case over 57 mm in diameter. Frederick Giles established an agency in Chaux-de-Fonds in 1860 that was to help make their firm an extensive importer of high grade Swiss watches. BELOW. 13. Beautifully enameled dust cover is inscribed "SELF WINDER, Independent Seconds, 43 Ruby Jewels, Giles, Wales & Co., Chaux-de-Fonds, Suisse, No. 10111." Note buttons in the band for setting the watch and operating the chronograph. Frederick Giles may have used these high grade Swiss products to help develop his own ideas for stem winding and button setting watches.

14. Movement view, large 49 mm nickel movement is fully jewelled in gold settings. Gold train wheels. Both the time train and independent chronograph train are wound by turning the crown right and left. Watches of this quality were probably among the highest priced products sold by Giles, Wales & Co.

to report to Julia, "I have become completely disgusted with trying to do business. No one is at home, or if by chance we find a man at home his head is so full of the Festival that it is impossible to get him to talk of business. If you will just imagine for a moment a fourth of July Celebration in a New England Country village to last ten days without interruption and you will have a faint idea (on a small scale) of what they are doing here now. This shooting is attended by people from all parts of Europe. It is estimated that the daily arrivals and departures at Chaux de Fonds will reach the number of 25,000 persons. On Sunday, the opening day, there was sold at the "Cantine" (a large building open at the sides, fitted up with tables and seats for dining, accommodating about 5,000 persons at a time) between the hours of 10 AM & 5 PM 26,000 bottles of wine!"[20]

It had been a good spring and early summer at 13 Maiden Lane. Frederick, commenting from Switzerland, estimated that their gross since March 1st had been eighty to ninety thousand dollars. He further felt that this would have increased their worth from the original thirty thousand to over forty thousand dollars.[21] Alas, neither his ebullience nor growing wealth was shared by all of his fellow New Yorkers. James D. McCabe, Jr., was to report in 1872 that, "The Civil War checked the growth and trade of the city, which languished during the entire struggle."[22] particularly hard hit by the war were the poor and working classes. Although their wages rose, the inflation rose faster.[23] Then in March, 1863, Congress passed the "Conscription Act." For many this was the last injustice of a war they felt no part in. Most galling was the fact that any man rich enough to spare three hundred dollars could literally buy his way out of the draft.

Conscription began in New York on Saturday, July 11, 1863. At first there were only angry mutters, then on the hot Monday which followed, things literally exploded. A volunteer fire company, known as the "Black Joke," enraged by the drafting of their leader on Saturday led an assault on the Draft Office at 46th Street and 3rd Avenue. For the next four days roving mobs laid siege to the city. It was a ghastly affair fought with clubs, guns and finally with cannon and grape shot. There were times when the ferocity of the mobs broke battle hardened, but not street wise, infantry. In the end it was the police and their murderous locust clubs that destroyed the mobs' will.[24]

The riot spared Maiden Lane, but just barely. On Monday night the mobs surged down to Printing House Square, just four blocks north, to lay siege to Horace Greeley and his hated Tribune. For the next three days and nights the battle surged around Maiden Lane, never quite reaching it or the neighboring financial district. It was a period of terror when the clerks and owners of the various jewelry houses sat armed, barricaded in their stores, awaiting the arrival of the burners and looters. In the end Maiden Lane's loss was largely a business one.[25]

After the Draft Riots an uneasy peace returned to the city. Only once, November 25, 1864, was there a serious attempt to destroy this peace. A highly organized band of arsonists (newspapers of the time felt they were Confederate agents) attempted to set fire to a number of hotels, businesses and factories.[26] This time Maiden Lane did not escape. Frederick's residence, the Howard Hotel, was the local target. At 3:30 in the morning the night watchman made an unpleasant discovery in the fourth floor room of a guest registered under the name of S. M. Harner of Philadelphia. Mr. Harner was gone and the contents of the room had been piled on the bed, where they had been saturated with an inflammable substance. Luckily, the watchman had arrived before the fire could get a start.[27] Much to New York's good fortune none of the fires accomplished anything more than burning out a room or two — Atlanta could not be avenged on fire proof buildings.

Despite New York's civil unrest Giles, Wales & Co. managed to prosper. Their stock, which now included the best that Switzerland could provide, also boasted of the Jurgensen Watch, the most expensive then sold in America.[28] Each year their gross sales increased. This burgeoning growth necessitated the establishment of a permanent representative in Switzerland. It seemed only logical to choose Fayette for this position.[29] He was the last member to join the firm and was less experienced in its operations than the other possible choice, George Wright. It never occurred to the senior partners, especially Frederick, that sending a footloose young man on an independent mission might be a mistake. Alas, the decision to send Fayette Stratton Giles to Chaux-de-Fonds was, for him, the beginning of a way of life that was to give him almost expatriate status and was for the firm the first of a series of errors that would ultimately figure in the destruction of Giles, Wales & Co.

CHAPTER 2 REFERENCES

1. Charles W. Moore, *Timing a Century* (Cambridge, 1945), pp. 44-7.
2. Albert Ulmann, *Maiden Lane* (New York), p. 34. Also: Fletcher Pratt, *Civil War in Pictures* (New York, 1955). This work contains an illustration by Winslow Homer of a sutler's store on p. 84.
3. Edward Chase Kirkland, *Industry Comes of Age* (New York, 1961), pp. 14-16. James K. Kindahl in his article "Economic Factors in Specie Resumption: The United States 1865-79," *Journal of Political Economy*, 59 (February, 1961), pp. 30-48, gives a slightly different version of the details of Specie suspension, however the fact of suspension is for our purpose more important than its details.
4. Moore, *Timing a Century*, p. 42.
5. Frederick Giles to Julia M. Wright, November 28, 1861, Burns Collection.
6. Ulmann, *Maiden Lane*, p. 84.
7. Kirkland, *Industry Comes of Age*, pp. 20-22.
8. Harold G. Vatter, *The Drive to Industrial Maturity* (Westport, Conn., 1975), p. 54.
9. Article of Agreement, Giles, Wales & Co., March 1, 1863, New York, N.Y., copy in Burns Collection.
10. Ulman, *Maiden Lane*, p. 86.
11. Ibid., pp. 73-77, 85, 86, 96, 97. Much of this information may also be found in the various trade journals after 1870.
12. "Putting New Watches in Order," *American Horological Journal*, September, 1871, III:63-4; also: Henry Abbott, *Watches and Men* (New York, 1933), p. 5. Further, trade catalogs usually contained several pages of parts for jobbers.
13. Frederick A. Giles to Mess. Giles, Wales & Co., July 1, 1863, Burns Collection.
14. Ibid.
15. "Fred. A. Giles — Obituary," Burns Collection.
16. James Dugan, *The Great Iron Ship* (New York, 1953), Chapters 9-10.
17. Frederick Giles to Julia Giles, July 18, 1863, Burns Collection.
18. Frederick Giles to Mess. Giles, Wales & Co., July 1, 1863, Burns Collection.
19. Ibid.
20. Frederick Giles to Julia Giles, July 18, 1863, Burns Collection.
21. Frederick Giles to Mess. Giles, Wales & Co., July 1, 1863, Burns Collection.
22. James D. McCabe, Jr., *Lights and Shadows of New York Life or, the Sights and Sensations of the Great City* (Philadelphia, 1872), p. 49.
23. Harold G. Vatter, *The Drive to Maturity, The U. S. Economy 1860-1914* (Westport, Conn., 1975), pp. 48-49.
24. Irving Werstein, *July 1863* (New York, 1957); also Herbert Asbury, *The Gangs of New York* (Garden City, 1927), Chapters VII & VIII.
25. Ibid.
26. Asbury, *The Gangs of New York*, pp. 171-3.
27. *New York Times*, November 27, 1864, pp. 1-2.
28. "Fred. A. Giles — Obituary," Burns Collection.
29. "F. S. Giles is at Rest," unidentified clipping from a Chicago newspaper published at the time of Fayette's death in 1897, Burns Collection.

Chapter 3
The Watch Factory Scheme

In February of 1865, after toying with the idea for several years, Giles, Wales & Co. decided to enter the business of watch manufacturing. While the origins of this scheme are unknown, it has already been pointed out that Frederick's former employers, the Gendar brothers, had once considered the idea. The germ of the belief that New York would be a desirable location for a watch factory even antedates them; in 1845 a Swiss, Pierre-Frederic Ingold, had attempted to found one there. Nothing came of it, however Ingold was later to claim that the American Watch Company made use of some of this machinery in its early days.[1] Of course, whether or not Giles was aware of Ingold's efforts is a rather speculative question.* In fact the first hard evidence that Giles, Wales & Co. was contemplating entering the manufacturing business appears in the co-partnership agreement of 1863. This document stated that the business of the firm was to be "Importers, Manufacturers and Jobbers of Fine Watches, Diamonds and Jewelry, American Watches, Silver Ware and Fancy Goods."[2]

The original Charter of the United States Watch Company was drawn in New York City on February 22, 1865. Although Giles, Wales & Co. could have entered the manufacturing business on their own, the large scale of the project they envisioned made a separate corporation a necessity. Not only would financing beyond their resources be needed, but it also seems likely that the firm chose not to gamble its security on so speculative a venture. Curiously, Fayette Stratton Giles was not included among the trustees of the watch company. The five trustees named in the Charter were Frederick A. Giles, William A. Wales, George C. F. Wright, Frank H. Page, and Daniel M. Wells. While it might be plausible to suggest that Fayette was omitted because of his absence in Switzerland, this is open to question inasmuch as neither Page nor Wells were signatories to the document. It seems more likely that the omission of Fayette was an additional safeguard in the event that the factory should fail. The total capitalization of the United States Watch Company was at this point five hundred thousand dollars; of this total two hundred fifty thousand dollars was in cash and the remainder was in manufacturing interests and future patents. Interestingly enough no mention is made of the future home of the watch factory, Bergen, New Jersey. Only New York City and Newark, New Jersey, are named as places of business. Frederick and his colleagues were gambling heavily on the success of a watch which may not have been totally designed as yet, whose key features were most certainly not protected by patent and whose place of manufacture was yet to be settled.[3]

By 1865 the concept of producing watches from interchangeable parts was almost ninety years old. As early as 1776 a Frenchman, Frederic Japy, was experimenting with machinery to produce ebauches (a rough or unfinished watch movement). In 1799 he patented a series of machines for this purpose.[4] From that time on, the dream of using unskilled or semi-skilled workers to operate machinery which would in turn produce the parts necessary for the construction of watches intrigued would-be promoters. Ingold, it is said, attempted to start factories in Switzerland in 1830, Paris in 1835 and England in 1840, as well as in New York.[5] While Ingold's attempts created far more machinery than watches, the first American venture achieved at least partial success. In 1838 Henry and James Pitkin of Hartford, Connecticut began producing watches with machinery. After manufacturing about eight hundred watches they were forced to discontinue production due to the fact that they could not compete with foreign hand produced timepieces. A key aspect of their failure was their inability to achieve interchangeability. This, however, was in some respects to remain an elusive goal for decades to come.[6] It remained for Aaron L. Dennison and Edward Howard to found the first truly successful enterprise, the Warren Manufacturing Company of Roxbury, Massachusetts, in 1850. After a move to Waltham, Massachusetts, and a change of name to the Boston Watch Company, failure in the form of bankruptcy caught up with this firm in 1857. The fates, which were seldom kind to American watch factories, smiled on this occasion and the enterprise, instead of dying, multiplied. Within two years the factory at Waltham, with new backing, became the American Watch Company. At the same time Edward Howard, on whose clock factory grounds stood the original Roxbury factory, began manufacturing watches on his own.[7] Both ventures, however, were to endure years of financial difficulty. The American Watch Company catering primarily to the low and medium grade markets profited greatly from the war and at last achieved success. Unfortunately, E. Howard & Co., which produced high grade watches, found the going extremely difficult. It survived for many years, but only marginally — saved by the fact that no matter what else happened Howard clocks were always enough in demand to support the watch enterprise.[8]

The difficulties of the pioneer watch factories should have stood as a warning to future promoters, however

---
*The timing of the Ingold and Gendar ventures is interesting — Ingold 1845 and the Gendars 1848-50.

the lessons were seldom understood until it was too late. Survival meant that four interrelated problems had to be overcome. These were scarcity of capital, lack of trained managerial personnel, insufficient technological development and foreign competition — to which may be added domestic competition as future American factories appeared. Since Giles, Wales & Co. had chosen to move, at least in part, from the category of foreign to domestic competition, it seems reasonable to examine the state of European watch manufacturing in 1865. At that time the bulk of such industry was concentrated in England and Switzerland. For the past quarter century the Swiss had, by means of sheer effort and a plentiful supply of skilled labor, managed to gain an ascendancy in world markets. Further, although they were behind the United States in developing centralized factories, it was a lag which had resulted from strength rather than weakness. Key to this power was their technique of operating with a decentralized industry utilizing small shops of specialists working on a subcontracting basis.* This approach, as any student of mass production methods will realize, is generally the alternative to the centralized factory. Properly managed, it possesses certain advantages in terms of labor relations, capitalization and flexibility. If there was any real weakness in the Swiss front, it was that they had concentrated the bulk of their efforts in the high and low priced fields. The English, on the other hand, made largely high and medium grade watches. Unfortunately, they had grown fat on their successes of the late 18th and early 19th Centuries. Having pioneered many of the key aspects of the modern watch it was difficult for them to accept that by the 1860's they were marketing outmoded designs produced by inefficient cottage methods.† Thus it was England which suffered first when the Americans began to build watch factories — fortified from 1862 on by a wall of protective tariffs.

Hardly two years elapsed between that fateful sheriff's sale in 1857 that had brought forth the American and Howard companies and the birth of another would be competitor, the Nashua Watch Company of Nashua, New Hampshire. The timing was, of course, impossible. So depressed was the market for high grade American-made watches that it could not absorb the few that the American Watch Company manufactured — the Howard Company was nearing bankruptcy.[9] Aside from the economic situation, Nashua had also had technical problems, particularly with its escapement design.[10] Not an altogether surprising difficulty for a new company, but American watch manufacturers seldom got

---

*Even well after the turn of the 20th Century one is awed by the number of small specialized shops that appear in directories such as Fellhauer-Calame's *Grand Indicateur Complet De L'Industrie Horlogere* which was published at Bienne beginning in 1905.

†*Timing a Century*, pp. 64-66. The reader should not be confused between the Swiss method of small subcontractors and the English cottage approach to manufacturing. Although there is some similarity between these methods, the Swiss approach is much more akin to the jobbing shop approach.

American Watches

For Soldiers

AT REDUCED PRICES.

American Watches for Americans!

THE AMERICAN WATCH COMPANY give notice that they have lately issued a new style of Watch, expressly designed for Soldiers and others who desire a good watch at a moderate price. These watches are intended to displace the worthless, cheap watches of British and Swiss manufacture with which the country is flooded, and which were never expected to keep time when they were made, being refuse manufactures sent to this country because unsalable at home, and used here only for *jockeying* and *swindling* purposes.

We offer to sell our Watch, which is of THE MOST SUBSTANTIAL MANUFACTURE, AN ACCURATE AND DURABLE TIME-KEEPER, and in Sterling Silver Cases, Hunting pattern, at as low a price as is asked for the fancy-named *Ancres* and *Lepines* of foreign make, already referred to.

We have named the new series of Watches, WM. ELLERY, Boston, Mass., which name will be found on the plate of every watch of this manufacture, and is one of our trade-marks.

Sold by all respectable watch dealers in the loyal States. Wholesale orders should be addressed to

ROBBINS & APPLETON,

Agents of the American Watch Company,

182 BROADWAY, N. Y.

15. American Watch Company ad, February 28, 1863, "Harper's Weekly." Growth in demand for watches that accompanied the Civil War helped produce a four-fold increase in sales for Waltham in 1863.

time to eliminate the bugs from their products before the money ran out. Nashua's promoters were, however, more fortunate than most, the American Watch Company bought them out in 1862.

Despite the long and bloody siege that gripped the Union and Confederate armies at Petersburg, it was evident to the movers and shakers of the young American watch manufacturing community that times were ripe for fresh ventures. 1864 saw no less than three new firms start up, the National Watch Company at Elgin, Illinois, the Mozart Watch Company at Providence, Rhode Island, and the Tremont Watch Company at Boston.[11] A year later, of course, came the United States Watch Company, followed by Robert Schell & Co.'s Newark Watch Works at Newark, New Jersey, in 1866.* While the factors of inflation, tariffs, population growth, expansion in the rail industry and rising technology played their parts in this sudden flowering, the locations of these companies are strong evidence that other states were beginning to challenge Massachusetts' ascendancy in the watch manufacturing industry. It is quite demonstrable, however, that these new firms were extremely dependent upon that state's established watch manufactories for a cadre of upper echelon managers and technicians. This supply of talent unfortunately was so limited that only the Elgin Company was born without some serious defect. The survival of the remaining firms was dependent upon the amenability of their difficulties to correction.†

Students of the history of watch manufacturing have often cited the American Watch Company's immense profits for the years 1864 and 1865 as the principal motivational factor for the incorporation of new factories during the 1860's. Dating, however, would tend to suggest that this generalization is open to question. In the instance of the incorporation of the United States Watch Company, the economic condition of other domestic producers seems to have been of minimal influence. It has previously been noted that Frederick Giles and his associates were considering watch manufacturing at least as early as February, 1863. Yet, at that date Waltham's balance sheets were at about the break-even point and the Howard Company was in great financial difficulty. In fact, Giles, Wales & Co. did better in gross sales and profits than the American Watch Company until well into 1864. It seems, thus,

---

*There is some debate about this firm's history. Crossman states that it was organized in 1863, however, Dr. Percy Livingston Small suggests in the NAWCC BULLETIN 4: 257, that the date of 1866 is more reasonable.

†Only two firms, Elgin and the New York Watch Company, would have extended corporate histories.

Table I
Sales & Profits
Giles, Wales & Co. vs. Waltham Watch Co.
1862 - 1865

	Waltham Watch Co.		*Giles, Wales & Co.*	
	Sales	*Profits*	*Sales*	*Profits**
1862	$ 61,465.00	$ -4,010.00	$117,104.72	$18,068.42
1863	$242,778.00	$ 36,050.00	$306,797.57	$47,310.48
1864	$576,885.00	$124,577.00	$390,106.52	$60,157.34
1865	$838,534.00	$491,573.00	$507,027.55	$78,187.42

*These profits are approximate since Giles gives as only a total ($741,729.86) for 1862-73.

Sources: Moore, *Timing a Century*, pg. 50 and Giles, "Memorandum — Sales & Profits, Giles, Wales & Co., 1862-73."

---

more reasonable to accept at face value Frederick's statement that he had planned to ultimately enter the watch manufacturing field as early as his years with the Platts.[12] This would, of course, infer that the primary reason for the incorporation of the United States Watch Company was simply the desire to possess or control a watch factory. The timing of the implementation of this wish was controlled by Giles, Wales' ability to provide sufficient working capital and their evaluation of economic conditions. Unfortunately, Frederick seems to have assumed that the wartime inflation would continue and thus everything had to be done in haste to keep costs within reason.[13]

Frederick had always worked long hours, but the pace he adopted after February was brutal even for a thirty-one-year-old man in excellent health. The watch factory had gotten off to a late start and now there were others racing ahead in the field. His timetable required that everything had to be done at once. A marketable watch had to be designed and patented, the land for a factory and housing for the employees had to be acquired, men had to be hired from Waltham to assist in the tooling operations, and still there were the demanding responsibilities at the store which he had to carry on.[14] There were, of course, Billy Wales and George Wright, but Frederick needed their assistance at the store. What he really wanted at this point was someone who could be trusted to act as agent for the tooling operation at Newark. Fortunately, Isaac Henry Wright, usually called Henry, was available. Frederick, who often had trouble in delegating authority, felt differently about Julia's older brother. There was a bond between these men that would last no matter to what shocks it might be subjected.

There are a number of unsolved riddles contained in the history of the Marion venture and Henry's total role in our story is one of these. It is known that he and his brother George, were ultimately to become Frederick's most trusted aides. Yet details about Henry's activities are almost totally lacking. What is known is that he was Dr. George and Julia B. Wright's eldest surviving child, born March 19, 1829.[15] It is suspected that he first entered the jewelry trade in Boston during the 1850's. In August of 1853 he was seriously injured in a driving accident. This incident was to leave him a partial cripple and in constant pain for the rest of his life.[16] Pain, however, did not deter him from pursuing an active business career. By the mid-1860's he had risen to the position of head salesman in the wholesale department of the jewelry firm of Jordan & Marsh Co. of Boston.[17] Further, the fact that on September 11, 1865, he married Georgeanna Baker would suggest that he did not allow his physical problems to interfere with his family life either.[18] Thus one critical fact about the man Frederick selected to manage the Newark tooling operation does emerge, and that is that Henry Wright shared his brother-in-law's ability to stand up to the most difficult odds.[19]

The reason that Newark, New Jersey, had been mentioned in the United States Watch Company's charter was that it was there the company proposed to construct their machinery. While there were a number of machine shops within reasonable commuting distance of Maiden

Lane where this could have been carried out, Newark possessed a facility uniquely suited to this purpose. Charles W. Dickinson and George Rowden had established a shop to build geometrical and engraving lathes at 54-56 Hamilton in 1863.[20] Newark with its growing jewelry manufacturing industry had need for such a firm. By 1865, the young machine shop was beginning to achieve a reputation for its ability; in fact during the previous December Dickinson had received a patent for an engine turning lathe (#45,455, December 13, 1864). Now they were about to take part in one of the largest watch tool building ventures ever attempted — the United States Watch Factory.

In 1865, experience with the building of watchmaking tools and machinery was limited to a small number of men in America. Only fourteen years earlier James M. Bottum had recieved his patent for a foot-powered wax lathe (#8216, July 15, 1851). Although this was basically a watch repairman's machine, watch factory lathes were only slightly more advanced.[21] When Dennison and Howard began their venture in 1850, they found that practically everything they needed had to be invented. The knowledge which Ingold and the Pitkins had acquired at great cost had largely been lost.[22] Edward Howard was later to state, "We did not know how to make a jewel, or a dial, or a tempered hairspring, or to do proper watch-gilding or produce a mirror finish on steel."[23] To which one might add neither could they at first achieve accuracy to the necessary ten thousandth of an inch and maintain high production as well.

Despite the fact that Maudslay and others had developed that basic machine, the lathe, to a high level of perfection, it was by no means adequate to the watch factory's demand for high speed precision work. Its primary weakness was that no one had yet devised a workable method for holding small cylindrical objects. One of the two available techniques required that the stock be either center drilled or pointed on both ends and then revolved between dead centers by means of a dog or small pulley. This method worked well enough, but the several hand operations involved placed severe restrictions on production. The only other suitable approach was to secure the stock to a live center by means of a small daub of shellac. Again the problem was set up time. A year after the failure of Howard and Dennison's enterprise, a Waltham, Massachusetts, manufacturer of small tools, George W. Daniels, achieved a break-through in this area.* On October 19, 1858, Daniels received a patent for a "Lathe for Cutting Screws from Wire" (#21,864). Actually the patent did not cover an entire lathe, but merely the spindle and its fittings. The spindle had been bored lengthwise, terminating in a conical opening at its front. Inserted into the spindle and running the length of the bore was a hollow steel tube. At its front the tube, called a spring collet, was shaped to match the conical mouth of the spindle and was sliced to form four equal lips. The rear of the tube was made slightly longer than the spindle and was threaded to accept a hand wheel. In operation, tightening the hand wheel would pull the tube into the spindle causing the lips to be forced inward so that they would grip any item of stock that had been placed in the mouth of the tube. Loosening the hand wheel would allow the lips to spring outward freeing the stock. The only real limitation of the spring collet is, as every machinist knows, that the stock must be relatively close fitting to the bore of the collet. This hardly represents a problem in a factory setting inasmuch as special collets to handle problem diameters could be produced at will.

As yet it is uncertain how Daniels' invention metamorphosed from its original form into the first modern instrument and jeweler's lathe, the American Patent Combination Lathe. Perhaps as some suggest this transition took place in the machine shop of the American Watch Company;[24] or it may be that John Stark, the first independent builder of watch factory tooling, was correct when he claimed that he had built the first of these lathes.* It is also possible that Daniels himself may have seen this application of his invention and produced the first of these machines. Certainly, all known examples of these lathes are stamped with the date of his patent and at least one of his own manufacture has surfaced in recent years. Whatever, during this process of change the collet was likewise altered to its modern form. The number of lips was reduced from four to three. Also the original hand wheel was extended so that it entered the spindle to form the now familiar hollow drawbar. This in turn permitted the collet, formed from expensive carbon steel, to be shortened to more practical proportions. There was still, however, one important flaw in this key invention in the history of high speed precision machines. The operation of closing the collet caused it to retract into the spindle and in so doing the stock was pulled backward out of position making it difficult to use any type of automatic cutting set up. In 1865 John Stark was the first to receive a patent on the concept that called for a segment of the spindle to go forward to close the collet, thus leaving the stock undisturbed (#47,997, May 30, 1865).† Later inventors merely concentrated on making Stark's plan more efficient. The invention of the spring collet and the modern instrument lathe by no means solved all of the watch manufacturers' problems. Rather, it was the beginning of efficient tooling, but years of costly experimentation lay ahead.

The process of hiring a crew of experts to assist in the tooling of the factory had gone quickly and before

---

*Biographical information about George W. Daniels is scant; see: Edmund L. Sanderson, *Waltham Industries* (Waltham, 1957) p. 44. A few tools of his manufacture are known to exist in collection, including: Two eyelet setting pliers, a hand vise, dividers, an American Patent Combination Lathe, and possibly an incomplete headstock for another.

---

*These claims are to be found in older ca. 1918 Stark Tool Company catalogues.

†This patent is also interesting because it acknowledges the Daniels Patent.

the beginning of May, 1865, work was underway at Dickinson & Rowden's machine shop.[25] Charles S. Crossman in his somewhat flawed account of this phase of the United States Watch Company's history states that this team consisted of James H. Gerry, Master Mechanic and ten machinists from the Waltham machine shop.[26] Of this group only three can be identified with certainty: David B. Gerry, Emery J. Gerry and John Pray.[27] There is a great deal more about the tooling phase than the names of seven men, however, that has slipped into oblivion. We do not know, for instance, whether the United States Watch Company built or purchased its general machines such as punch presses and lathes. While either course could have been followed, the latter seems most likely. The factors of time and cost, and the fact that it was becoming common practice for watch factories to purchase this classification of machinery tend to support this belief.

There is still another question about the approximately twenty-seven months of the tooling phase which remains unanswered. The jigs, fixtures, dies and special machines of mass production are and were in 1865, each designed to perform a specific operation or series of operations on a specific part. None of these tools can be built until an acurate set of parts dimensions exist. There are, no doubt, readers who will question the applicability of this notion of process to watch manufacture of the late 1860's. Certainly, the lack of primary source material has made this view somewhat conjectural until recently. Fortunately, the inventories of the New York Watch Company for December 31, 1872 and 1875, were recently unearthed by Richard Ziebell.[28] The specific tool, jig, die and quill listings found in these documents clearly support the view that watch factory engineering of the period was totally orthodox by modern standards. These inventories also indicate that most operations were conducted on standard equipment, such as Stark lathes, that had been adapted to perform specific operations in a semi-automatic fashion. Although the records for the United States factory have never been located, every known source indicates that this factory followed contemporary standard practice in its tooling.[29] Which thus brings us down to this question: If the watch which the Company placed on the market in July of 1867 was not patented until March 13, 1866 (Design Patent #2281), then just what watch was tooling being constructed for in the Spring of 1865?[30]

One might argue that Frederick Giles and the United States Watch Company simply did not patent his design for almost a year or that the Patent Office delayed the patent. The latter notion will not work, however, because there is a Patent Office receipt stamp on the original drawing giving the date of February 15, 1866. As for the former argument, it would require one to believe that a company which did its research behind locked doors would not seek some way to protect an invention in which it intended to invest a great deal of money. Further, inasmuch as the model for this design was constructed by Oliver J. Baldwin at George W. Platt & Co., 20 Maiden Lane, no real security short of patenting could have been achieved.[31] If it is thus illogical to assume that the full plate model of 1866 was the watch for which the Gerry team was building tooling in 1865, then there must have been another design.*

Although no example of a watch that might be considered to be the United States Watch Company's model of 1865 has been located, the evidence that such a design once existed is extremely strong. It seems most likely that this model would have been a three-quarter plate stem-wind movement based on Frederick Giles' patents of April 25 (#47,412) and May 9, 1865 (Design Patent #2055). Covered by the first patent was a design for a stem-winding mechanism which was essentially a button-set variation of Adrien Philippe's of 1842.[31] Unfortunately, the Giles' version retained the weakness of Philippe's design of allowing the stem to turn in only one direction, and added a few faults of its own. The other patent was for a three-quarter plate arrangement that has to be unique. Basically, the intent was to produce a design where the plate and cock are joined by the regulator index. This would reduce the number of supporting pillars and screws for the plate and cock to three. There are at least two other reasons beyond the patents for the belief that the company attempted to place this decorative and airy movement into manufacture. First, United States Watch Company stock

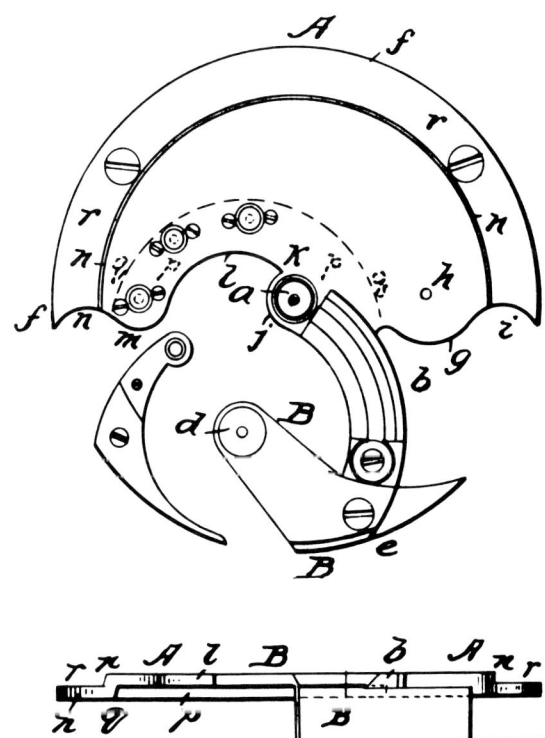

16. F. A. Giles Design Patent 2055 Drawing.

Harper's Weekly, July 25, 1870, p. 414, also *The Gazette,* June 18, 1870, p. 2. The United States Watch Company was hardly alone in its penchant for secrecy; in his revised version of "The Watch Factories of America," Henry G. Abbott relates how the National Watch Company suppressed an unauthorized newspaper article about the model for its first watch.

17. USWC stock certificate with movement illustration based on Giles design patent 2055.

certificates were decorated with an engraving of this design in completed form. Further, this illustration shows evidence of production modifications. Most noticeable of these is that the curved edge which sweeps from the train through the regulator index to the cock has been considerably smoothed out to form a regular elipse. Unhappily, in this process of change the tail of the cock became somewhat awkward in appearance. Worse still, the primary weakness of the design — the balance wheel could not be removed without disturbing the train — had not been attended to.* The other reason for supposing that construction of these movements was attempted is that both United States Watch Company advertisements and surviving watches indicate that the 1866 full plate serial numbers begin at 1001 — thus leaving a tantalizing gap of 1000 numbers.

---

*There are, of course, several solutions to this problem, such as a sub-plate on the pillar plate or a reversal of the cock to regulator index relationship coupled with large resting surfaces on the pillar or backplate, but until or unless a movement is found we will never know if the problem worked out.

The fate of the model of 1865 was sealed even before it was patented, for time was against Giles. John C. Adams, Patton S. Bartlett and Ira G. Blake had already skimmed the cream of the available men at Waltham for their National Watch Company at Elgin, Illinois.[33] The men that the directors of the United States Watch Company were able to employ were excellent, but generally lacked the seasoning for such an undertaking. One has only to compare the two master mechanics; National acquired the talents of the extremely experienced and highly skilled watch and machine designer, Charles S. Moseley, whereas United States got James H. Gerry, who had never held the post of master mechanic before and who was still several years away from his best work as a designer.[34] Further, Frederick Giles could be as demanding as he was inexperienced in the business of operating a watch factory. It is therefore not altogether surprising that problems developed almost immediately at Newark. The first design for the stem-winding mechanism, which was to be one of the main selling points of the United States watch, had to be discarded. By August of 1865 a rocking bar type stem-wind design, a variant on Louis Audemars' classic pattern, was patented and adopted (#49,397, August 15, 1865).[35] Once again, however, they had attempted to cheap it out by using fewer intermediate gears than normal and the operating characteristics were not optimum. Work on the rest of the three-quarter plate was not going well either; its complex shape was proving difficult to manufacture. Considering the interlocking plate and cock, assembly of the movement must have taken the patience of Job. By this time, it has to be assumed that matters between the directors of the Company and James Gerry must have become rather strained, hence the letting out of the task of building the model for the 1866 movement design.

CHAPTER 3 REFERENCES

1. Paul M. Chamberlain, *It's About Time* (New York, 1941), pp. 416-17.
2. Article of Agreement, Giles, Wales & Co., March 1, 1863, New York, N.Y., copy in the Burns Collection.
3. Articles of Incorporation, *United States Watch Company*, New York, N.Y., February 22, 1865.
4. Cecil Clutton and George Daniels, *Watches* (New York, 1965), p. 137, and G. H. Baillie et al, *Britten's Old Clocks and Watches and Their Makers* (New York, 1956), p. 412.
5. Dr. Leonard Waldo, "The Mechanical Art of American Watch Making," *Van Nostrand's Engineering Magazine* 35:52-53; and Chamberlain, pp. 416-417.
6. Crossman, Adams Brown Reprint, pp. 4-7.
7. Moore, *Timing a Century*, Chapters 1 & 2; also Chauncey M. Depew, ed., *One Hundred Years of American Commerce* (New York, 1968), Chapter 82.
8. Dana Blackwell, the former vice-president of E. Howard Clock Co., is the source of this information. There is, as well, a great deal of data from other sources to suggest this view.
9. Moore, *Timing a Century*, Chapter IV, also p. 66.
10. Crossman, p. 68. Also: Frederick Mudge Selchow, "1812 — Belding Dart Bingham — 1878, 1859 — The Nashua Watch Company — 1862," *NAWCC* BULLETIN 17: 550-553.
11. Crossman, pp. 91, 104 and 109.
12. *The Gazette*, June 18, 1870, p. 2.
13. Frederick Giles to Julia M. Giles, August 6, 1865, Burns Collection.
14. Frederick Giles to Julia Giles, August 8, 1865, Burns Collection.
15. Wright Family Bible; Isaac Henry was preceded by another Isaac Henry, born July 4, 1826, died December 15, 1827.
16. From an obituary, original source unknown, information supplied by Carolyn Burns.
17. Ibid.
18. Wright Family Bible.
19. James Gopsill, *Newark City Directory 1865-6*.
20. Charles H. Folwell, *Newark City Directory 1863-4*; Gopsill, *Newark City Directories 1864-8*.
21. E. A. Marsh, *The Evolution of Automatic Machinery as Applied to the Manufacture of Watches* (Chicago, 1896), p. 29.
22. Waldo, *Van Nostrand's Engineering Magazine* 35, pp. 53-54.
23. Chauncey M. Depew, *One Hundred Years of American Commerce*, p. 542.
24. Edmund L. Sanderson, *Waltham Industries* (Waltham, 1957), pp. 45-6.
25. Gopsill, *Newark City Directory 1865-6*.
26. Crossman, Adams Brown Reprint, p. 83.
27. Gopsill, *Newark City Directory 1865-6*. The reason that Crossman and Abbott's lists of foremen are not used is that city directory entries suggest that many of these men served the company at different times and were not necessarily contemporary with one another.
28. R. J. Ziebell, *Serial List, New York Watch Company, Springfield, Mass.*, Old Post Office Clock Shop (Ipswich, Mass, 1972).
29. Crossman, p. 85, as well as contemporary articles about the factory are quite clear on this point.
30. Crossman, p. 86, says late summer 1867; company advertisements indicate that July, 1867, is the exact time.
31. Ibid. Also see: W. Barclay Stephens, "The United States Watch Company, Marion, New Jersey," *NAWCC* BULLETIN 14:154.
32. Elsworth H. Goldsmith, "Keyless Watches," *NAWCC* BULLETIN, 59:497-8.
33. Henry G. Abbott, "The Watch Factories of America —Past and Present" (Revised), *American Jeweler* 24:85-86.
34. Crossman, p. 67; Abbott, 1888 edition, pp. 33-36; Robert S. Woodbury, *History of the Grinding Machine* (Cambridge, 1959), pp. 51-53.
35. Goldsmith, "Keyless Watches," pp. 496-7.

WORKS OF THE UNITED STATES WATCH COMPANY
GILES, WALES & CO.

18. One of the most elegant watch works ever constructed in the United States, the Giles, Wales & Co. structure in Marion, New Jersey, was built using iron and glass with a frontage of 253 feet.

Chapter 4
Marion

Bergen Town is one of those ancient place names that have now been retired to a few respectful lines in history books. Its origins can be traced to the days when the Dutch inhabited the lands which surround New York's Upper Bay. There had been even older settlements in New Jersey, but the Indians, fearing for the neighborhood, had ended these intrusions in 1643. Alas, the Dutch, a stubborn lot, returned in 1660 to found the fortified settlements of Communipaw and Bergen. Inasmuch as it had become apparent that 1643 could not be repeated, the red men soon decided that the time had come to concern themselves with the possibilities of suburban living. As the years passed by, although it grew and prospered, Bergen remained essentially what it had always been, a Dutch agricultural community. In later years the arrival of new settlers from across the Hudson led to a process of dismemberment to form new towns. By 1840 the last remnant of the town stood astride the heights gazing down at the lands that were once its proud possessions. And so for the last thirty years of its existence this remaining fragment was bounded by the growing industries of Jersey City on the east, by Greenville on the south, by the Hackensack River and its tidal marshes on the west, and by Shanley's Cut and the Newark Railroad on the north.[1]

Frederick's surviving letters do not tell us why the site at the northwest corner of Bergen Town adjacent to the New Jersey Railroad was selected for the factory, so we must indulge in a certain amount of conjecture. While his first acquaintance with New Jersey dated back to his days with the Gendars, more recent matters may have led to Frederick's interest in this particular property. Giles, Wales & Co. purchased a portion of their jewelry and fancy goods from various manufacturers in Newark.* Thus, Frederick in the normal course of business would have had to travel there. Even today the most expedient methods of getting from Maiden Lane to Newark is by rail. True, the leg from lower Manhattan to Journal Square in Jersey is now underground; the trip by ferry across the Hudson and the climb of the cars up the east slope of Bergen Hill are long gone. But after that the route is the same, only the name of the railroad has changed. Today, if one sits on the left hand side of the Hudson and Manhattan Tube Train as it moves west in Shanley's Cut, the vision of what Frederick Giles saw still passes through the mind.

---

*At least two firms with which Giles, Wales & Co. did business, Thomas G. Brown, and Dodd & Hedges, had factories in Newark. In the matter of Frederick A. Giles et al., U.S. District Court, District of New Jersey, Case No. 1124, December 21, 1874, National Archives Record Group No. 21.

Nor was Frederick a solitary holder of the vision of a watch factory and a commercial complex on the western slope of Bergen Hill. Alexander Hamilton Wallis, whose name first appears as a director of the Marion Building Company, seems to have been the other moving force in this dream. Wallis, a prominent lawyer, was born in New York City in 1819 and moved to Jersey City in 1846. Jersey City owes much to Wallis, who, besides being one of that city's staunchest boosters, founded its public school system and was influential in securing for the community a modern system of water supply based on the Passaic River.[2] While the choice of a factory site in or near Jersey City was not uncommon in the New York jewelry trade, the location advocated by Giles and Wallis was more ambitious than desirable. Their arguments for acquiring the large tract of land from Bergen Junction to the Hackensack River were based on the knowledge that it could be purchased cheaply, on the belief that the Junction would retain its

19. Alexander Hamilton Wallis

importance, on the promise that another railroad, the New Jersey Midland, would be constructed through the tract, on the speculation that Jersey City would eventually annex Bergen Town, and on the dream that easy money could be made by converting salt marsh to dry land. Unfortunately, besides containing a great many "ifs" these reasons did not outweigh the costly means that would be necessary to reclaim the wet land that made up about half the area.

The marsh land which Frederick Giles and his associates proposed to reclaim was but a small part of a vast topographical feature locally referred to as the "Jersey Meadows." Meadows is a pleasant-sounding euphemism for a fifteen-mile stretch of tidal marsh which is formed by the Hackensack and Passaic Rivers as they travel through the low lying ground on the western side of the Palisades. (Bergen Hill is a small section of this range.) Since the early 1800's this area, which was then covered with salt grass and cedar trees, has excited the energies of land reclaimers and real estate speculators. Unfortunately, until recently the all-pervasive swamp muck, the mosquitoes and the tunneling habits of the area's teeming population of muskrats have proved to be economic quicksand for the unwary investor. The boom of the late 1860's lulled more than one group into investing in dreams of immense profits from reclaimed salt meadow. Directly across the Hackensack from the tract that Giles and Wallis had selected, the Iron Dike and Land Reclamation Company headed by one Samuel N. Pike was attempting to reclaim six thousand acres of marsh.

In order to implement their plans for the tract, which Frederick had named Marion, after his wife, the directors of the United States Watch Company had two bills placed before the New Jersey Senate on February 5, 1866.* One of these bills, Senate Bill number 111 of 1866, would permit the watch company to establish works in New Jersey, "For the purpose of manufacturing, selling, and dealing in watches, watch movements, watch cases, and timepieces of every description." The capital stock of the corporation was to be a maximum of one million dollars in one-hundred-dollar shares. It is interesting to note that the bill was amended several times in both houses of the legislature before it became law on March 9, 1866. While the exact nature of the amendments of the United States Watch Company Act are not known, there is reason to suspect that article three, which permitted the firm to become a New Jersey corporation at any time the board of directors so desired, was the chief amendment.[3] The other act, Senate Bill number 110, was for the incorporation of the Marion Building Company. Since the preamble to this act is both important and interesting, the author wishes to present it in toto: "Whereas, it is represented that a factory is about to be established in the town of Bergen, in the county of Hudson, near the Hackensack River, for the manufacture of watches, and extensive

---

*Frederick Giles to Julia M. Giles, August 6, 1865, "We call the company the 'Marion Building Company' after the middle name of my darling wife ('Maria' changed 'a' to 'o' and added an 'n')."

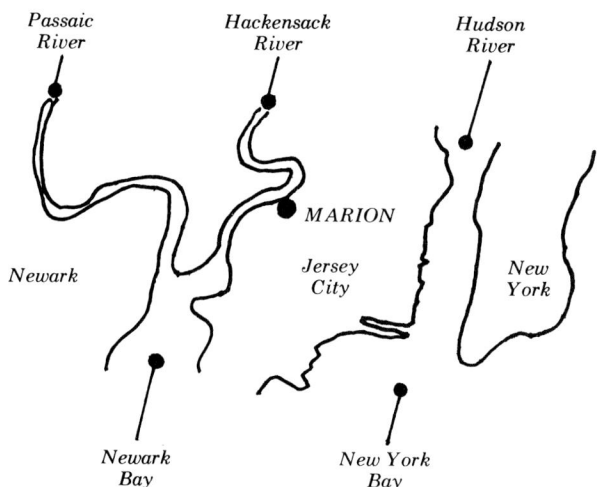

Chart II
Sketch illustrating location of MARION and the surrounding area.

buildings are now in the course of erection for the purposes of said manufacture, and that the persons interested in said manufacture have purchased certain lands and salt meadows situated and being in said town, bounded on the northeast by the lands of the New Jersey Railroad and Transportation Company, on the northwest by the Hackensack River, on the southwest by the lands now or late of H. Van Wagenen and J. D. Van Winkle, and on the southeast by West Side Avenue, leading from the Paterson Depot (Bergen Junction) to the Newark Plank Road and containing about seventy-seven acres of land, near which the said factory is located and being erected, and may desire to acquire other tracts of land and salt meadow adjacent thereto, which salt meadow they desire to fill in and reclaim, and that they desire to divide said lands and salt meadow they now have and may acquire into suitable building and other lots, and to erect dwellings and other buildings upon some of said lots, for the accommodation of the employees in said factory, and to rent, lease, sell and dispose of said lots and buildings, or said lots without buildings thereon, for the benefit of the said association, with a view of accommodating the said employees of the said factory, and to erect a town with buildings, docks, wharves and bulkheads: Whereas, it is necessary for the success of the said undertaking that they should be able to lease or convey said premises, from time to time, notwithstanding any cause affecting the individuals owning or who may own said land and salt meadow, and in order to give greater efficiency and facility to their effort for improvement

---

OPPOSITE PAGE. 20. G. M. Hopkins 1873 map of main part of Marion showing the factory facing on NJRR and the surrounding Marion Building Company land.

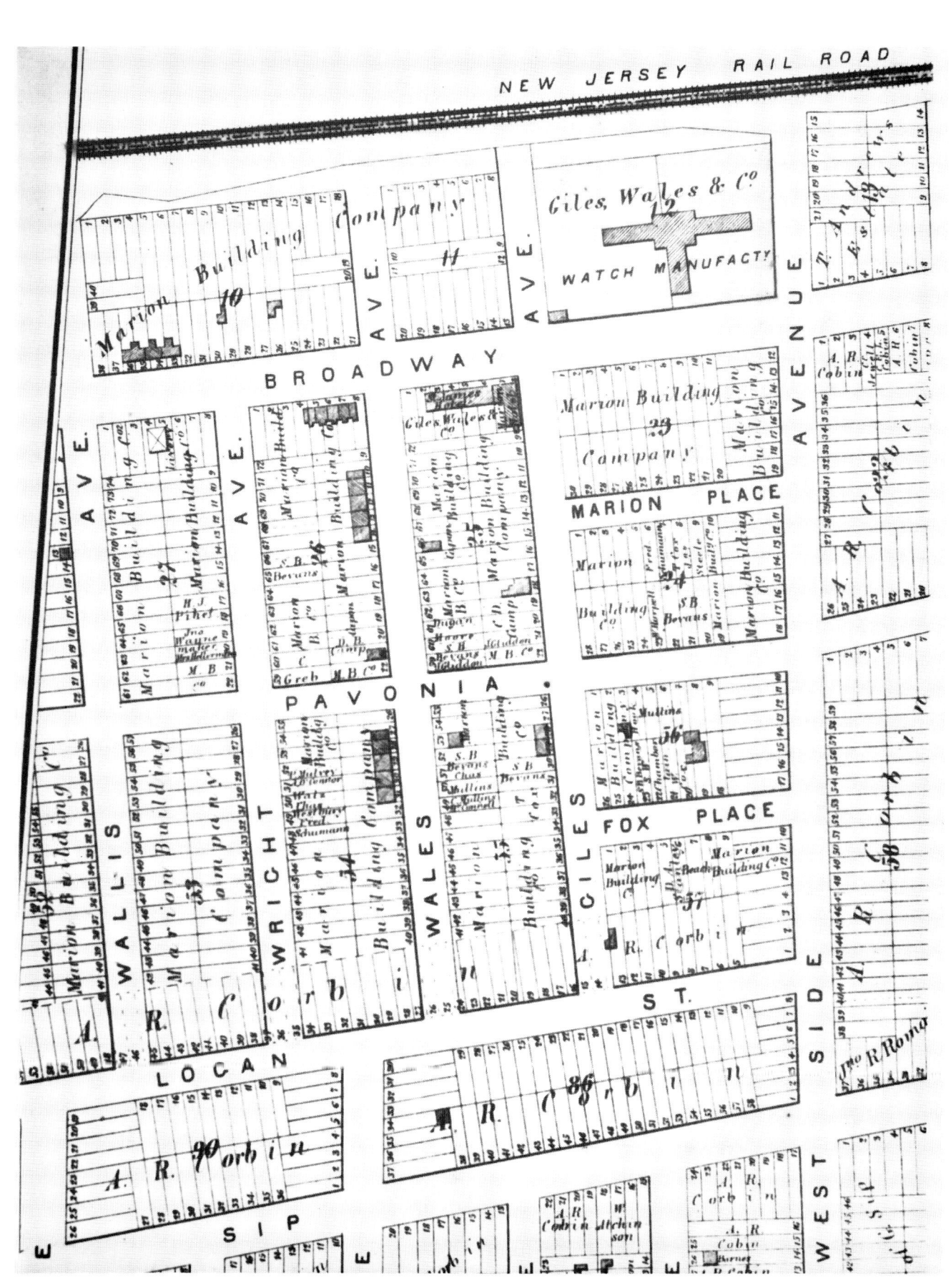

of the said land and salt meadows." The bill further provided for a seven-man board of directors consisting of Frederick Giles, William Wales, George C. F. Wright, George F. Pratt, Daniel M. Wells, Benjamin G. Clarke, and Alexander H. Wallis. While the act set the land company's maximum capital at five hundred thousand dollars, the initial capitalization figure of twenty-five thousand dollars presented in article four leads one to the conclusion that either the directors were in no hurry to implement their plans or that they were extremely naive as to the possible cost of the project.[4] It must be said, however, that experience has proven that although costly, the Marion Company's plan of using land fill in the meadows was feasible. On February 27, 1866, the Marion Building Company Act became law and thus, at least on paper, Marion, New Jersey was created.

The acquisition of the land for Marion had taken place during the previous summer. Frederick had hoped to close the matter in June of 1865, but it was August 4th before the transaction could be completed. This delay had greatly disturbed him since the beginning of construction of the factory was contingent on the land deal. In the end the loss of the two precious months caused him to do an uncharacteristic thing: he turned the entire management of the Marion Building Company over to its directors. He knew that this was not altogether a wise move, but he had spread himself too thinly and thus had to trust that the directors could carry out their promises.[5]

The factory of the United States Watch Company was to outlast the company by almost fifty years. This impressive cast iron structure, begun in August of 1865, is so well documented that there is almost no disagreement about it whatever. Among the documents relating to the building are at least two photographs, and a survey made in September of 1925, just prior to the structure's dismantling. The survey, made by James V. Hogan of Hoboken, New Jersey, and the photographs leave little doubt as to the accuracy of the early descriptions of the building. Students of cast iron and American industrial architecture have suffered a severe loss with the wanton destruction of this building, as the following lines, written in 1868, will indicate: "This magnificent structure is a striking object to the traveler who passes by in the cars, and is probably the most elegant building used for manufacturing purposes in the world, having a front of two hundred fifty-three feet on the Newark Turnpike and New Jersey Railroad, is composed entirely of iron and glass, main building four stories, with a lofty spire rising from the center, windows five feet wide, and iron columns between them, one foot, being five feet of glass to one foot of iron, thus giving the operatives the benefit of light and ventilation almost equal to the open air; the building stands by itself on the western slope of Bergen Hill, in the outskirts of the city."[6] And in another description which appeared two years later in *Harper's Weekly* one finds further details: "The building is in the form of a 'T' (The third or rear wing was of brick.), the design

21. Actual surviving photograph of the factory testifies to the accuracy of the C. Wright engraving shown in Figure 18. The iron railed fencing shown in the photo was added to subsequent engravings used in USWC promotion.

of which excites the admiration of all who are so fortunate as to see it, and the interior arrangements are perfect for the convenience and comfort of all concerned. Sets of pipes, of which there are a number arranged throughout the building, are used respectively for light, heat and the supplying of water. Great care has been exercised in guarding against accident by fire, every room and hallway being supplied with fire-hose, so that every floor can be drenched almost immediately, if necessary (a very real possibility as the histories of the New York and Freeport Watch Companies demonstrate). The toilet arrangements for the vast number of employees (approximately four hundred men and women) are perfect: hot and cold water, with abundance of room and light, serve to make this one of the pleasantest rooms in the building, and could be copied with advantage by many employers throughout the country. The sanitary arrangements are complete; the consequences of which are health and cheerfulness expressed by every one engaged in the establishment."[7] And if you were not cheerful — well, you cannot have everything! Aside from the jarring note that each employee had one four-hundredth of a lavatory, it must have been a very fine building; even Charles Higginbotham, who had few pleasant things to say about the company, refers to the building as "attractive."[8]

In his article about the United States Watch Company, Charles S. Crossman mentions several things about the factory which, although they are quite possibly true, cannot be totally supported by original sources. For instance, he tells us that not only did the factory cost one hundred twenty-five thousand dollars to construct, but also that the west wing was the property of Giles, Wales & Co.[9] We are further informed that Frederick had intended to sell this space to the United States Watch Company when the factory had expanded its operations sufficiently to have need of it. This sale, for some reason that Crossman does not explain, never took place. In this same paragraph we are told that the factory was completed during the summer of 1865 and that the machinery was moved in at once — an interesting fiction. As for the west wing, while Crossman may be correct about its ownership, he is in error on the question of space utilization. Although the approximately fifty-nine thousand square feet of floor space in the factory seems to have been underutilized, the numerous descriptions of the factory's interior suggest no distinct pattern or location for this unused space.* On the first floor, for instance, both the east and west wings are known to have been in use, however, the south wing is something of a question mark. And so it goes from floor to floor, a seemingly vacant wing here and full floor in use there.

---

*Crossman gives the following floor areas: "The center building 40' x 53' and four stories high; the east and west wings 40' x 100' and three stories high; and the south wing 53' x 75' and three stories high." This is correct, but he fails to count in the basement, which was also used for manufacturing purposes. There are three general descriptions of the factory's interior: *The Gazette*, June 18, 1870; *Harper's Weekly*, June 25, 1870; and *Moore's Rural New Yorker*, December 17, 1871.

There is much more support for Crossman's comments about the early activities of the Marion Building Company. Agreement is not total, however; for instance there is some dispute about the age of the trees that were planted in an attempt to beautify the district. While Crossman says these were old and often did not survive the transplant, John W. Barber, who visited Marion in 1867 or 68, states that they were young.[10] Aside from this, the notion that the company originally had grandiose dreams about wharves and cutting a canal are clearly supported by the incorporation document. Both Crossman and Barber agree that a great deal was done in the way of constructing walks, drives, a park, and a fountain. Crossman notes that the Marion Building Company erected the Marion House in 1866 for seventeen thousand dollars. This four-story wooden structure stood on the southwest corner of Giles Avenue and Broadway, catty corner from the factory. About two years later a five-story wooden center section and another four-story wooden wing were added to the Marion House. These additions, costing sixty-six thousand dollars, were referred to as the St. James Hotel, however the Marion House seems to have remained separate in name at least. The whole structure did have a semblance of architectural unity, taking on the appearance of a mansard summer hotel surmounted by a flagstaff-topped cupola. A block away, on the south side of Broadway between Wright and Wallis Avenues, the company also built a livery stable which housed a few employees on its upper floor. Crossman also refers to a superintendent's house which was supposedly adjacent to the Marion House, however no such building appears on plat maps or in the sole surviving cut of the hotel.[11] It is possible that an unidentified structure, which appears on one plat map on the Broadway and Giles Avenue corner of the factory site, may have served this function.

Once construction began on the factory, it proceeded at a reasonable if not rapid pace. By the late spring of 1866 the exterior of the building had begun to take on a finished appearance. It would be several months, however, before it was much more than a vast hollow shell. A great deal of millwright work installing the line shafting and numerous benches had to take place before a single machine could be moved in or set into motion. The Marion factory, in common with all other mills of the period, was designed to be powered by a central source. In this instance, the source was an eighty horsepower steam engine located in the basement of the factory. This engine, constructed by the Putnam Machine Company of Fitchburg, Massachusetts, received its energy from a pair of nearby boilers.[12] Unfortunately, as the United States Watch Company was soon to learn, the boilers of the 1860's were not totally dependable. That difficulty, however, was to appear later and did not affect the process of getting the factory into production.

Although a certain amount of conjecture is involved, it seems reasonably safe to assume that the operations at Newark were terminated by the end of 1866. By that time the factory's own machine shop at Marion should have been finished and in operation. Its availability was an absolute prerequisite to severing the relationship

22. *C. Wright engraving of St. James Hotel and Marion House as seen from the factory's lawn.*

with Dickinson and Rowden. The inability of the United States Watch Company to achieve an absolutely acceptable design meant that a great deal of major tool building remained. Curiously, the nearer the company got to being able to market a watch, the closer it got to a managerial crisis. The first crack appeared when David Gerry left the firm early in 1866, however this gap was filled by the addition of a new machinist, George E. Hart.[13] While the difficulties with the three-quarter plate design and the stem-winding mechanism had assisted in destroying the relationship between Giles and the Gerrys, Frederick's inability to delegate authority was a much more important factor. He simply was not capable of stepping back and allowing his staff to get on with their work. Ultimately, James H. Gerry resigned, taking with him several members of the tooling crew.[14] Soon after, these men reappeared as part of the management of the recently reorganized New York Watch Company at Springfield, Massachusetts.[15]

CHAPTER 4 REFERENCES

1. Daniel Van Winkle, ed., *History of the Municipalities of Hudson County, New Jersey* (New York, 1924), pp. 130-34.
2. *New York Times*, July 23, 1879, 4:5.
3. *Laws of New Jersey 1866*, pp. 277-279; also see minutes for New Jersey Senate and Assembly, 1866.
4. *Laws of New Jersey 1866*, pp. 137-140.
5. Frederick Giles to Julia M. Giles, August 6, 1865, Burns Collection.
6. John W. Barber and Henry Howe, *Historical Collections of New Jersey* (New Haven, 1868), p. 28.
7. *Harper's Weekly*, June 25, 1870, p. 414.
8. Charles T. Higginbotham, "Incidents in the American Watchmaking Industry," *National Jeweler and Optician*, March, 1912, p. 182.
9. Crossman, Adams Brown Reprint, p. 84. Abbott also gives $125,000 as the factory's cost, see p. 51.
10. Crossman, p. 85, and Barber and Howe, p. 28.
11. G. M. Hopkins, *Atlas of the County of Hudson and the State of New Jersey* (Philadelphia, 1873), p. 95.
12. J. Leander Bishop, *A History of American Manufacturers 1608-1860* (Philadelphia, 1868), III:372.
13. Gopsill, *Newark City Directory 1866-7*.
14. Crossman, p. 86.
15. Ibid., p. 112.

Chapter 5
First Fruits

The appointment of William B. Learned to the superintendency of the United States Watch factory marked the beginning of the production phase of the company's history.[1] Not that the tooling phase was completed — in truth it never was. The difference in emphasis which began with the production of the company's first watches in June of 1867 was, however, quite suitable to Learned's abilities, which were in the production area.[2] It seems probable that he was a member of the original group of machinists that Gerry had brought from Waltham and that he had gained his early experience in the factory there.[3] Hardly as adventuresome an individual as Gerry, Learned nonetheless seems to have moved about a bit. Certainly, on at least one occasion, at the Howard factory, he would follow Gerry as superintendent.[4]

James H. Gerry had stayed at the United States Watch factory only until the task of fitting out was done. It thus fell to William B. Learned to conduct this leviathan's shake down cruise. Charles S. Crossman's comment that the firm had only fifty employees at this time points out one of the serious difficulties that commanded Learned's immediate attention. The factory at Marion was large, so large that the small staff must have been totally swallowed up by its vast interior. Efficient operation demanded a staff of at least two hundred fifty.* Added to this burden of having to hire and train four-fifths of the company's personnel was the fact that Learned had not, despite a five-year contract, gained Frederick Giles' absolute trust. While there were several reasons for this situation, an important factor was that Giles and his directors had not grasped the lessons of the Gerry affair. Not that they were unique in their lack of understanding of the proper role of a superintendent; more than one watch factory suffered badly at the hands of an administration which considered itself more skilled than its technical experts. Frederick's role as the factory's principal designer and backer, however, tended to exacerbate the problem. Regardless of what standard design practice might indicate, Giles' notions had to take precedence. This is nowhere more evident than in the instance when Oliver Baldwin submitted his model for the full plate watch only to find it rejected because the standard size balance wheel he had used masked Frederick's patented visible pallet action.[5]

Baldwin's difficulty with the model for the full plate watch is the logical place to begin a serious critical look at this design. Disguised by various model names and details, this watch, which first appeared under the label "Frederic Atherton & Co.," was to constitute a large part of the firm's production during its eleven-year manufacturing history. From the outset it should be noted that despite its faults, this design was the first mass produced stem-winding watch. Later it was the first American watch to have nickel plates and damaskeening, and when in good repair it was generally a first-rate timekeeper. Its shortcomings, however, at least in the eyes of the watch trade, seemed to outweigh its various forward-looking features.[6] In appearance Frederick's design of 1866 was a piece of pure idiosyncratic elegance. There was, despite the eccentricities, a plausible rationale for his design. From the outset Frederick had wished to follow the current trends and manufacture a three-quarter plate watch. The ill-fated design of 1865 had proved this to be beyond his immediate capabilities. Rather than settle for just a full plate watch, he produced a movement which would retain some of the mechanical visibility of a three-quarter plate. His means for accomplishing this end was a butterfly-shaped opening situated on the top plate, longitudinally centered on and located between the pallet and escape pivots.[7] The intent of the design was, of course, different from skeletonized watches; there the plan is to expose as much of the gearing as possible for aesthetic purposes. Giles wished to use the aperture to simplify the task of adjusting by allowing observation of the pallet action without removal of the dial.* Whether or not Frederick was the first designer to attempt this is unknown, however he certainly was not the last. A year after Giles received his patents, James Ferguson Cole produced a pocket chronometer which used a semi-circular opening to expose its escapement.[8] Later, Edwin Bourquin,† the Rockford Watch Company‡ and the New York Standard Watch Company§ also used visible escapement designs.

Through the years various sources, including Charles Crossman, criticized the United States Watch Company's full plate design, alleging that the balance wheel was inordinately large.[9] While there is some validity to

---

*This figure is approximate. It is based on the number of employees at the factory when large scale production was achieved in 1869.

*The standard method of viewing the pallet action is through two inspection holes in the pillar plate. This, of course, requires removal of the hands and dial.

†The relationship between Bourquin and Giles will be discussed later.

‡The Rockford Watch Company used an exposed escapement on the 6th model of their 18 size watches.

§The New York Standard Company used a star-shaped opening to show their worm drive escapement.

this criticism, it ought not to be assumed that the difference between their balances and those by other American manufacturers was particularly great. Comparative measurements generally indicate a difference of less than one mm.* Marion-made balances do have a rather large and imposing appearance, but this is largely a visual effect produced by the layout of the rear surface of the watch. Curiously enough, the size of the balance wheel has little practical effect on the value of Giles' visible escapement — the correct way to adjust a watch's banking is with the balance removed.† The real weaknesses in Giles' 1866 design can all be attributed to his quest for elegance. Elegance and nothing else led to the adoption of a long, thin and all too bendable balance bridge. Elegance led to the use of fine pitch plate screws that stripped all too easily. And the result of this quest for elegance was a watch all too quickly brutalized in the inexpert hands of the "practical watchmakers" that infested the countryside.

The achievement of production and the settlement of the managerial crisis should have given Frederick some relief from the pressures under which he had been working, but in truth new problems immediately arose. Tooling the factory had proven to be extremely costly and a great deal of time had to be spent seeking out and wooing new investors.‡ Further, 1867 marked the low point in business at Giles, Wales & Co. and Frederick was compelled to devote more of his time to the store.[10] A bright note in this confusion — Julia was pregnant and a child, their first, was expected in January. Julia was in Montague and just as her time neared one of the almost new boilers in the factory had to be replaced.[11] The result was that Frederick, unwilling to trust that Learned could do the job right, sat in Marion fuming and fussing about Julia and the boiler that never seemed to arrive.

Despite Frederick's worries, December and January had had their bright spots. Business during the Christmas season had been good at the store. Giles absolutely chortled over the fact that Giles, Wales & Co. had been filled with customers throughout the season while other stores on Maiden Lane had been relatively empty.[12] The hectic pace at the store continued until late New Year's Eve. Although there was a ball at the Marion House, Frederick did not arrive until near midnight

---

*This whole discussion of balance wheels begs the real question of fast versus slow train watches, and concentrates on only one aspect — wheel size. Had Frederick been clever enough to foresee this revolution, the subsequent history of the United States Watch Co. might have been much different.

†*The Joseph Bulova School of Watchmaking,* USA, 1945, p. 255. Anyone who feels it is possible to adjust a watch's banking with the balance wheel on is welcome to try, but *not* on our watches!

‡In his letter of December 29, 1867, to Julia, Frederick mentions potential investors that he is wooing. Later in other letters to Julia, January 5 and May 31, 1868, we find him still trying to interest the same individuals, Mr. Barber and Mr. Shawe. The December and January letters also speak of Frederick trying to sell stock to a Mr. Scribner — this is probably the publisher who was known to have invested in the venture. Burns Collection.

23 and 24. Dial and movement views SN 1130, 18s, full plate, "Frederic Atherton," 19j, stem wind, button set. Note this example fitted to accommodate both types of setting and winding. Typical flat hairspring. Gold banking pins, balance screws, settings on plate for escape and pallet arbor plus gold arbor cup in polished steel, chamfered disc. Typical double recessed dials on early "Frederic Atherton" grades. This was the first stem wind, button set, mass produced watch in U.S.

25

27

26

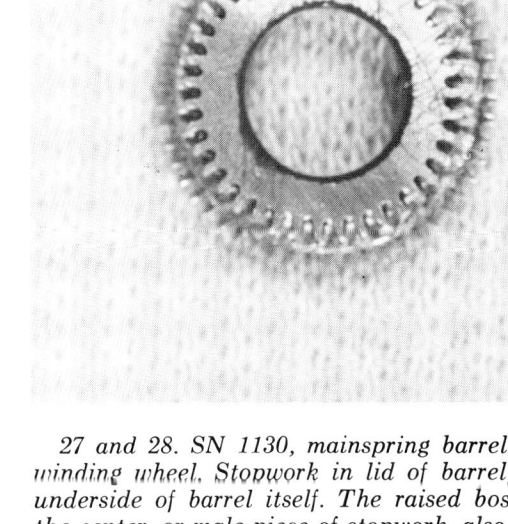

28

25 and 26. SN 1130, under dial and yoke views showing August 15, 1865, patent winding mechanism. No intermediate wheel on plate between cannon pinion and yoke wheels. Because yoke sits underneath the double main winding wheel, it must have some clearance around the wheel and winding pinion. It will shift around under the wheel without the steadying screws, causing binding or skipping of teeth during setting and winding respectively. A very early, but poor design.

27 and 28. SN 1130, mainspring barrel and double winding wheel. Stopwork in lid of barrel, not on the underside of barrel itself. The raised boss portion of the center, or male piece of stopwork, also functions as the upper arbor pivot turning in a larger than normal hole in the barrel bridge. The double winding wheel shows signs of problems over the years. It should also be noted that balance wheel on SN 1130 measures 17.6 mm in diameter.

and stayed only long enough to get a few chuckles over some of the crew who had a bit too much to drink. The weather made New Year's Day something of a social disaster. Henry and George Wright, Mr. and Mrs. Learned and Henry Lowe, a foreman, and his daughters had been invited to dinner, but the storm kept the Learneds and Lowes away.[13] As for Julia, there is no way of knowing whether Frederick was at last able to disentangle himself long enough to go to Montague, but anyway on January 22, Hattie Chenery Giles arrived.

Hattie's birth had little or no effect on Frederick and Julia's preference for pursuing their independent ways.

29. *Hattie Chenery Giles*

Julia remained in Montague, while Frederick maintained an elegant, if solitary, existence in Marion at the St. James Hotel. His arrival there had been recent. In 1866, he and William Wales had moved their residences to the Astor Hotel.[14] In some respects the Astor, which always numbered among its guests men of firm Republican tendencies and excellent business connections, was a better location than the St. James. The loss of the immediate elbow rubbing, day to day contacts that were essential for the recruitment of capital proved to be a real hardship. When it comes to doing business, New York is much like a vast club where friendship and acquaintance often prove to be the strongest assets one can possess.

Although Giles found it increasingly difficult to recruit major investors after the move to New Jersey, he had previously managed to acquire an impressive bevy of stockholders. Aside from Alexander H. Wallis, whose role in the venture has already been mentioned, there were: G. A. Read, whose own firm Read, Pratt & Co. of Deep River, Connecticut, was among the principal manufacturers of ivory combs and piano keys;[15] Sylvester M. Beard, whose firm S. M. Beard Sons & Co. was a prominent importer of tea, coffee and spices;[16] Henry Randel of Randel, Baremore and Billings, diamond importers;[17] A. S. Hatch of the banking firm of Fiske & Hatch;[18] J. Abner Harper and William S. Wyse of Harper Brothers publishers;[19] James A. Alexander, who had been the New York representative of the Aetna Insurance Company since 1851;[20] and there was also Sylvanus Sawyer of Fitchburg, Massachusetts. J. Leander Bishop, that brilliant early historian of industry in the United States, described Sawyer as one of the most distinguished of American inventors. While his inventions covered a number of fields, his principal fame rests on his total mechanization of the processing of cane. Before Sawyer's work all cane used in the manufacture of furniture had to be laboriously hand processed. Inasmuch as caned chair seats were de rigueur, it should come as no surprise that his success in this field permitted his retirement from management to full time inventing at age thirty. From this point on, 1852, he spent much of his time experimenting with rifled cannon, and later tools.[21]

Assembling a complete list of major stockholders presents certain problems. We have already cited the individuals mentioned by Charles Crossman and Henry G. Abbott. Unfortunately, although their lists are seemingly accurate, they are somewhat incomplete. As Abbott points out, the United States Watch Company named a number of their movements after stockholders. But while the list of these movements is much more extensive than he assumed, the company's inconsistent practices thwart the acquisition of a complete roster by this means. One has to be curious as to why neither Frank H. Page nor Daniel M. Wells, the early trustees of the firm, ever had watches named after them. Perhaps they dropped out of the company before production began. Of course, a number of important investors never had movements named after them. Certainly the omission of Sylvanus Sawyer in this regard is inexplicable. Another important omission was Frank H. Richardson and the other partners in the jewelry house of Enos Richardson and Company. Neither was a movement named for Thomas G. Brown, an important manufacturing jeweler from Newark, New Jersey. Despite these omissions, the court records from the bankruptcy period of the company's history indicate that both the Richardson house and Brown were longtime major investors in the firm.[22] There were at least three important stockholders who did have movements named after them that both Crossman and Abbott fail to mention. These were Andrew J. Wood of Brooklyn, New York, and later of Brick Church, New Jersey, a merchant in fats and oils who supported the company from its earliest days to its end;[23] Cyrus H. Loutrel, a partner in the firm of Francis and Loutrel, stationers, as well as a commissioner of Castle Garden;[24] and William S. Wyse.[25] Court records also suggest that there were several banks and insurance companies that held the factory's stocks and bonds. Among these were the First National Bank of Jersey City, the Continental Life Insurance Co., the Hanover Bank and the Aetna Insurance Company.[26] The complexities of the financial arrangements between the various firms owned by the Giles brothers leads to the assumption that both Giles Brothers of Chicago and E. A. Giles & Co. of Dubuque, Iowa, had returned Frederick's earlier favors by making investments in both the factory and the Marion Building Company.

The full plate model of 1866 was never manufactured in large quantities. Various problems at the factory conspired to limit combined production for its two grades, "Frederic Atherton and Co." and "Fayette Stratton," to fewer than thirty watches per week. Besides, the company was not totally satisfied with the design. Although they had rejected Baldwin's original model, the thought that the balance wheel might be too large obviously troubled the directors. Further, and more importantly, the stem-winding mechanism, despite its handsome blued steel appearance, was deemed to be less than satisfactory.* Fayette Giles, upon his return from Switzerland in early 1867, had patented a new and interesting design which utilized a large internal gear (#65208, May 28, 1867). Cutting gear teeth on the inside of the ring was not, however, considered economically practical at this time. Since no other solution was in the offing, it was finally decided to rework the original design. The new stem-wind adhered much more closely to Louis Audemars' original pattern using a longer rocking bar (yoke), an improved ratio between the contrate (crown) wheel and its intermediates, as well as an additional wheel in the setting train. As for the balance wheel — well, Frederick and the directors at last resolved the question in favor of a slightly smaller wheel.

The introduction of the improved stem-wind model of December 22, 1868, marked the end of the company's fumbling and uncertainty over designs.† Future models may have had their faults, but there would be no more failures or indecision. This new attitude meant that for now no further attempts would be made to redesign the full plate. Henceforth, every effort would be made to increase production and the number of grades available. Unfortunately, the United States Watch Company soon found it far easier to accomplish the latter. There was

---

*A number of years ago Bill Muir and the late F. H. McMillan gathered together a great horde of Marion-built watches and parts. This collection included well over a hundred examples spanning the entire history of the factory's existence. It was this research collection, now dispersed, which provided the foundation and basis for much of the data that appears in the Historical section.

†No records have been located that would indicate whether this improvement was registered with the Patent Office.

30 and 31. Dial and movement views, SN 1034, 18s, full plate, 19j "Frederic Atherton," stem wind, button set. Cannon pinion extends up only enough to hold hands. No disc around barrel arbor and arbor is flush with plate indicating this early product was exclusively stem wind. Later, USWC products were equipped for both types of systems. Note double recessed dial. USWC was one of the first American watch companies to use this double-sunk type of dial.

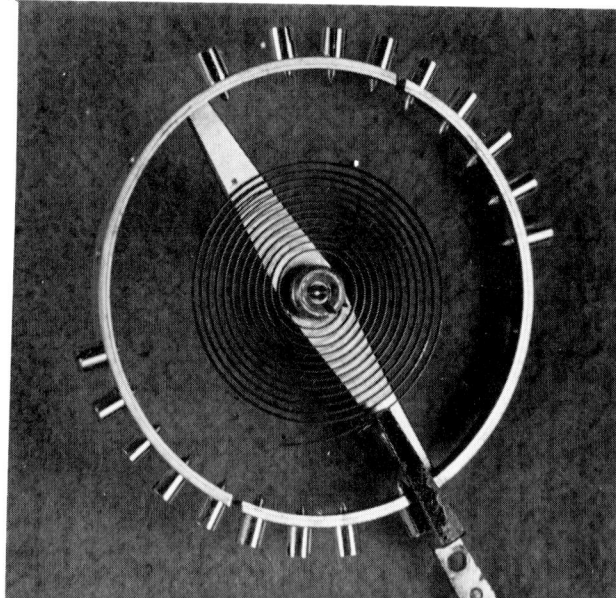

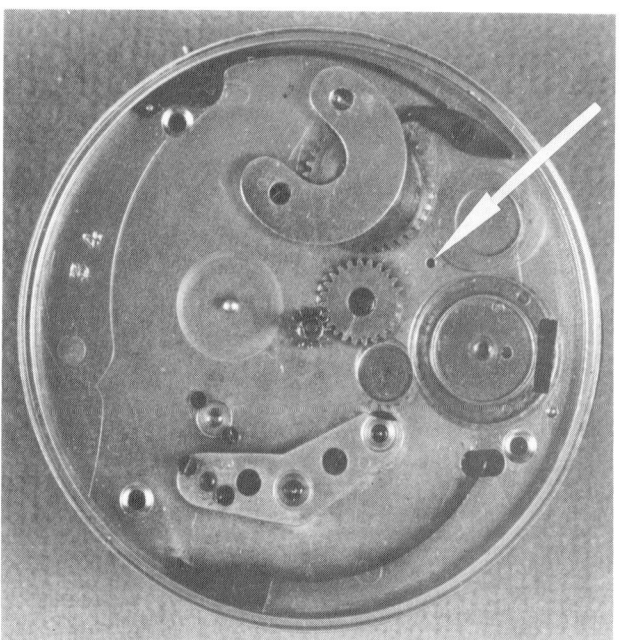

32 and 33. SN 1034, under dial views showing movement has been fitted, or retrofitted, with the December 22, 1868, improved winding mechanism. Arrow shows threaded hole originally used on earlier '65 patent as a limit of travel for yoke. The "less than satisfactory" 1865 system obviously has been replaced with the improvement. This is the earliest example to surface with the 1868 mechanism.

34 and 35. SN 1034, balance wheel and mainspring barrel. This balance wheel measures 18.25 mm and is believed the largest size used by USWC (SN 1108 also has same outside diameter). Other USWC products in 18s have balances which measure 17.6 mm, 16.85 mm, and as small as 15.4 mm. Note the stopwork and arbor on top of the mainspring barrel typical of early USWC full plate models.

some flaw, deeply seated in the organization of the factory, that allowed them to excell at almost everything but getting out production.* It is true, of course, that the company was a trifle slow in getting out the additional three new grades of the '68 model, but this can largely be attributed to the energy expended preparing "The United States Watch Co.," a very high grade version of the full plate. The other two grades, "George Channing" and "Edwin Rollo," were medium quality and involved little that was new outside of a flat regulater for the Rollo.

Charles Higginbotham tells a story, albeit an amusing one, in his "Incidents in the American Watchmaking Industry" that sheds some light on the efforts that went into engraving "The United States Watch Co., Marion, N.J." in Old English letters. In those days the lettering and ornamentation on watches was entirely cut by hand. The work cost upwards of two dollars each to cut and scroll United States watches; it would be done today for a few cents — and better done at that. So much for the difference between hand and machine work. In connection with the trade mark the following incident was related by the engraver — Jacob Sanborn. The style of the initial letter in the word 'Watch' did not suit the president [Frederick] or someone else in authority, and was ordered changed. Sanborn tried repeatedly to make a letter to suit, but was unable to do so without destroying its identity. In sheer desperation and with a daring characteristic of the man, he cut a letter 'M'. It suited exactly, the fraud not being detected. The result was that a number of watches went out engraved "United States Match Co."[27]

---

*Although it is hard to document this statement, the company's advertisements giving testimonials and every modern statistical study of the company's production tend to indicate very low output.

LEFT. 36. Movement view, SN 12017, 18s, full plate, 19j, "United States Watch Co." grade, stem wind and button set. First run, gilt, produced early in 1869 while William B. Learned was superintendent. Flush gold settings throughout back of movement, gold arbor cap in machined finish steel ring, blued screws, heavy regulator. First use of Breguet hairspring by USWC. Balance wheel measures 17.6 mm, same as SN 1130.

ABOVE. 37 and 38. SN 12017, under dial and side views showing the December 22, 1868, improved winding system. Winding yoke polished with chamfered edges. 3rd and 4th jewels in pressed settings. Jewel screws for escape and pallet arbor not in counter-bored holes. No gold jewel settings on pillar plate. Side angle shows winding pinion block and setting push piece between plates. Plate screws nicely chamfered.

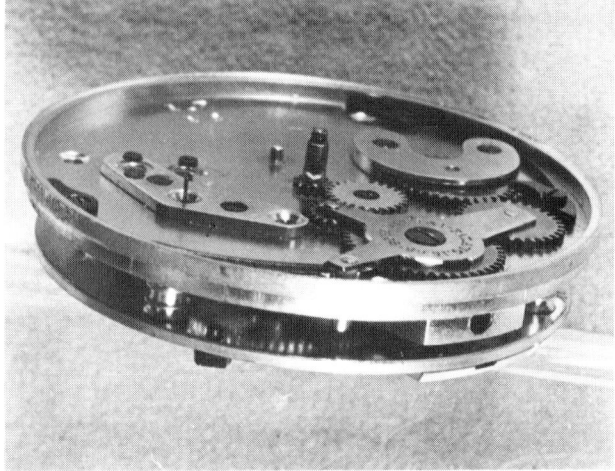

39, 40, and 41. SN 12017, pillar plate side view, yoke and hour wheel. Side view shows chamfered yoke with attached push piece for setting between plates. Both '65 patent and '68 improvement dates on yoke. Full serial number on hour wheel.

42. SN 12017, double recessed, two-color dial with large block Roman numerals. "Giles, Wales & Co" in black and red. Large USWC monogram in seconds bit also in black with red outline. An early example of the fine dials produced by the company.

Neither Sanborn's attitude nor his difficulties were unique at the factory. It has been noted that the company suffered from personnel problems from the beginning and they would persist until its demise. Frederick's elegance and charm do not seem to have carried over to his relationships with his employees. Following the Sanborn incident, one of a far more serious nature rocked the factory. Time had not improved the situation between Frederick and William B. Learned; it remained a cold, formal relationship. Further, Frederick had long dreamed of giving his brother, Fayette an important role in the factory.[28] Now the factory's inability to get out production presented the perfect occasion for an administrative reorganization.

CHAPTER 5 REFERENCES
1. Crossman, Adams Brown Reprint, p. 86.
2. *Watchmaker and Jeweler*, November, 1869, p.41.
3. Crossman, p. 86.
4. Ibid., p. 62.
5. C. B. Garrett to Charles S. Crossman, April 10, 1887, quoted by W. Barclay Stephens, "The United States Watch Company," NAWCC BULLETIN, 4:154.
6. Higginbotham, March, 1912, p. 182.
7. Giles patented designs for both straight line escapements, U. S. Design Patent 2281, March 13, 1866 (as described), and right angle escapements, U. S. Design Patent 2266, February 27, 1866.
8. Cecil Clutton and George Daniels, *Watches* (New York, 1965), illustrations 551-2. Giles, Wales & Co. handled Cole's watches, however whether there is some sort of connection between Giles' and Cole's use of the aperture is unknown.

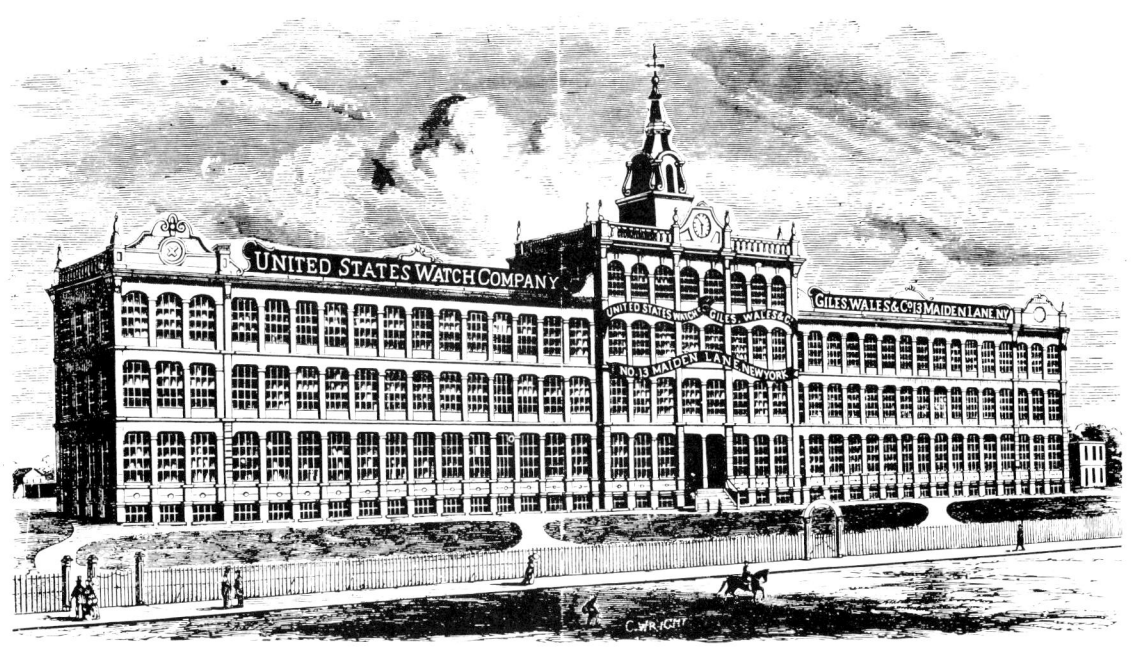

MANUFACTORY OF THE UNITED STATES WATCH COMPANY.

43. Advertising cut showing the use of "Frederic Atherton & Co" and "United States Watch Co" grades. Note this C. Wright engraving has the added iron railed fencing around the factory.

9. Crossman, pp. 88-89. Also see: Stephens, *NAWCC* BULLETIN, 4:154.
10. Memorandum, "Giles, Wales & Co., Sales and Profit 1862-73," sales were $366,041.90 for 1867, Burns Collection.
11. Letters of December 29, 1867, and January 5, 1868.
12. Letter of December 29, 1867.
13. Letter of January 5, 1868.
14. Trow, *New York City Directories 1866-7, 1867-8.*
15. Crossman, p. 84; Abbott, p. 54; Bishop, III:426.
16. Crossman, p.84; Abbott, p. 53; Dun and Barlow, *Mercantile Agency Reference Book,* 1879.
17. Crossman, p. 84; Abbott, p. 53; *Jewelers' Circular,* August, 1880, p. XXVIII.
18. Crossman, p. 84.
19. Ibid.

20. Crossman, p. 84; Abbott, pp. 53-4; *New York Times,* December 4, 1913, p. 9:6.
21. Crossman, p. 84; Bishop II, pp. 579-589.
22. Petition, December 22, 1874, In the Matter of Frederick A. Giles, William A. Wales and George C. F. Wright, U. S. District Court, District of New Jersey, Case No. 1124, National Archives, Record Group No. 21.
23. United States Watch Co., *Price List,* September 2, 1872.
24. Empire City Watch Co., *Price List,* undated.
25. Ibid.
26. Case No. 1197, List of Creditors.
27. Higginbotham, March, 1912, p. 182.
28. *Harper's Weekly,* June 25, 1870.

Chapter 6

William Alexander

Giles family members of later generations were often given to comment that Frederick and his brothers did not act like brothers — they would not support each other in times of crisis. One must suspect, however, that this view resulted from the fact that after Frederick's generation the family took to producing fewer children. Large families are seldom deceived by pleasant terms such as "brotherly love" and "band of brothers." They know from sad experience that siblings can indulge in some of the hardest and most protracted warfare ever to hit our species. No, Cain and Abel set the style a long time ago and fortunate is the family that arrives at adulthood with all of its members speaking with one another.

While quarrels among the Giles brothers seldom reached the level of open warfare, nonetheless these were strong-willed and intelligent men willing to take a stand and defend it regardless. Further, the special circumstances of their childhood had produced an unusual series of relationships. The most important of these was the bipolar effect of the two elder brothers — Frederick, the brother who ruled from afar, and William, who was on the scene. Perhaps even if Lemira and Prescott had lived, the elder sons would have struggled against each other, but the reasons and thus the effects would have been different.

After the deaths of Prescott and Lemira, William, then eight, came under the rule of his aged grandfather. This did not last for long, however, as the old man was as worn down as his hilltop farm. Within three years the elder Giles had died and William was left to carry on as best he could. Unfortunately, labor as he might his efforts and the small help that Frederick could send home were not sufficient to save the place from the debts that Daniel had acquired. At fourteen William was thus forced to hire out as a farm laborer. Yet, despite these circumstances he managed to acquire an education at the local high school.[1]

At the age of fifteen William had his first brush with the jewelry business. Frederick, in hopes that his younger brother would follow the path he had cut, apprenticed William to a jeweler named Cook in Northampton, Massachusetts. The young man, who was quite academically inclined at this point, soon found the tedium of retail business not to his liking. Despite his disappointment, Frederick was forced to consent to the termination of the apprenticeship and William's return to school. After a semester at the high school in Athol, Massachusetts, William then moved on to New Salem Academy. While studying there he began teaching at a local common school. In 1854 he opened his own high school at South Royalton, Massachusetts. Later he returned to his studies at Thetford Academy. All of this moving from school to school was necessitated, of course, by William's chronic lack of funds. He was perpetually forced to teach one semester so that he could study during the next. In the end he hoped to be able to attend college, but this was a dream beyond his means. Ultimately, his health broke under the strain and young Giles was forced to seek other goals.[2]

While Frederick was the earliest of the Gileses to visit the midwest, it is uncertain whether Anna or William was the first to settle there. What is known is that friends of the family had moved to the Prairie du Chien area by the mid-1850's. In 1857 William traveled there in search of his health and a small business in which to invest his few hundred dollars of savings. Not long after, he did what had been unthinkable a few years earlier; he opened a jewelry store. Despite his previous abhorrence of retail business the store proved a success and he soon opened a branch on the opposite side of the Mississippi at McGregor, Iowa. Since he needed assistance in operating this second venture he brought his brother, Charles, west. By 1861 William and Charles were ready for bigger and better things, but the old problem appeared — a lack of sufficient funds.[3]

Although William was now a resident of the midwest he was hardly independent of the east in the late fifties and early sixties. First there was the matter of a wife. While William claimed it was money and health that led to his abandonment of an academic career for business, romance may well have played a role. In 1858 he returned east to marry a Miss Mary Harper of Enfield, Connecticut.[4] It is also suspected that at this time he arranged with his elder brother, who was by then a financial success, for aid in opening the McGregor store and for Charles' trip west. Three years later, in September of 1861, he, Mary and their baby journeyed east once more, this time to secure Frederick's assistance in William's plan to open a wholesale jewelry house in Chicago.[5] Times were still difficult from the recession and Frederick was struggling to make a success of Giles, Wales & Co., but his sense of responsibility made him contribute everything he could to his brother's venture.[6]

From the moment that Giles, Bros. & Co. first opened their doors at 142 Lake Street they were a total success. They had chosen the perfect moment to enter the whole-

OPPOSITE PAGE. 44. Giles, Brothers & Co. (William Alexander and Charles Giles) original gift box with an 18k USWC "William Alexander" grade stem wind watch. Giles, Brothers & Co. was organized in 1862 at 142 Lake Street in Chicago.

45. "Wm. Alexander" movement of watch illustrated in Figure 44. SN 20293, 18s, 15j, nickel damaskeen finish, stem wind and button set. Case is marked G B & Co., SN 9684, initials for Giles Brothers & Company.

sale jewelry trade in Chicago. The midwest had been experiencing rapid growth even before the war and the Windy City stood at the center of things. Further, surrounded by timber, wheat, minerals, cattle, and as the rail hub and terminal port of a unique chain of inland seas and connecting rivers, the city had just begun to feel its future potential. All of this burgeoning economic activity created an ever expanding need for new businesses. In the instance of the jewelry trade the time for a first class wholesale house dealing in the highest class of merchandise had arrived and William and Charles were prepared to meet this demand.

Once he had committed himself to a life of business William became totally immersed. He was, however, an active and somewhat nervous person whose energies could never be soaked up by a single enterprise. When the strain of opening the store had passed he began to involve himself in other ventures. The first of these sidelines seems almost a trifle strange for a man who would, as time passed, gain a reputation for being a hard-headed businessman — he opened an art gallery. This excursion into the arts proved successful and soon the gallery was a gathering place for artists and patrons.[7] Interestingly, Giles' next venture, which was in a field more closely allied with the jewelry business and much less romantic, proved to be a total failure.

Fate has a curious way of dealing with things. If Bob and Carolyn Burns had not decided to leave Duxbury, Massachusetts, they might never have discovered the biography which explains William Giles' connection with the United States Clock and Brass Company. Certainly those who studied the matter for so many years were never able to ascertain a link between the United States Watch Company and the United States Clock and Brass Company. Samuel W. Jennings, who has spent over thirty years diligently sifting the meager data and who undoubtedly discovered most of what we know about the Austin, Illinois, firm, was unable to progress much beyond the names of Henry W. Austin and Chauncey Jerome.[8] Even with the new-found knowledge of the Giles connection we know far less than we would like, and were it not for the role that William's illfated venture plays in our story the affair would rate only brief mention. Necessity, however, must cause us to probe as deeply as we dare, even if some conjecture is required.

According to the article about William Giles in *Biographical Sketches of the Leading Men of Chicago*, the United States Clock and Brass Company was exclusively his conception. He had, after careful study of the matter, concluded that a Chicago based factory could produce clocks equal to those of any eastern manufactory. Although there is no explanation of the rationale for the rolling aspect of the factory, it seems probable that he hoped to capitalize on the fact that the midwest had become a major copper mining center.* Armed with every expectation of success, Giles and several friends launched the new company as a joint stock venture with two hundred thousand dollars in capitalization. The stock offering was soon subscribed and Henry W. Austin donated forty acres of land for the factory in the village which he had founded on the Galena Division of the Chicago Northwestern Railroad. By mid-1866 the United States Clock and Brass Company had constructed three large factory buildings as well as a village of tenement houses for their employees.[9] Equally important, they had managed to hire the illustrious Chauncey Jerome as their superintendent. This latter triumph, however, was not without flaw. Jerome, who had done so much to develop inexpensive brass clock movements, was well past his prime.[10]

At first everything seems to have gone well. It is recorded that the company rolled its first brass in the autumn of 1866. While it is uncertain when the first clocks were finished, it seems probable that this took place about the same time. By late 1867 the factory had grown to the point where it was employing about two hundred people. Curiously, the prime mover of all this, William Giles, was not the president of the company;

---

*The Lake Superior copper mines began to produce ore in quantity by the late 1840's.

46. *William Alexander Giles probably about the time he had formed the United States Clock and Brass Co. at Austin, Illinois.*

47. *Nice example of U.S. Clock & Brass Co. (Chauncey Jerome, Superintendent) 30-hour Ogee, Austin, Illinois.*

this honor was held by a C. N. Holden. There is no way of knowing what William's biography meant when it referred to him as "chief practical manager" of the company, but it is believed that his role was largely financial.[11]

Why the United States Clock and Brass Company did not succeed is not known. The truth is that no one even seems to be certain about the date of the company's demise. All that can be gleaned is that William's biography states unequivocally that things were going well when it was written ca. late 1867, yet every other known source seems equally certain that the factory was out of business by 1868.[12] Obviously something must have suddenly taken place and the young company was unable to resolve the difficulty. This notion of suddenness is underlined by the information that the factory was able to manufacture only one style of movement, a thirty-hour weight driven model. There was another model, but as Samuel Jennings points out this was almost certainly produced by the Gilbert Manufacturing Company.[13]

Two details are known that may suggest possible reasons for the sudden collapse of the United States Clock and Brass Company. First, Chauncey Jerome died at New Haven, Connecticut, on April 20, 1868.[14] Second, a local newspaper, *Oak Leaves*, stated in 1936 that, "It was a three or four story building. Later it became a shoe factory and then a tannery. . . . It was destroyed by fire in 1868."[15] This article is, of course, quite obviously flawed, but not necessarily beyond redemption. If the fact that there was actually more than one building in the factory complex is taken into account, then this apparently garbled oral history might make some sense. The remaining difficulty is whether or not the reader is willing to accept an interim solution. If so, then it becomes reasonable to suggest that the demise of the young company may well have been caused by the twin catastrophe of the death of its superintendent and then the destruction of a part of its plant by fire.

Regardless of the exact causes for the end of the United States Clock and Brass Company, it seems improbable that their investors escaped without financial loss. Fortunately, times were prosperous enough that it is doubtful that any of these would-be capitalists suffered unduly. Certainly the continuing success of Giles, Bros. & Co. left William only with the embarrassment of a side venture that did not quite work out. Still, the damage to his pride left a permanent mark on his psyche; in the future he would be unwilling to enter into any financial scheme that presented an element of risk.

CHAPTER 6 REFERENCES

1. *Biographical Sketches of the Leading Men of Chicago*, pp. 511-12.
2. Ibid., pp. 512-13.
3. Ibid., p. 513.
4. Ibid., p. 515.
5. Frederick Giles to Julia Wright, September 29, 1861, Burns Collection.
6. *Jersey City Journal*, May 27, 1871.
7. *Biographical Sketches*, pp. 513 and 515.
8. Samuel W. Jennings, "Chauncey Jerome, Austin, Illinois," NAWCC BULLETIN, December, 1947, 3:351:3; Samuel W. Jennings, "Some Further Notes on Chauncey Jerome, Austin, Illinois," NAWCC BULLETIN, April, 1978, 20:136:9.
9. *Biographical Sketches*, p. 514.
10. Samuel W. Jennings, NAWCC BULLETIN, 3:351:2.
11. *Biographical Sketches*, p. 514.
12. Mrs. Edward G. Snodgrass, *Oak Leaves*, December 31, 1936, as quoted in Samuel W. Jennings, NAWCC BULLETIN, 3:352.
13. Samuel W. Jennings, NAWCC BULLETIN, 20:138; Although Jennings is unsure about the place of manufacture of the weight driven model, it now seems reasonably safe to assume that a factory complex with 200 employees and a rolling mill produced at least this one type.
14. Jennings, NAWCC BULLETIN, 3:351.
15. Snodgrass as quoted by Jennings, NAWCC BULLETIN, 3:352.

Chapter 7

Fayette Stratton

By the late sixties, time had altered Frederick's role in the family from juvenile patriarch to revered elder brother. Of course William was an exception to this, but his attitude was unique. The other brothers and sisters were well aware that as they had matured Frederick had underwritten their entry into the world. For example Edwin, upon his return from the war, had settled briefly in New York.* He had, however, in common with many other veterans, become an independent, self-reliant, but restless individual. While his loyalty to Frederick never swayed, his preference for standing alone meant that he could not fit into Giles, Wales & Co. Thus when he wanted to go west to start his own business, Frederick wisely gave the venture in Dubuque, Iowa, his blessing and wholehearted financial support.† The girls, Sue and Anna, by this time had married and moved to Iowa, but when the need arose they still called upon Frederick for aid.[1]

---

*Although the exact dates are not known, Edwin stayed long enough to patent a stem-winding watch case, U.S. Patent 57,495, August 28, 1866.

†The first reference to Edwin's firm was in the *Jersey City Journal* article of May 27, 1871.

Fayette seems to have always been Frederick's favorite. Even before the other brothers and sisters had gone west, Frederick had lavished his special hopes and plans on him. It was Fayette that he took into partnership. When a regular agent was needed in Switzerland, it was the younger brother that was sent to supervise the house at 13 Rue Leopold Robert in Chaux-de-Fonds.[2] Later, the watch factory would name its second grade of the full plate watch "Fayette Stratton." Further, it was Frederick's ultimate plan to turn over the supervision of the factory to him.[3]

While Fayette had genuine affection for his elder brother and was to contribute brilliantly to his business schemes, he had his own plans for his future. This independence first appeared after he had arrived in Switzerland in 1866. Not suffering from Frederick's tendency to become totally absorbed in business affairs, Fayette found time to court Bertha Faigaux, the daughter of a local hotel owner. It soon transpired that the young couple fell deeply in love and Fayette proposed marriage. Although Bertha's parents had some misgivings about the separation the marriage of their daughter would produce, they gave their consent. Until this point, Fayette had said nothing to Frederick about Bertha. Worse still, just before leaving for Europe he

48. E. A. "Edwin" Giles & Co. corner location in Dubuque, Iowa. This engraving features their Howard, Illinois and Elgin agency, but Edwin was also a distributor for the USWC.

49 and 50. Fayette Stratton Giles and Bertha Faigaux were married at Chaux-de-Fonds early in 1867.

had promised to avoid romantic entanglements. One can thus imagine the elder brother's shock when he received the letter requesting his blessing upon the impending marriage.[4] Despite his reservations there was little that Frederick could do but give the young couple his blessing.

Although it is known that Fayette and Bertha were married at Chaux-de-Fonds early in 1867, a slight mystery surrounds their whereabouts for the rest of the year. On May 28th Fayette's stem-winding mechanism was patented. Whether or not he returned to New York to submit the patent application is open to some conjecture. It does not seem altogether unlikely, however, that the young couple may have journeyed to New York. Aside from the patent application and other business matters, Frederick would have undoubtedly wanted to meet the young woman who had turned his brother's head. Whatever, it is fairly certain that Fayette was to remain in charge of the Swiss branch of the business until 1869.*

In the beginning Fayette had not wanted to go to Europe. At first, Switzerland was a place totally alien to his New England ways. Bertha changed all of that. Soon Fayette had become as much a French-Swiss as an American, even to the point of adopting the name Lafayette.† While in the long run this conversion would cause serious problems for Frederick's plans, in the short run it was to prove a positive benefit. It opened Swiss doors that might normally have remained closed to a foreigner.

As a material for the manufacture of watch movements, nickel possesses several important advantages. Among these are: first, nickel is superior to gilded brass in long term resistance to the corrosive effects of watch oils and moisture; second, it is physically stronger than brass. Of course its strength does present some additional machining difficulty, but since the strength improves the dependability of the watch, the additional effort in manufacturing is not overly important. Finally, with the proper decorative treatment, nickel is perhaps the most elegant of all plate-making materials.

By the middle of the 19th Century the Swiss, after a long process of experimentation, had not only adopted nickel as the standard material for much of their watch manufacture, but had also perfected methods of improving its rather harsh appearance. This technique of embellishment, which utilized geometric designs produced by small wheels or laps, was called damaskeening. While American manufacturers were well aware of the Swiss use of nickel and damaskeening, for one

---

*The data concerning the date of Fayette's return to the United States is conflicting. If New York City Directories are one's source, then the date of his return would be before May 1, 1867; however, the *Harper's* article says mid 1869. There is, unfortunately, no way at present of proving either date. It seems probable that the correct date is early in 1869 or late in 1868.

†This use of the name Lafayette may have been Frederick's doing; it would seem that he got a certain pleasure from playing with names.

51. 18s, 15j, "Fayette Stratton" grade, SN 10033, with gilt damaskeening. With the help of Fayette, Mr. F. Wilmot of St. Imier agreed to train USWC personnel in the art of both gilt and nickel damaskeening and introduced this technique to the American watch industry.

reason or another they were unable to duplicate their results.* Further, all attempts to acquire the necessary machinery and information from the Swiss had proven futile. This is hardly surprising, considering that the young American watch factories were already making serious inroads into the world market.[5]

Fayette, during his stay in Switzerland, had been able to make the acquaintance of a Mr. F. Wilmot of St. Imier. Wilmot, who is perhaps best known for his invention of a patent regulator used by the Tiffany and Patek Philippe companies, was by all accounts extremely knowledgeable about the field of nickel finishing (#129,197, July 16, 1872). Precisely how Fayette was able to lure Wilmot to the United States is not known, however it is entirely possible that Wilmot, being aware that several American companies were willing to pay handsomely for the process, needed little

---

*Henry G. Abbott points out in his book, *The American Watchmaker and Jeweler*, that this application of the term was erroneous. He states, "The embellishment of the surface of metals with rings or bars is snailing and is not damaskeening, although improperly called so by watch makers and watch factory employees particularly." Incorrect or not, the term damaskeening was used by every watch factory starting with the United States Watch Company.

persuading. By the end of Mr. Wilmot's one-year contract with the United States Watch Company, they had not only designed and constructed their own damaskeening machinery, but had become totally self-sufficient in the art for both gilt and nickel movements. Crossman notes that Charles Berlin of the company developed the wet process and introduced the ivory lap as a replacement for the wooden ones that had previously been used.[6] While Wilmot was later to launch the damaskeening process at both the Waltham and Elgin companies, the work produced by these firms is quite different from that of the United States Watch Company. The Marion work was far more restrained and generally avoided the sin of over-decoration.

There was one difficulty in all of this; production at the factory still lagged so badly that full advantage could not be taken of the company's technical lead. Watch manufacturing in the early days was not much different from modern baseball; when things went wrong it was time for a change of management, and the manager of a watch factory was the superintendent. On the face of things, then, William B. Learned's removal from the superintendency of the United States Watch Company on June 14, 1869, appeared to be nothing more than the demotion of an individual for less than satisfactory performance of his duties. Certainly this was the position taken by Frederick Giles and the board of directors. The factory's low production figures and the promotion of Henry J. Lowe, foreman of the finishing room and a friend of Learned's, to the superintendency seemed to suggest that Frederick's only concern was for the business. Learned, whose five-year contract still had several years to run, felt, however, that his exile to the factory's setting up room resulted from factors quite separate from his handling of the superintendency. In August he severed his connection with the company and began a law suit to recover both the balance of his contract and his reputation.[7]

Evidence that Learned's feelings were not totally unjustified was almost immediately forthcoming. Despite first appearances it soon became obvious that Giles and the directors had not merely dismissed another superintendent, but that they had altered the nature of the position as well. True, Henry Lowe had been given the title, but his duties were in fact limited to the direct supervision of manufacturing operations. Moreover, the factory had suddenly acquired two other de facto superintendents.* George E. Hart, in charge of machinery, was even being referred to as superintendent in the press.[8] The other was Fayette, returned from Switzerland and now nominally in charge of research and development.[9] In this new structure the activities of the superintendents were coordinated by the cashier, Henry Wright.[10] It must be remembered, however, that Wright was in reality Frederick Giles' shadow, thus real, total and direct control was now in Giles' hands.

The primary advantages of this new arrangement, at least from Frederick's point of view, were that on one hand he now had direct control over the factory, and on the other he now had an executive staff that he could trust. Unfortunately, the fact that Fayette was Frederick's brother as well as business partner created a serious flaw in the structure. At the outset, however, the younger brother's abilities far outweighed the liabilities of nepotism.

Upon his return to the United States Fayette had settled Bertha and their baby son in Westport, Connecticut. Why he had chosen to locate his family so far from the factory is hard to understand. Even if New York, particularly in the summer, seemed too crowded and unhealthy, certainly there were many desirable locations nearby in New Jersey. Whatever, the distance meant that Fayette would be away from home for many days at a time. The problems of distance became all too evident when, in August, illness struck Fayette's household. It was Saturday the 28th when Fayette received a letter from Bertha, dated the day before, stating that she had been so ill that she had the doctor twice that day. Then later, while he was at the factory, another dispatch arrived: "Baby very sick come soon." Fayette dropped everything and was soon entrained for Westport. Sadly, it was all to no avail. Frederick received a message from his brother — "Baby dying. Poisoned to death. Will one of you come up."[11]

Precisely what had happened we do not know, however one strong possibility suggests itself. It was an age when childhood illnesses were treated with opiates. All too often lethal dosages were administered by parent and doctor.

Shortly after Fayette's message was received William Wales left for Westport. Frederick would have gone as well, however the legal battle between the watch company and Learned had just opened. Since he was due to give testimony in court on Monday there was no possibility he could leave until after his appearance. By then it was too late.[12] Circumstances had prevented Frederick from being in the one place at the one time he was absolutely needed.

After the death of Fayette's son matters returned to normal as best they could.[13] Learned's lawsuit against the company dragged on. Fayette and Bertha moved into the Merchant's Hotel near the store on Maiden Lane.[14] Fayette buried himself in his work and in the months immediately following the tragedy produced a set of brilliant designs for a small woman's watch.[15]

The United States Watch Company had lagged behind its competitors in introducing a watch designed for the growing women's market. Fayette's design would have immediately eliminated this disadvantage, however there was one slight problem. Getting it into manufacture would have entailed setting up an entirely new production line. Since the company was still having far too much difficulty getting out production of their rapidly expanding 18 size line, producing a new watch of unique design was out of the question. There was, fortunately, one possible solution; the 10 size could be manufactured in some other factory.

Once again Giles, Wales & Co.'s overseas connections proved useful. The 10 size watch was jobbed out to the Bourquins of Bienne, Switzerland.* Inasmuch as these movements, labeled "Chas. G. Knapp" and "R. F. Pratt," were imported in the grey — without dials or cases — they could easily be classed as parts, requiring the payment of only ten per cent duty.[16] As far as Giles, Wales was concerned, profits were further enhanced since Fayette's patents were assigned to them rather than to the United States Watch Company.†

There was, unfortunately, a negative side to things. Besides their Swiss operations the Bourquins also had a store on Maiden Lane. Their establishment at number 20, upstairs from George W. Platt and Co., was a competitor in the import trade with Giles, Wales & Co.[17] Also it would seem that the arrangements with the Bourquins required certain patent concessions on the part of Giles, Wales. The Bourquins were not only per-

---

*The question of the line of succession of superintendents is one of several tacky points which delayed completion of this work for many years. While the present solution may still leave unresolved questions, it is supported by several sources including an unidentified clipping, "The President at a Watch Factory," October 1, 1870, Burns Collection.

*The fact that these 10 size watches were not manufactured in the United States has long been suspected by collectors, however proof was difficult to come by. This problem was at last resolved in the late 1960's when an example of the 10 size complete with Bourquin markings was found on the Bowery in New York City.

†As noted on U.S. Design Patents 3885 and 3886.

52. *10s, 15j, gilt "Chas. G. Knapp" grade, SN 60362. Designed by Fayette Stratton, both Knapp and the companion size "R. F. Pratt" grades were manufactured for USWC by the Swiss firm of Bourquin. The USWC called this style ¼ plate, cock and bridge.*

53. *18s, 15j, nickel "Ls. A. Bourquin" SN 15154 with a modified USWC type cut out in the plate. It is likely that they were granted this variation as part of some arrangement with the USWC and Giles.*

mitted to manufacture watches for European distribution based on Fayette's patents, but also watches utilizing a modified form of Frederick's "butterfly opening." The watches incorporating the latter feature were 18 size and may have been sold on the American market.* Aside from the Bourquin concessions there was another problem. Fayette's designs had been for key-wind watches and did not lend themselves easily to conversion to stem-winders. These factors, plus the passage of a law in 1871 that required all imported movements to be marked with their country of origin, forced Giles, Wales & Co. to reconsider the entire 10 size matter.[18] Much to Fayette's chagrin, it was decided to discontinue the Swiss-produced model. When the next series of 10 size watches were manufactured they would be built at Marion and designed by Frederick Giles.

---

*Inasmuch as neither Giles, Wales & Co. nor the United States Watch Co. brought patent infringement action against the Bourquins, a *de facto* arrangement must be assumed.

CHAPTER 7 REFERENCES

1. Sue Hubbard to Julia Giles, February 21, 1882, Burns Collection.
2. Dunn and Bradstreet, *1867 National Business Directory.*
3. *Harper's Weekly*, June 25, 1870; also *The Gazette*, June 18, 1870.
4. Fayette Stratton Giles to Frederick Giles, July 4, 1866, Burns Collection.
5. Crossman, Adams Brown Reprint, p. 87; also *American Horological Journal*, September, 1870, II:71.
6. Crossman, p. 87.
7. Ibid.
8. *American Horological Journal*, January, 1873, IV:158. Also: *American Jeweler*, January, 1906, p. 7 and *Jewelers' Circular Weekly*, January 3, 1906, pp. 42-3.
9. *Harper's Weekly*, June 25, 1870. Also: *The Gazette*, June 18, 1870.
10. *The Gazette*, June 18, 1870. Also: "A Gold Watch Added to his Collection of Testimonials," October 1, 1870, unidentified clipping, Burns Collection, and note on back of business card, dated June 24, Burns Collection.
11. Frederick Giles to Julia Giles, August 29, 1869, Burns Collection.
12. Ibid.
13. "F. S. Giles is at Rest."
14. Trow, *New York City Directory 1870-71.*
15. U. S. Design Patents 3885 and 3886, March 8, 1870.
16. Henry Abbott, *Watches and Men* (New York, 1933), p. 5. Abbott also points out that the duty on complete watches was 25%, thus it was cheaper to import them as parts.
17. Ulmann, facing p. 97.
18. H. F. Piaget, *The Watch* (New York, 1868), quoted in Wesley R. Hauptman, "Swiss Imitations of Early American Watches," NAWCC BULLETIN 9:272-4.

MARION

UNITED STATES WATCH CO.'S WATCHES.

REPORT OF JUDGES
AT THE SECOND
INTERNATIONAL INDUSTRIAL EXPOSITION OF THE
MECHANICS' INSTITUTE:

TO THE EXHIBITION COMMITTEE:

GENTLEMEN,—The undersigned, Judges in Department 1, Group 7, report that they have carefully and impartially examined, according to the "Instructions to Judges" transmitted to them, the several competing articles submitted for their judgment, and that the following are their conclusions:

No. 795. United States Watches. Entered by   T. & E. DICKINSON,
Buffalo, N. Y.

"T. & E. Dickinson exhibit American Watches, manufactured by the United States Watch Co. (Giles, Wales, & Co.), Marion, N. J.
"These Watches are, without doubt, the best manufactured in this country."

ALLEN CHURCH,
GEO. I. BENTLEY, } Judges.
B. S. BENTLEY,

I hereby certify that the above is a correct copy of the report of Judges on Entry No. 795, and that the same was awarded First Premium. (Large Medal.)

Buffalo, N. Y., Dec. 29th, 1871.   D. B. McNISH, Secretary.

Price-List furnished the Trade on application, inclosing business card. For sale by the trade generally.

Ask your Jeweler to see the MARION

WATCHES.

WHOLESALE ROOMS OF THE **United States Watch Co.,**

GILES, BRO., & CO.,   **GILES, WALES, & CO.,**

384 Wabash Ave., Chicago, Ill.   No. 13 Maiden Lane, New York.

54. "Harper's Weekly" ad, March 30, 1872, featured the "United States" grade and first-place results at the 1871 Buffalo Industrial Exposition.

Chapter 8
The Apex

While there had been a small setback or two, the national economy had improved steadily since the war.* By 1870 business in general and the jewelry industry in particular were booming. Giles, Wales & Co.'s gross sales had shown steady growth since their low point in 1867. When the books were closed at the end of 1870 they showed sales within fifty thousand dollars of the company's peak wartime year of 1865.[1] 1870 also marked a high point for the factory. The ninth federal census, taken on July 1st, indicated that the works now employed 315 hands and paid out one hundred eighty-three thousand six hundred fifty dollars in wages during the year. In the same period the company had five hundred thousand dollars in working capital, spent forty-three thousand three hundred eighty-five dollars on materials and produced five hundred thousand dollars worth of watches.[2] Although Giles, Wales & Co. would continue to improve on this rosy picture for a few years and the factory's capital and number of employees would also grow, there were clouds on the horizon. Both the United States Watch Company's production figures and their ability to dispose of that production were less than they should have been.†

It should never be assumed that the United States Watch Company's sales difficulties resulted from a lack of merchandising efforts in the media. Shortly after the administrative reorganization of 1869, Giles, Wales & Co., Giles Brothers & Co. and their jobbers began an advertising campaign that was, for scale, without precedent in the watch manufacturing industry. While the extensive use of advertising was hardly new to the watch factory entrepreneurs of the 1870's, it was Frederick Giles who first made use of saturation techniques. Pictures of the factory, pictures of the company's watches, articles about the company and straight advertising by the company began to appear in copious quantities in the major magazines and newspapers of the day. In fact both the frequency at which material appeared and the number of periodicals in which it could be found would be considered respectable by even today's standards. Not only does one find material in such trade journals as the *Watchmaker and Jeweler*, but also in important farm papers such as *Moore's Rural New Yorker* and prominent mass circulation periodicals such as *Harper's Weekly*.* Nor did the ballyhoo stop there. Somehow the company managed to gain mention even in books of the period, such as watchmaker's manuals and state and industrial histories. In statistical terms there are at present about one hundred advertisements which are known to have appeared during the period 1869-71.† After that time the number and frequency diminishes greatly, but by one means or another the company used advertising until its demise.

United States Watch Company, or more correctly Giles, Wales & Co., advertisements fall into two distinct categories, the first being articles about various facets of the company or its products. Typical of the period, only a few of these were identified as advertising, however this is, of course, exactly what they were. Perhaps it is not unreasonable to go further and suggest that more than one of these works smacks of the hand of Frederick Giles. The first known article about the company appeared in the 1868 edition of J. Barber and Henry Howe's *Historical Collections of New Jersey*. This friendly article must have been particularly gratifying to Frederick since Howe was noted for his biographies of eminent mechanics. In 1869 brief articles appeared in Kenelo's *Watch Repairers Handbook* and in the September edition of E. Albert's *Watchmaker and Jeweler* magazine. Mid-1870 witnessed the publication of two major articles. The best known of these appeared in *Harper's Weekly* on June 25th; no less important, however, was a long article published on June 18th in a Jersey City newspaper, *The Gazette*. Once more (on May 27, 1871) a local newspaper, *The Jersey City Journal*, published a long article about the factory. This item is noteworthy for its biographies of William Wales and Frederick Giles. Later in the year, December 17 to be precise, *Moore's Rural New Yorker* ran a substantial piece about the company's machinery. 1873 saw the first publication of an article entitled "Marion and

---

*One of the best known of these occurred with the "Black Friday" disaster on the gold market in September of 1869. Although all should have learned something about the dangers of speculation, subsequent events were to prove otherwise.

†The lack of production records has pretty much left the various scholars who have approached the problem totally adrift. In earlier years it was assumed that the solution lay in locating the highest serial number; modern research, however, has demonstrated that large gaps in the numbering make this approach useless. Thus, current thinking suggests that rather than a total production figure of almost 300,000, the pitiful result of a decade's work was a mere 50,000. (See Chapter 24.)

Harper's Weekly seems to have the longest string of advertisements, twenty-six in all, running from the December 18, 1869, edition to November 5, 1870.

†It is suspected that there are more, however there comes a point when further research does little to add to the result.

SCIENCE AND ART.

Marion and its Temple of Labor.

While the world-traveled stranger is being whirled out of Jersey City, just beyond the western verge of that expanding hive of industry, the first object of beauty that strikes his eye is a vast but airy structure, chiefly of glass and iron, lifting its chaste proportions from a garden lawn of several acres.

It seems as it built for a colossal conservatory, or a summer palace for a prince of fairy tale; which it were our traveler could not guess. But what would be his astonishment if told that this structure of such wondrous beauty, held six hundred artisans of both sexes, moving intelligently among nearly as many curious and tireless machines, doing the work of one hundred thousand human hands, and that there were sent from this scene of enchanted labor a quarter of a thousand perfect watches in silver, and gold, and precious stones, every twenty-four hours. The first thought of our traveler would be about horology in former ages, and American watches made by perfected machinery in 1874.

The difficulties which attended the final establishment of a fixed standard of time are inconceivable in our day, just as the importance of accurate time-keeping is incalculable. Without a reliable measure of time there could be no union of action among men. The vast net-work of railways, steam lines and telegraphs, which are now making a brotherhood of men in business and feeling, would beget nothing but inconceivable disaster. Nor could the triumphs of modern science over the stubborn forces of nature on earth, or in the stellar universe, have been won. And yet man began his observations and experiments in this unknown field without a single ascertained landmark, wholly unconscious even of the period when he commenced his own existence—with no idea of time, or natural or mechanical means for fixing its beginning, or measuring its flight. He had to make an alphabet from the sun, moon and stars, and create a language, harmonious and universal, by which to read, arrest and fix the hurrying movements of the whole physical creation.

Shepherds watching the stars, five thousand years ago, on the plains of Chaldea, traced imaginary forms in the sky, and named them after the animals in their various herds, and, by the aid of the hand and the stars, fixed the first rude measure of time. Then came the noon-mark, and the dial and the flowing tide of sand or water; and finally, at the close of the tenth century, the discovery of the balance clock with weights, and eight hundred years later the improved pendulum clock, and the balance spring clock, or watch for the pocket. The mechanical principle of time-keeping consists of the uniform division of a constant force in a given time, so that a vibratory or reciprocating movement is continually changed into rotary motion. The agent for doing this is called an escapement, and when this delicate point was reached the great invention was made.

There could be only one thing lacking, and that, like so many things, was to be achieved by American genius—viz: making watches by machinery. England, France and Switzerland chiefly supplied the world with watches until the American discovery. Since 1812 we imported up to last year probably over $300,000,000 worth, nine-tenths worthless as time-keepers and costing more than the original outlay in trying to make them go in some sort of correspondence with the motions of the solar system.

The American Watch is an abolute triumph of science and art. A thousand from the same model could be distinguished from each other only by the number. Being made with the precision of unvarying machinery, perfect uniformity is secured, and if the first is made right, all the rest must be.

This grand result—this new miracle—is performed nearly a hundred thousand fresh times every year in that fairy Palace of Labor in Marion. Hitherto the magnetic needle has been considered the only perfect type of unchangeable fidelity. But we must give up the comparison, for the variations of the magnet are endless. Its main fidelity is towards the North Star. But it yields to so many other attractions that, if followed implicitly, it would sweep all navigation from ocean. But the Creator's latest and grandest work is the genius of man, which atones for the otherwise imperfect reliability of the magnet. The Marion pocket watch seems to us a near approach to perfection. Scores of these watches will follow their wearers around the sun—a circuit of over four hundred million miles—and not make a variation of five seconds; while on the same long journey the magnet may vary a million of times. Never did a devout Heathen or an illuminated Christian, worship so unwaveringly his idol or his God.

The extraordinary record of these Watches as the time-keepers in the pockets of the American public, including many of the leading men of our country, R. R. Engineers, Conductors, Expressmen, and others whose occupation renders it absolutely necessary they should have the most accurate time, speaks volumes.

It has remained for this Company to supply two serious wants, long felt by both dealer and consumer, but which have heretofore been unattained. One of these is a Patent Reversible Barrel, to prevent damage to the train in case of breaking the mainspring. This (which they now put in all of their watches) is quite an ingenious contrivance, extremely simple, and so arranged that it is always free and ready to act, even though the watch may have run for years, unlike other appliances for similar purposes, which, after the watch has been wound a few times, become set and entirely useless, so far as accomplishing the purpose intended. The other is a Patent Double Index Regulator, beautiful in design and finish, while in novelty of construction and the results obtained, it is a little Wonder, and we are sure will be hailed with delight, not only by watch wearers, who appreciate fine time, but by the seller, who with the old style of regulator, is often exceedingly annoyed (even with watches that are otherwise fine), and, in fact, is often unable, after repeated trials and consuming much time, to touch the regulator fine enough for a small variation, getting it first too far one way, and then to far the other, backwards and forwards, sometimes for months.

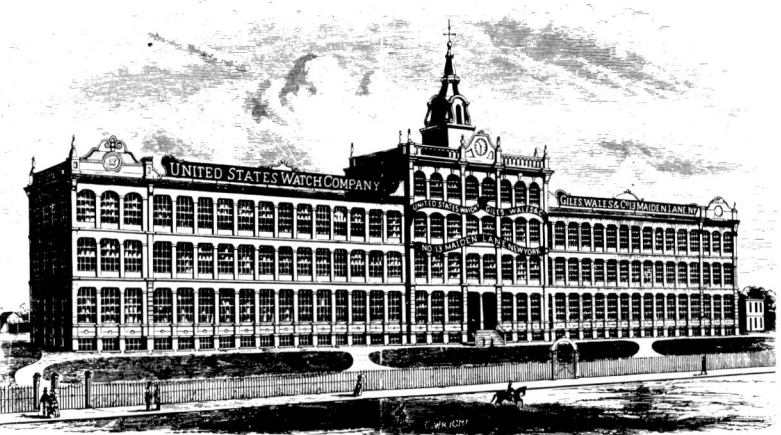

MANUFACTORY OF THE UNITED STATES WATCH COMPANY.

This house makes a specialty of Stem Winders, for which they own patents, and the mechanism of which, for simplicity, durability, strength and smoothness of action, excels anything yet produced, either at home or abroad, and is becoming quite a feature in the trade—so much so that prominent dealers predict that in less than five years there will be none but stem-winding watches sold. This Company, we observe, has taken the front rank among American manufacturers. At the first Fairs held in different parts of the country, where there has been great competition in this line, the Marion (Giles, Wales & Co.) United States Watches have been regarded as greatly superior in every particular to any on exhibition, and have been awarded the first premiums, over all competitors, at a Fair of Cincinnati Industrial Exposition, Ohio; Ohio Mechanics' Institute, Cincinnati, Ohio; at "Louisiana State Fair," New Orleans, La.; at "Texas State Fair," Houston, Texas, 1871; at New Jersey State Fair, 1872, and Iowa State Fair, 1872.

In conclusion we will introduce the individual members of this model Establishment, that our readers may appreciate the skill, perseverance and manly qualities possessed in so large a degree by each individual Member as a Citizen.

Mr. Giles, being left an orphan at the age of eight years, and being the oldest male member of a family of seven children, it became incumbent on him, with his eldest sister, to provide for and superintend the education of the rest. Feeling his responsibilities as a son and a brother, he determined on learning the manufacturing of watch-work, with the idea, at some future day, of controlling the destinies of a watch manufactury. The manufactory, now completed, stands to-day a monument to the skill and determination of its worthy head.

W. A. Wales, Esq., is a gentleman who, once met, is not easily forgotten. Having experienced all the disadvantages possible in his early efforts to rise, he is always ready to cheer and encourage the young man, striving to win himself position.

In G. C. F. Wright, Esq., the junior member of the firm, we find all the elements of the true business man. Bold and determined in the prosecution of trade, genial, generous and decided in character, he has won for himself a host of friends, both in active and private life.

55. Complete text of "Marion and its Temple of Labor" article used in promotion between 1873 and 1876.

its Temple of Labor."* This work was used again a year later in Asher and Adams' *Pictorial Album of American Industry, 1876*. Throughout the period, 1869-73, the *American Horological Journal* carried numerous articles about the factory and its machinery. There were, of course, many other newspaper articles about the United States Watch Company; unfortunately, all too often these were straight news and sometimes they were anything but flattering.

The second category of advertising used by the company was the standard variety that one might expect. Except for quantity, the United States Watch Company restrained their usual creative impulses and followed normal practices in the industry. This meant that the bulk of their ads appeared in magazines rather than newspapers. In terms of style they relied, as did other manufacturers of the third quarter of the 19th Century, upon the testimonial to tout their products. Whatever creativity one might find in watch advertisements was

---

*Unfortunately, the original 1873 version of this article, "Marion and its Temple of Labor," is one of those unmarked clippings from the Burns Collection and thus the periodical which carried it cannot be identified.

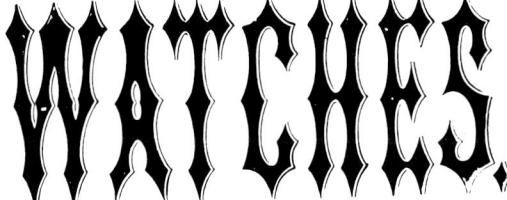

56. *"Watchmaker & Jeweler" ad, March 1872, with the testimonial of Sam'l Merrill, Governor of Iowa.*

in the claims. Perhaps it would be better to hear from an eyewitness, Charles T. Higginbotham: "Certificates from owners of the various makes of watches as to the time-keeping qualities were eagerly sought for by the makers, and many were the schemes resorted to for the purpose of obtaining them. Presenting watches to individuals, thus placing them under obligation to the company, was a common occurrence. One of the cleverest schemes was that practiced by a certain large wholesaler doing business at that time. In the establishment of this wholesaler were three timepieces located in separate parts of the establishment. One was a fine Howard Regulator near the front of the establishment; another was a Bliss Chronometer placed in the watch repairing department; a third was a Negus Chronometer, in the private office. These timepieces were purposely kept ten seconds apart from each other. The chronometer in the repair department was correct, the clock in the front of the store ten seconds fast, that in the private office ten seconds slow. This gave a range of twenty seconds. Thus when the carrier of a watch came in a glance was given by some member of the firm at the watch and at the Howard clock; this gave the cue as to whether the time should be taken from the clock or from one or the other of the two chronometers."[3]

On October 1, 1870, Frederick Giles accomplished the ultimate coup in the testimonial field when he was able to get President Ulysses S. Grant to visit the factory and accept the gift of a watch from the company. The watch was described by the press as a "heavy double-cased gold watch, with stem-winder and worth about five hundred dollars" and one might add, all neatly enclosed in a rosewood box to boot. It would seem, further, that the President and the press enjoyed their visit which included champagne at the St. James.[4]

It was also quite common for watch factories of the period to enter their products in industrial and trade fairs. The United States Watch Company was hardly a laggard in this area of advertising either. In September of 1870 the company had a large exhibition of its watches at the fair of the American Institute of New York. Their display included over fifty different styles, among which were some very expensively cased examples of the 10 size model. Another aspect of this display centered upon the company's prowess at dial manufacture. For their efforts at this fair the United States Watch Company was awarded a gold medal.[5] Shortly thereafter the company was awarded first premium at the Cincinnati Industrial Exposition of the Ohio Mechanics' Institute. The company made great use of this award in some of their advertisements. Surrounded by no less than sixteen different kinds and sizes of type, a facsimile of a handwritten note announced, "No. 1650 United States Watch Company, Marion, New Jersey. These watch movements, the result of American capital, skill and perserverance, as specimens of this branch of science and manufacture rival those of European make. The watches exhibited by this company are of all grades, from that of the most exact timekeeper, to the cheapest kind for the million, all are excellent of the kind, and are too well known and appreciated by the people to need further comment, *First Premium* is awarded to the *United States Watch Company*. E. L. Wayne, James Powell, William M. Davis, Judges."[6] Following the Cincinnati Fair the company was awarded first premium at the Louisiana State Fair, New Orleans; the Texas State Fair, Houston; the New Jersey State Fair and the Iowa State Fair.[7] (See Chapter 20 for summary of all known exhibitions participated in by the USWC.)

Although 1869-71 was the period when, all aspects considered, the United States Watch Company was at its apex, there were areas which peaked before or after this time. Mechanically, the highest level was achieved between late 1872 and early 1873. Unfortunately, since the company's ability to advertise had declined before then, there are several aspects about this late period that must remain conjecture. This, however, is not the case for 1869-71. There is almost a surfeit of information about the machinery and its various locations in the factory during this period. The articles which appear in *Harper's Weekly, Moore's Rural New Yorker, The Gazette,* and *The American Horological Journal*

57. Watch factories of the early 1870's were constructed in long wings to afford maximum windows and sunlight. Everything had to be connected to the central power source by means of belts, pulleys, and shafting suspended from the ceiling. The visual effect is illustrated in this engraving from "Watchmaking in America," first printed in 1870 by Robbins, Appleton & Company.

represent one of the largest bodies of specific details we have about any watch factory.[8] Its only real rivals are E. A. Marsh's *Watches by Automatic Machinery at Waltham*, George H. A. Hazlitt's (Henry G. Abbott) "The Lancashire Watch Company, Prescott, England" and the recently discovered inventories of the New York Watch Company, Springfield, Massachusetts.

It is a trifle difficult, even for those familiar with the modern equivalent, to visualize the factory of a hundred years ago. The contemporary manufactory is the product of cheap electric motors, the power demands of heavy, high speed machinery and the advantages of the horizontal production line. In old time mills the central power source was everything; it was seldom possible or practical to provide individual machines with their own mill wheel or steam engine. Everything had to be connected with everything else by means of belts, pulleys, and endless lines of shafting suspended from the ceiling. This centralization of power produced a totally different type of plant organization than is now practiced. In New York City all of the newspapers had their buildings at a single location, Printing House Square. The cause of this phenomenon was the fact that one power plant with two steam engines, one of one hundred fifty horsepower and another of seventy-five horsepower, powered all one hundred twenty-five of the local presses. All being nicely belted together by long strands of rubber or leather which ran under the nearby streets and up through the various newspaper buildings.[9] Factories of the period were multi-storied not because land was expensive, but simply because a vertical arrangement allowed shorter and more efficient power transmission. Belts were omni-present in these mills; not only were there the heavy main drive belts, but also a forest of smaller ones. The total visual effect was a dense jungle of leather and rubber strands that slapped and vibrated as they turned the wheels of industry.

Marion's source of power was, of course, the eighty horsepower Putnam located in the factory's basement. The power plant was not the only thing, however, that was located on this level. Weight and heavy power requirements also necessitated that the company's extensive press, rolling, and slitting equipment be located in the basement. The United States Watch Company used punch presses in many phases of their operations; among these were the stamping out of dial blanks, bridges, and rough plates. Included in the process of punching the plates was the piercing of the butterfly opening. Also situated on the basement level were the blacksmith's shop, the hardening room and storage for the company's rough materials.[10] The latter function, dictated by the location of the press shop, proved to be not altogether desirable, since the local dampness was constantly attacking the company's costly stocks of high grade American steel.[11]

In the central section of the factory's first floor were located the reception room and a hallway, on one side of which were the business offices and on the other the director's office. Flanking the central section were the wings containing the machine shop and the forwarding room. All of the tools, dies and special machines required by the factory were serviced and constructed in the machine shop.[12] The machine shop had, ever since the move from Newark, been the special province of George Hart. His skill as a machinery builder is attested to by some of the extremely well constructed and accurate machines in the shop. One of these, a master screw cutting machine (lathe) came to the attention of *The American Horological Journal*. This early trade paper's editor reported that the mechanical wonder could cut threads as fine as four hundred per inch and further that it produced all of the taps used within the factory.[13] Unfortunately, this machine may well have been a factor in the mistaken choice of fine pitch over coarse pitch plate screws. That error, however, cannot be attributed to Hart or the machine shop. Their monument was the nearly five hundred machines which filled the factory.

Perhaps the busiest and most important area in the factory, the plate and forwarding room, was also located on the first floor. Here the United States Watch Company machined all of its watch plates, flat work, most of its screws and numerous other components. This facility was over a hundred feet long and contained four long benches that ran its entire length. On these benches were one hundred and twenty-five lathes and other machines.[14] E. A. Marsh indicated in his *Watches by Automatic Machinery at Waltham* that the lathes used in machining watch plates were extremely inefficient. Despite the fact that these machines required the operator to use his hands, feet and knees, each lathe could perform only a limited set of operations. The watch factories of the period were thus forced to employ large numbers of these cumbersome machines.[15]

By way of contrast to the awkward plate lathes a brief comment about the steady pin machine is in order. This machine was a precursor of the marvels of mechanical automation that were to fill later watch factories. In American-made watches steady (banking) pins were oval head brass screws which had a small pin machined eccentrically to the line of centers at the threaded end. Imagine then, if you will, that one machine beginning with raw stock performed all of the necessary operations of heading, slotting, threading, turning, and as a final flourish screwed the finished piece into the back plate.[16] By our contemporary standards not a wonder, but in 1870 it had to inspire a little awe.

The forwarding room's regular screw manufacturing lathes were identical to those used at the Waltham factory during the same period, thus again we can turn to Marsh's book for accurate pictures and information.[17] Generally speaking these machines were little more than regular watchmaker's and instrument lathes equipped with special fixtures to convert them into mass production machines. An example of this type of arrangement would be the head slotting lathes. *The American Horological Journal* described the Marion machines in this fashion: "This chuck is simply two steel disks, . . ., of perhaps an inch and a half diameter, held together by a screw nut, . . ., and the line of junction of its two surfaces in contact, pierced and tapped with screw holes as near together as the head will permit. This chuck, when fitted in the lathe to which it belongs has a slow revolving motion, while the rotary slitting saw, . . ., revolves rapidly, making in the head of each screw that passes under it a diametrical slit and on removal from the lathe, opening the disk allows the screws to fall out."[18]

The June 25, 1870, *Harper's Weekly* article about the factory stated that "In the centre of the second floor the office of the superintendent is situated, and in which the private drawings are kept. The drawings of all parts of the movements are on an enlarged scale and upon which are marked the measurements of the private standard gauge that is used by this company. So accurate is this machine that the very smallest possible fraction of difference can be divined, even to the twenty-fifth thousandth part of an inch. To make this more plain, and to show the exactness required on the part of every movement, we will say that the finest hair from the human head will show by this unerring finger a difference of one half inch." As well as the superintendent's office the second floor contained the carpenter's shop, the escapement shop and the lapidary. The lapidary, of course, housed the equipment used to manufacture watch jewels. There is reason to believe that this facility may have been moved to the third floor in 1871. Of the escapement shop, it can be said that this room contained almost as many machines as the forwarding room. While it is a certainty that balances, pallet forks and other escapement parts were manufactured there, some confusion exists in the various

58. *Escapement wheel machine, circa 1870, illustrated in this engraving from "Watchmaking in America."*

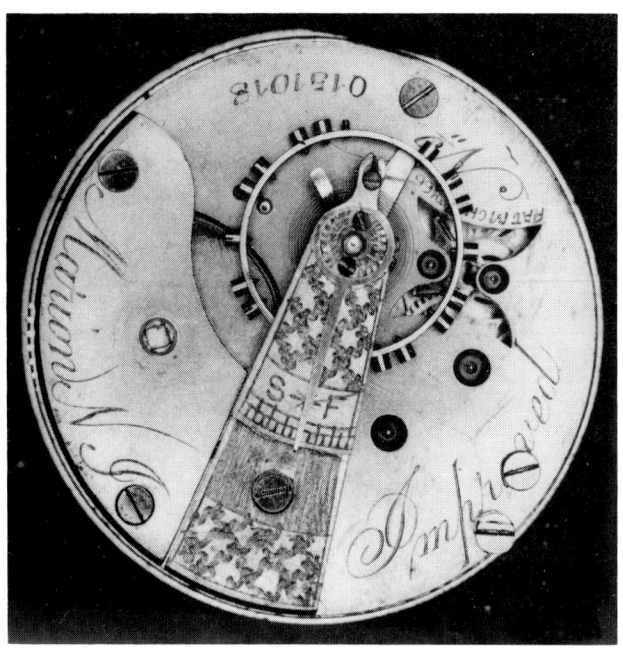

ABOVE. 59 and 60. Two probable products of USWC experimental department. The 18s full plate marked "Improved," SN 0151018, would appear to be a basic "G. A. Read" grade that has been upgraded from 7j to 11j. The SN 151018 is from a run of Reads; the 0 in front of the SN is a different engraving style and was probably added later. The unmarked 18s movement with rounded edge cut out is a variation probably evaluated by the USWC. The "Independence America" Centennial product shown in Figure 210 used almost exactly the same type of cut out. F. A. Giles patent No. 2266 of February 27, 1866, covers the design for a top plate using a type of rounded edge cut out. BELOW. 61 and 62. Two examples of fine dials produced by USWC at their apex. Both are 18s. The Masonic dial has the full Mason's emblem below XII in blue, gold, and black and a USWC monogram in seconds bit done in black. The Symbol of Peace (dove with an olive branch) dial is done in black and green with black USWC monogram in the seconds bit.

63. The USWC was one of the earliest, if not the earliest, American watch companies to develop a fine dial making reputation. Custom dials were offered at an additional price ranging between three and fifty dollars. This grouping includes a bold numeral conventional double recessed dial, one USWC monogram, two banner dials, and a unique presentation dial with the "Lord's Prayer" in the seconds bit.

sources as to whether pinions may have been produced there also. It seems more probable that this was done in the motion room on the third floor.[19]

A large number of departments were located on the third floor. Among these were the closely guarded and highly secret damaskeening and experimental rooms. Another third-floor department was the company's exceptionally fine dial making operation.[20] The United States Watch Company seems to have had great success in hiring dial painters and as a result was able to produce quality dials including many fine special ones, particularly for presentation watches. Masonic and custom dials were offered by the company for an additional fee, which ran between three and fifty dollars. It ought to be noted that the interest in Masonic business stemmed from the fact that Frederick, his brothers and the management were all members of that fraternity.[21] For display and other special purposes the company's dial painters produced a series of delightful tours de force including a seconds bit containing the "Lord's Prayer" and one dial with a painting of the factory upon it. Sadly, only a few of the factory's fancier dials seem to have survived the passage of time. They, however, give ample evidence as to the company's abilities in this area.[22]

By way of contrast, the smallest third-floor operation was gilding while the largest was the motion room. Inasmuch as only a fraction of the company's movements and some of its watch parts required gilding, little space was required for plating operations.[23] The motion room's situation was, on the other hand, quite different; this topmost of the factory's major machinery rooms produced the spring barrels and gearing for all of their watches. It rivaled both the forwarding and

OPPOSITE PAGE. 64. Enlarged view to show detail of "Lord's Prayer" in seconds bit. Note entire prayer occupies less than half of seconds bit, with block letters that average .008" height. Prayer begins at 60 and ends at 45 with break in words as artist progressed to next smaller circle in clockwise direction. The prayer is painted in black with bit chapter ring and numerals trimmed out in gold, red, and green.   BELOW. 65. Enlarged view of "Lord's Prayer" presentation dial showing "CKL" monogram and fancy Old English numerals in black trimmed out in gold, red, and green.

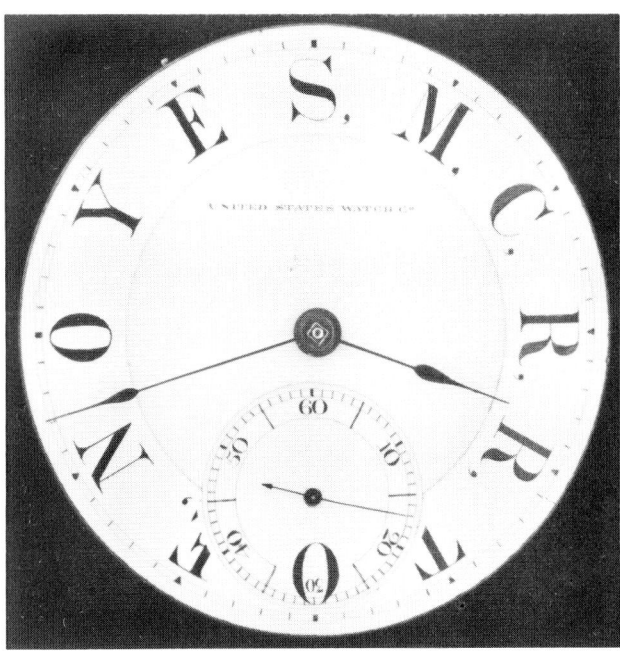

66. Another presentation dial by USWC with bold capital letters in place of numerals. These letters are believed to read "To E. Noyes, M.C.R.R."; possibly Mississippi (or Maine) Central Rail Road. It could have been "O. E. Noyes, M.C.R.R.T."

67. USWC monogram dial. It is interesting to note that the dial painters used several different variations of the USWC monogram over the years. The USWC fancy dials are usually double recessed and found on the nicer "United States" and "A. H. Wallis" grades.

escapement shops in size and contained nearly as much machinery as either.[24] The history of this facility is, however, far more complex than the other two major rooms. Basically the direction of production flow in the factory was upwards; that is, new materials entered at basement level and then proceeded floor by floor upward until the finished watches emerged at the upper levels. As new models were added and segments of existing ones were modified it became necessary to reorganize various rooms in the factory. For one reason or another the third floor, and in particular the motion room, bore the brunt of these changes. The full extent of these alterations becomes apparent when one realizes that the assembly aspect of the motion room, which

68. 18s, 19j, SN 24042 "United States Watch Co." grade, nickel frosted with damaskeening. Prestige item of the USWC, these products in nickel with Breguet hair spring were first introduced in 1869. Note fine engraving on the balance cock and damaskeening pattern; no two examples are the same.

gave it its name, was moved from there to the fourth floor in 1871.[25]

One series of production operations of particular interest was located in the motion room. These were the facilities for pinion cutting and polishing. The Marion company's setup was undoubtedly one of the most advanced then in use. At the time when the factory's machinery was constructed most, if not all, of the other American watch manufacturers were using drawn English pinion wire as stock for their machines. This material was purchased rough drawn in twelve-inch lengths and the production operation which converted its fluted shape into precise pinion leaves was by rounding up rather than gear cutting. Marion procedures, however, dispensed with the costly imported pinion stock and instead utilized round drawn wire. In brief, the United States sequence of operations were cutting to length, pointing, arbor cutting, pinion roughing, hardening, grinding, and finally polishing. After a while other factories caught up with and passed Marion, but while it lasted the United States Watch Company's semi-automatic pinion machinery was the wonder of the industry.[26]

The last floor in the factory was the fourth. This level

69. 18s, 19j, SN 19210 "A. H. Wallis" grade, nickel frosted with damaskeening. One notch down from the United States grades, these products did not have a Breguet hair spring but were still sold at prestige prices. Cock engraving and damaskeening patterns also vary on each example. Circa 1869.

was limited, of course, to the central section of the building. Originally this space was occupied by the adjusting department, however the room was really quite a bit larger than was necessary for this purpose. When the factory was reorganized in 1871 watch assembly (motion works) was moved to this room and the adjusting operation was sent to less commodious quarters on the third floor.[27] One of the reasons that assembly required more room was its need for storage space.* Back-logged movements and parts were a constant headache for all of the early American factories. The New York Watch Company inventories indicate that that company tinkered with some models for years and that parts for the Mozart watch were retained long after the ill-fated design had passed into the status of legend. As subsequent events will show, the United States Watch Company not only conformed to this pattern, but set something of a record of its own for string saving.

*"The President at a Watch Factory" states that "The pieces are bottled up and labeled and ready to be counted out, like the parts of a Springfield musket, and put together, . . ."

CHAPTER 8 REFERENCES

1. Frederick's Memorandum "Giles, Wales & Co. Sales and Profit 1862-73" gives the following sales figures:
    $378,540.39 — 1866
    $366,041.90 — 1867
    $387,482.53 — 1868
    $408,006.48 — 1869
    $450,335.66 — 1870
This would in turn suggest the following approximate profits:
    $58,373.76 — 1866
    $56,446.40 — 1867
    $59,752.71 — 1868
    $62,917.65 — 1869
    $69,445.13 — 1870.
2. *Ninth U. S. Census* (1870) Vol. III:694.
3. Charles T. Higginbotham, "Incidents in the American Watchmaking Industry," *National Jeweler and Optician*, March, 1912, p. 184.
4. "The President at a Watch Factory," October 1, 1870, unidentified clipping, Burns Collection. Also: "A Gold Watch Added to his Collection of Testimonials," October 1, 1870, unidentified clipping, Burns Collection.
5. "Fair of the American Institute of New York," *American Horological Journal*, November, 1870, II:116-117. Most of the medals which the United States Watch Company won are extant, but the passing of time is beginning to scatter them.
6. *Moore's Rural New Yorker*, September 9, 1871.
7. "Marion and its Temple of Labor," unidentified clipping, Burns Collection.
8. The complete known list of articles containing information about the factory's machinery is as follows:
    Harper's Weekly, June 25, 1870
    The Gazette, June 18, 1870
    Moore's Rural New Yorker, December 17, 1871
    American Horological Journal, May 1872, III:251-4; June 1872, III:266-9; July 1872, IV:5-8; September 1872, IV:54-56
    "The President at a Watch Factory."
9. James D. McCabe, Jr., *Lights and Shadows of New York Life* (Philadelphia, 1872), p. 247.
10. *The Gazette* and *Harper's Weekly*.
11. Higginbotham, pp. 181-2.
12. *The Gazette* and *Harper's Weekly*.
13. *American Horological Journal*, III:252.
14. *Moore's Rural New Yorker* and *The Gazette*.
15. E. A. Marsh, *Watches by Automatic Machinery at Waltham* (Chicago, 1896), pp. 20-22.
16. *Harper's Weekly*; also, *American Horological Journal*, III:254.
17. *American Horological Journal*, III:251-254, and E. A. Marsh, pp. 95-100.
18. *American Horological Journal*, III:253.
19. *Harper's Weekly*, *The Gazette* and *Moore's Rural New Yorker*. There is an excellent description of the pinion cutting operations in *American Horological Journal*, IV:5-10, but the location of the machinery is not mentioned.
20. *Harper's Weekly*, *The Gazette* and *Moore's Rural New Yorker*.
21. This fact was gleaned from the obituaries of various members of the management and oral information from the Burns family.
22. *American Horological Journal*, IV:54-6 has a complete description of the factory's dial making operations.
23. *The Gazette*.
24. *Harper's Weekly*, *The Gazette* and *Moore's Rural New Yorker*.
25. *Moore's Rural New Yorker*. "The President at a Watch Factory" suggests that this change may have taken place in the last half of 1870.
26. *American Horological Journal*, IV: 5-8.
27. *Moore's Rural New Yorker*.

Chapter 9

The People of Marion

There was never a day or even an hour when the United States Watch Company experienced total success. No matter how bright things seemed there was always some problem that hovered ominously. How then can we determine when Frederick Giles' dream began to drift into failure? It could be said that the low production figures or the difficulties with superintendents or Frederick's inability to support his brother in time of need were portents of things to come. But these events had been lost amid many successes. By 1871, however, the glow of optimism had begun to wear a bit thin, and the failures began to appear more frequently. Although more serious than the earlier problems, the new ones were still not individually significant enough to cause irreparable harm, but collectively they began to sap the vitality of the company. Nor were these difficulties caused by new or unique sources. They were in fact extensions of the past, and as before the people of Marion were their common denominator.

Aside from being an intellectual, an inventor and a businessman, Fayette Stratton Giles was also an avid sportsman. How he was able to indulge his love for fishing and hunting and still keep up with the hectic pace at the factory and the store is an unanswered question. Somehow he managed to steal away every so often to pursue this interest for a few days. Frederick, if he was aware of his brother's pastime, must have considered it a frivolous aberration. No matter, early in 1870, about the time his 10 size designs were patented, Fayette conceived of a scheme to form a large game preserve.[1] At first, the plan lacked one important element, a suitable piece of land. Thus, until one could be located Fayette was forced to bide his time. Frederick, totally unaware of Fayette's intentions, was busily pursuing his own plans for his younger brother. He even went so far as to discuss with the press how Fayette would take over the management of the factory.[2]

By early 1871, Fayette had decided that a large tract of land which was for sale in Pike County in eastern Pennsylvania would suit his needs. The purchase of this property would, however, require that he strain every resource to acquire the necessary funds. Further, the scale of the enterprise, the Blooming Grove Park Association, demanded nothing less than a full time management. Faced with these considerations Fayette withdrew from his partnership in Giles, Wales & Co. and his association with the United States Watch Company.[3]

Frederick was thunderstruck! Once again Fayette had jarred his plan. This time, however, it was something far more serious than marrying an unknown Swiss girl. In an instant Fayette had deprived Giles, Wales & Co. of almost one quarter of its capital. Equally important, his services were lost to the firm. The blow, of course, would have been devastating at any time, but it came at an extremely critical moment at the factory. An ambitious attempt to totally revise the company's complete line of watches was in the early planning stages. This model change was an absolute necessity if the factory was to remain competitive. Now the sudden loss of funds and Fayette's skills would seriously delay this step. The managerial depth might allow for Fayette's replacement, but the loss of funds created problems for which there was no simple solution.

Things between Fayette and Frederick were never really patched up. At first Frederick struck back in anger. Fayette Street in Marion was suddenly renamed Frankfort, an obvious slur against Fayette's pro-French leanings.* Further, the Fayette Stratton model was dropped from the company's line. The 10 size model

70. The second grade introduced to the market. "Fayette Stratton" models were dropped by USWC in 1871.

---

*This change can be seen in the fact that Hopkins shows the street as Fayette in 1870, but Marion Building Co. maps of 1872 refer to it as Frankfort. Years later, after the factory had closed, the street reverted to its original name.

was also eliminated, but this was undoubtedly for other reasons. As for Fayette's game preserve venture, it proved remunerative although his active connection with it was short lived. After a couple of years he, Bertha and their second child, Emma, returned to Switzerland where he entered the watch exporting business.[4] In later years he would try to restore his relationship with Frederick, but that must wait until the end of our story.

In September of 1871, just a few months after Fayette's resignation, another disaster struck Marion. Again, it was a situation that had been several years in the making. This time it was not the management who created the problem, but the one group we have largely ignored so far — the workers.

Although statistics covering the makeup of personnel in 19th Century American watch factories are generally not available, there is enough information from the literature of the period to give us some reasonable understanding of this subject. One is immediately struck by the large numbers of women who seem to have been employed in all of these factories. As might be expected, the delicacy of the work and the fact that they could be employed for low piece work wages made their presence most welcome to management. Ordinarily, one might expect to find them running much of the lighter, more sensitive equipment, their smaller hands being ideally suited for this purpose. There were, of course, exceptions to this rule and in a few shops they operated heavier equipment as well. It should not be assumed that the situation was totally one of exploitation; working conditions in watch manufacturing were generally better than in other industries and skilled women operators were generally rewarded for their labors. Of course, modern day feminists would quite rightly point out that women were almost universally excluded from both management and highly paid jobs such as tool building. Conversely, however, they were not exiled from the board room and at least one firm, The Peoria Watch Company, had a female director of considerable power.* As for male employees, it can be simply said that for them the hierarchy of sex was very much in evidence and that they occupied most if not all of the highly skilled and leadership positions. The situation of child labor, however, was not so simple or clear. In general watch factory personnel were young, but only a few firms seemed to have employed many children.

In several respects the United States Watch Company was unique in its employment practices. While the ratio of male to female workers seems to have been normal, it is evident that contrary to standard procedure women operated many of the heavier machines such as plate lathes. Further, the United States factory made far greater use of child labor than did most of its competitors.† Whether this was a matter of labor costs or Frederick acting on the basis of his own childhood experience is difficult to say. It must be said, however, that if Frederick assumed that he was aiding the younger generation, he was ultimately to discover that the mores and imaginations of teenage boys were not the same as his own.

One of the critical factors in watch factory survival was employee loyalty. Certainly no greater evidence of the benefits of happy personnel can be found than in the case of the New York Watch Company. This firm endured a fire which totally destroyed its factory, a couple of bankruptcies and numerous other difficulties. Yet, in the end it was able to triumph because its employees cared about its future.[5] Their loyalty was such that when the successor, Hampden Watch Company, was moved to Canton, Ohio, John C. Dueber arranged transport for most of the staff to the new location.[6] Unfortunately, the United States Watch Company's experience with its employees was much different. Frederick and his managers not only fought with their superintendents, but they also found it extremely difficult to establish satisfactory relations with the rest of their personnel. This lack of cordiality seems to have been one important factor in the company's low production figures. A measure of the depth of this employee hostility can be seen in the tone of contempt which permeates Charles Higginbotham's comments about the company.[7] Even forty years were not sufficient to mellow the opinions of this former Giles, Wales & Co. watchmaker. The watch factory, however, did not need to wait for such later writings; during its lifetime the uglier aspects of its personnel relations had the unfortunate habit of finding their way into print.

On Friday, September 22, 1871, the readers of the *Jersey City Evening News* were jolted by the news that four boys employed at the factory had been arrested. They were charged in a complaint, filed by George E. Hart, with having stolen watch parts in the value of two hundred dollars. According to the newspaper it had been the intent of the boys to make watches for themselves out of the parts they had taken from the factory. The following day the *Journal* announced that the number of employees arrested had grown to fifteen and that the amount of recovered material in police custody now amounted to one thousand dollars worth. Worse still, it was reported that on Saturday before noon the police had locked the doors of the factory and searched the employees.*

One can imagine how Jersey City and the other surrounding small towns must have buzzed through the remainder of that early autumn weekend. The fact that ten of the young men had held a banquet, supplied by friends, in the town jail on Sunday evening could have only added to the shock of unreality.[8] In the cold light of the Monday that followed the shock became resentment, but not directed at the erring young men. It would seem that what had been taken from the factory

---

*This was Lydia Bradley; see: "A Modern Horological School," *American Jeweler*, October, 1892, Vol. XII, No. 10, p. 8, and "Some Reminiscences," *American Jeweler*, March, 1893, Vol. XIII, No. 3, pp. 11-12.

†This information was gleaned from the various articles about the factory. In the absence of actual statistics these comments are open to some question.

---

*The *New York Times* carried the story on the 23rd. They made up for the delay by increasing the amount taken to $2,000. The story also appeared in Higginbotham, but in a curious version.

was damaged material. Further, the company had required its employees to pay for such unfit parts by means of stoppages from their pay. In fact it had been customary for the company to demand a ten dollar reserve to cover such damages. Were this not issue enough, it would seem that the factory had indulged in extremely questionable if not unethical methods in apprehending its freelance salvage men. Certainly the idea of using promises of amnesty to entrap one's employees could hardly be expected to meet with public approval.[9]

By Tuesday, the 26th, when the first group of young men came to trial, both the press and the public had turned on the watch factory. All of its past mistakes and problems were dragged up. The fact that the various managers and foremen seemed to be unsure about just how much material was actually missing raised serious questions about the company's methods of inventory control. Far worse, however, were the charges of parsimonious management if not downright injustice to employees. Needless to say, the young men were let off with a lecture and in one or two flagrant cases a small fine.[10]

In the end the company was forced to back down. In truth there was simply no way it could reply to the various charges. Of course it was parsimonious; its always slender purse had never permitted it to be otherwise. Admission of this poverty, however, might frighten the very investors who could remedy the situation. Much to its chagrin, the company also discovered that pressing the issue would force it to discharge most of its employees. As it was, a number of key people were hired away by other watch factories.[11] Even if fewer workers had been involved, the senior staff, and in particular Frederick, would have found the issue of the damaged parts most embarrassing. It would not have been seemly to admit publicly that the United States Watch factory never threw out anything — ever! The idea that a first class high quality operation might have considered damaged parts of value would have raised far too many eyebrows. The choice they had was to accept the damage to their reputation as an ideal employer and to wish most fervently that the whole business of handling erring boys had been managed in some other way.

Another matter had also given Frederick pause in pressing charges in the theft incident. The case of William B. Learned, the demoted superintendent, was still far from closed. After resigning, Learned had sued the company and to add insult to injury he had made sure that the suit was an extremely public one. As the reader will undoubtedly remember, it was the opening of this case which had kept Frederick in Marion at the time of the death of Fayette's son. In September of 1870 a decision had been reached before a jury in the court of Judge Depue of the Essex County Circuit. Learned won and was awarded the balance of his five-year contract.[12] The case was, however, appealed to the New Jersey Supreme Court on a writ of error. In March of 1872 the court rendered its decision, and again the company lost.[13] Thus they were forced not only to pay Learned, but his four hundred five dollar lawyer's fee as well.[14]

Even then the dust did not settle and public recriminations continued long after the trial.[15]

There was yet another shock which shook Marion in March of 1872. While it was not caused by any immediate conflict at the factory, it was most certainly motivated by the economic consequences of the continuing personnel problems. As has already been stated, as early as 1871 the United States Watch Company began to work on a plan to remodel its line. Unfortunately, despite the fact that the books showed a profit, both Giles, Wales & Co. and the factory had been hurt financially by the various personnel conflicts. Thus, in order to raise the necessary retooling funds Frederick was forced to search out every possible source of additional capital. One aspect of this effort was a deal in which a number of surplus movements were sold to a New York City merchant, Ellis H. Elias.[16] Since Mr. Elias figures prominently in the last years of the watch company's history a full introduction will be left until then. Suffice it to say, however, that this gentleman was a genius at devising somewhat unethical business schemes.

Ellis Elias' current brain storm in early 1872 was the "Great Geneva Watch Company" sale. His first move in this scheme was to acquire a hoard of Swiss and American watches, mostly of low grade. Having accomplished this, he set up a relative, Henry P. Elias, in a store at 763 Broadway. As an inducement to the public, he then announced that the watches, now located at the store, represented the bankrupt stock of the "Great Geneva Watch Company" which had recently collapsed after thirty years of existence. No attempt was made to engrave the watches with a Geneva Company label; rather it was insinuated that they had

71. Waltham, Philadelphia or Swiss fake? Buying reject or surplus movements and engraving them with a variety of legitimate names was just one of many Elias "enterprises" during the 1870's.

72. *Even the dial is signed "American Watch Co." Waltham, MA. Case is a light weight, low karat gold; it is beautifully engraved and is convertible from hunting to open face ... the so-called "Magic Case" so popular during the 1870's. Many of the so-called Swiss fakes were actually the product of, or ordered by, American fraud merchants like Ellis Elias.*

manufactured timepieces for most of the other companies in the world. This cover story allowed the store to push its hodge podge of real and fake watch brands.[17] In connection with the matter of spurious brands, it should be noted that Ellis Elias and his engravers were responsible for a number of the so-called Swiss fakes, such as "James Russell & Co."* and "John Ellery."[18]

---

*The fact that Ellis H. Elias used Russell & Co. is noted on the official list of frauds in the *U.S. Postal Official Guide* of May, 1880, Anthony Comstock, *Frauds Exposed* (1880) pp. 516-17. Also: *New York Times*, December 13, 1877.

Although the Waltham Company seemingly had sold some watches to Elias, his use of fake Ellery watches in the Great Geneva operation was more than they could endure. Their lawyers attacked the Elias store on two fronts. First, they had the Great Geneva Watch Company enjoined from using the "John Ellery" or "William Ellery" labels.[19] Second, Wayne Litzenberg and Edward S. Newell, two employees of Howard & Co., Waltham distributors, brought suit against the Elias store claiming that it had sold them worthless watches. This, however, proved to be something of a mistake since the information brought out in court did little to harm Elias or anyone else in the store. It soon became evident that while his watches were of low grade, he had not misrepresented the metal content of their cases. Elias then evened the score by explaining from the witness stand that he had acquired his American made watches from Giles, Wales & Co. and the Waltham Co., and his Swiss watches from an unnamed source on Maiden Lane.[20] In the end it was Giles, Wales & Co. that suffered most. While the other parties were able to do a discreet fade, the already tarnished reputation of the folks from 13 Maiden Lane left them no alternative but embarrassed silence.

CHAPTER 9 REFERENCES

1. *New York Times*, December 7, 1897, p. 9.
2. *Harper's Weekly*, June 25, 1870, and *The Gazette*.
3. *New York Times*, December 7, 1897; also Thomas W. Herringshaw, *National Library of American Biography* (Chicago, 1909), Vol. 2.
4. *New York Times*, December 7, 1897.
5. Higginbotham, April, 1912, p. 270. One sour note: Higginbotham notes that some watches were taken during the New York Watch Company fire.
6. James W. Gibbs, *The Dueber-Hampden Story* (Philadelphia, 1954), pp. 8-9.
7. Higginbotham, March, 1912, pp. 181-182.
8. *Jersey City Journal*, September 26, 1871, p. 1.
9. "The Watch Factory Robbery — Further Developments," *Watchmaker and Jeweler*, September 25, 1871, III: 1:41.
10. *Jersey City Journal*, September 26, 1871, p. 4, and September 27, 1871.
11. *Watchmaker and Jeweler*, September 25, 1871.
12. Crossman, Adams Brown Reprint, pp. 87-88.
13. United States Watch Company v. Learned, 36 N. J. Law Reports, Vroom 7:429.
14. L. Zabriske, *Cash Book*, manuscript in Jersey City Library.
15. Crossman.
16. *New York Times*, March 27, 1872, 8-4.
17. *New York Times*, March 5, 1872, 8-4.
18. Ibid.
19. Ibid.
20. *New York Times*, March 27, 1872.

Chapter 10
Three-Quarter and Quarter Plate

By 1871 it had become increasingly evident that United States watches were losing favor with the trade. Unless action was taken soon to modernize the line the company would be in serious difficulty. Already the very jewelers who had adored the watches two years ago were now complaining that the "Mary Anns" were too outmoded to be sold.[1] Perhaps they were being unfair, but it was a period of rapid change. New designs and manufacturers had begun to crowd into the field. While the United States Watch Company offered a large complex line of watches in terms of finish and grade, they could no longer match the latest innovations of their competitors. The loss of Fayette's 10 size, perhaps the best basic design the company had, constricted their products exclusively to 18 size. Last modified in 1869, the 18 size design was now growing increasingly obsolete. By 1871, its marketability, except in the lower grades, was becoming problematic.* No longer could a slow train, full plate watch with a large balance wheel and peculiar setting mechanism be called high grade or modern.

Dealer complaints about the United States watch become understandable when one realizes how far the competition had moved in the brief space of two years. The American Watch Co. (Waltham) had recently introduced an extremely successful 14 size model and a new 18 size, the "Crescent Street," developed by the company's prestigious Nashua Division. Further, they were marketing watches in 20, 18, 16, and 10 size designs as well. Added to this was the fact that they were now producing stem-winding movements.[2] The New York Watch Co., perhaps a bit old fashioned, had just given up the production of 20 ligne watches for 18 and 16 size models.† Their workmanship, however, was good and their watches had a rugged dependability. They also had produced a few stem-winders, but as yet had had little real success in this field.‡ While the National Watch Company (Elgin), the strongest of the newer competition, had not as yet introduced a stem-winding model, they were producing both fast and slow train 18 size movements as well as a 10 size women's design. The fast train watches were of particular importance since they represented the wave of the future in the railway field.[3] E. Howard and Co. was in production with both "N" (18) and "L" (16) size movements; both were fast train and available in stem-wind. A "G" (6) size was nearing completion and would be introduced in 1872.* Nor did things stop there, for other American as well as Swiss firms were pouring both small and stem-winding watches into the market†

As one might expect, Frederick had devised an elaborate plan to improve the United States Watch Company's competitive position. First, the 18 size full plate models were to be substantially reworked. In particular, a new pillar plate utilizing a one-piece click spring and an improved stem-winding mechanism was to be the main feature of this line. Unfortunately, the plan did not include the conversion of the full plate design to quick train. A strange decision considering that the company did a respectable amount of business with railroad companies and that the eighteen thousand beat watches were becoming popular in these circles. Second, a new line of three-quarter plate watches in 18 and 14 size in both nickel and gilt was to be undertaken. Another new line, a quarter plate and bridge design in 16 and 10 size was also to be manufactured. Curiously, Frederick chose to use the quick train feature in these non-railroad models.‡ Another important new feature to be used on the higher grades of the three-quarter plate, the quarter plate and bridge and some full plate watches was a double index regulator (#100,511, March 8, 1870) patented by Julius Elson of Boston, Massachusetts.§ It would seem, at least from the available company records, that the right to use this feature was purchased outright. There was also another new

---

*This fact becomes obvious in the ratio of high to low grade watches manufactured by the company.

†Before the fire New York Company advertisments in *Watchmaker and Jeweler* indicate the firm's original use of Swiss sizes. One of these ads appears on the flyleaf of R. J. Ziebell's *Serial List, New York Watch Company* (Ipswich, Mass., 1972).

‡Their difficulty with the stem-wind model is hard to understand. Seemingly the design was good, but they were unable to achieve consistent production quality.

---

*Crossman, pp. 59-63. It should be noted as well that both Waltham and Howard were producing nickel watches at this point.

†Both the Cornell Co. in Chicago and its predecessor, the Newark Watch Works, had considerable stem-wind experience at this point. The Swiss, as might be expected, were manufacturing large numbers of small watches.

‡It would be many years before the railroads would accept anything other than an 18 size watch for their service.

§Elson also received another patent for a regulator, U.S. Patent No. 113,281, issued on April 4, 1871. This design, however, does not seem to have been adopted by any manufacturer.

feature, a patent reversible spring barrel designed by Frederick, that was to be used on many of the new models. This reversing barrel, a substitute for a safety center pinion, has a curious history. It was in use for over a year before it was patented (#145,939, application filed December 1, 1873, and granted December 30, 1873).

Some idea of what the new line of watches would be like can be grasped from the company's plans for their super watch, "The United States Watch Company." They intended to produce this model in all of the new sizes and configurations. The most splendid and expensive version of the "United States" grade was to be model 17, a quarter plate and bridge design in 10 and 16 sizes. For four hundred fifty dollars the purchaser would receive, without case, in either size, a stem-wind quick train movement with nickel damaskeened, frosted and enameled plates, gold index plate, gold jewel settings, gilded steel work, 19 ruby jewels, three pairs of conical pivots, Breguet hair spring, patented reversible barrel and patent double index regulator. For those who were pinched for cash or were just old fashioned a key-wind version of the same could be purchased for four hundred twenty dollars.[4] Other versions of the "United States" in three-quarter and full plate were almost as expensive. (See Chapter 18 for additional details.)

Text continued on page 90.

74. "American Watch Co." grade, SN 670086, 16s, ¾ plate, 18j Model 72 with interesting push setting device. Note the conservative damaskeening typical of early Model 72 products. Train wheels of these grades are typically gold.

73. "American Watch Co." grade, SN 670012, 16s, ¾ plate, 18j early Model 72 (circa 1872). Everybody's major competitor was Waltham; these finer grades were a product of the prestigious Charles Vander Woerd Nashua Department. Note this particular watch is button set similar to early USWC products. These grades sold at $250 to $275 in the 1872-73 period.

75. SN 670086, cocked view of dial side to show setting device. The setting mechanism is activated by pushing tab adjacent to bezel. Patent date by tab is June 11, 1872, for the mechanism patented by E. C. Fitch. A similar type of push setting may have been used on the "United States" ¾ plate, SN 54406 shown in next series of illustrations. Simple lever setting soon replaced both button and push tab setting mechanisms.

ABOVE. 76 and 77. Dial and movement, "United States Watch Co." grade, SN 54406, 18s, ¾ plate, 19j. Beautiful double recessed dial with USWC monogram in black. The monogram is repeated on movement. Stem wind, push set. This model sold for $400. Damaskeened, frosted, enameled, gold jewel settings throughout, gilded steel work. Breguet hairspring, patent double index regulator, patent applied for reversible barrel. Circa 1872-1873. BELOW. 78 and 79. SN 54406, top and bottom pillar plate views. Note the fine damaskeening, even on winding mechanism yoke. Yoke extends to outer rim of plate for push setting by either push tab device or case attached lever ("side set"). Patent date August 15, '65, and improved December 22, '68, on yoke. Intricate damaskeening on bottom side of pillar plate. Two dial feet held by screws instead of pins.

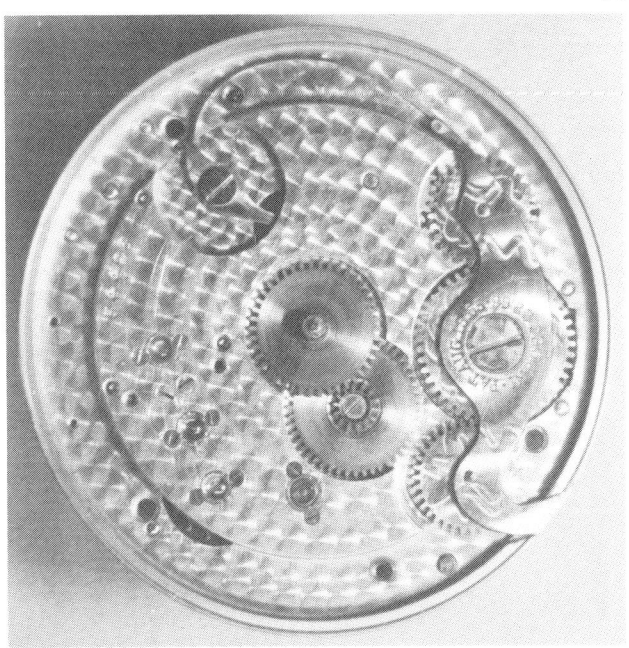
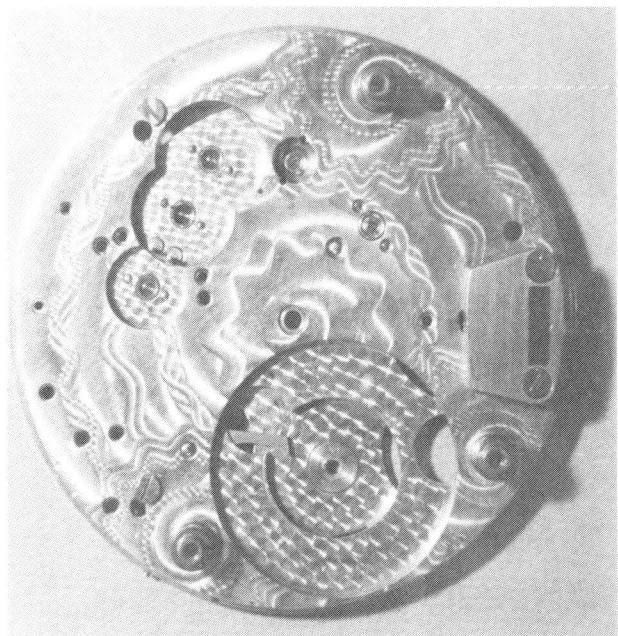

80

81

ABOVE. 80 and 81. SN 54406 miscellaneous components. Escape wheel is gold plated, but rest of train wheels are plain brass. This discrepancy from Crossman and Abbott references was originally reported by Wm. Barclay Stephens in his June 1950 NAWCC BULLETIN article on the USWC. Use of damaskeening on winding mechanism components, such as this example, is seldom seen on USWC products. Note damaskeened barrel cover with "Patent Applied For" abbreviation. This barrel cover provides one method for approximately dating USWC products. Later barrel covers show the Giles reversible barrel patent date of December 30, 1873. BELOW. 82 and 83. Dial and movement, "United States Watch Co." grade, SN 54415, 18s, ¾ plate, 15j. Double recessed dial with bold Roman numerals. Stem wind, lever set. Introduced in 1873, this model sold for $325. Damaskeened, frosted, enameled, flush gold settings on 3rd, 4th, escape and balance; gold setting arbor but brass winding arbor set in flush polished steel disc. Gilded steel work, flat hair spring, patent double index regulator, patent reversible barrel. Note chamfered plate screws. Dial feet screws and balance guard, typical of these ¾ plate models.

82

83

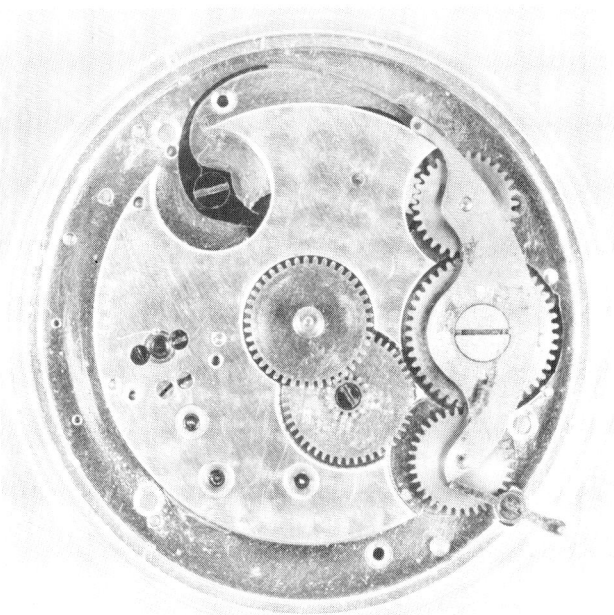 

ABOVE. 84 and 85. SN 54415, top and bottom pillar plate views. Finish of this 15j is not as nice as the 19j pillar plates illustrated in 78 and 79; winding mechanism yoke not damaskeened. Contrast this lever setting to the push setting shown in Figure 78. Typical gilt train, hollow center arbor. Two dial feet held by screws, one held by pin. Nice item, but not as nice as the 19j.   BELOW. 86 and 87. SN 54415, Giles reversible barrel views. Top view illustrates mainspring in barrel, wheel and top of barrel lid. Note patent date on this particular example is December 30, 1873. Side view shows bottom of barrel lid and spring tab in side of barrel. The nickel lid and gilt barrel sandwich the wheel when assembled. Spring tab in barrel fits notch in wheel permitting wheel to turn on barrel assembly in one direction only.

 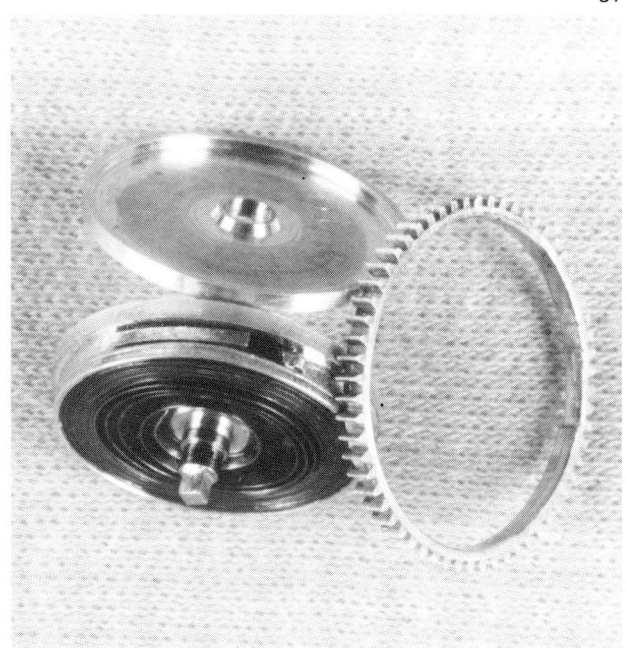

Implementation of this ambitious model change would require construction of two new sets of machines as well as extensive modification of the original equipment.* Conditions within the factory had totally reversed from the early days. Now the mechanical and design aspects of the problem had become minor in comparison to the financial ones. In a sense, Frederick and the United States Watch Company were paying the price of their education. A great deal of time and money had been lost getting the factory into production. During the intervening years the company had produced only modest profits. Tragically, their errors in learning had seriously depleted the funds now needed to apply their lessons. Further, little direct aid could be expected from the Marion Building Company. That firm had so far proven to be more of a source of loss than anything else. Its original serious under-capitalization had prevented recovery of the salt marsh.[5] Moreover, the remainder of its acreage had not sold well.* Although Giles, Wales & Co. was doing well, Fayette's defection had sapped its immedately available resources.

Whether money could be raised from Giles, Wales & Co. or from renewed pushing of Marion Building land or from new investors, the short fall of income at the factory made it imperative that the watch company issue more stock. Already the first million had been nearly exhausted. Further stock issue would, however, require modification of the company's charter. Accomplishment of this proved to be quite simple and on March 8, 1872, the New Jersey legislature authorized the United States Watch Company to increase its capitalization to two million dollars.[6]

---

*This was without doubt Frederick's most serious mistake. If he had concentrated on either the quarter plate or the three-quarter plate and a total rework of the full plate, then the company's chances in later years would have been greatly improved. In no way however, does this reflect on these two fine designs.

*Both Hopkin's atlas and the Marion Building Company's maps of 1872 show much unsold land.

88. *This map produced for the National Board of Health in 1880 clearly shows the extent of the salt marsh. The Hackensack River is in lower right-hand corner of the map, actually west of Marion factory.*

89. Cover pages from the Marion Building Company's last desperate attempt to sell a large block of its land by auction on June 20, 1872. The fact that they hired Grafulla's 7th Regiment Band and the "celebrated improvisatore vocalist," Mr. W. B. Harrison, both quite well known at that time, indicates the importance that management placed on the event.

The Great Geneva affair, the public nature of the Learned suit, the even more sensational difficulties with employees, Fayette's resignation and the declining popularity of their watches made the raising of money from new sources a difficult business at best. Frederick Giles' reputation as a wunderkind had become a bit too tarnished to attract new investors. Attempts to auction off Marion Building Company land had been colorful, but no more remunerative. Neither Grafulla's Famous 7th Regiment Band, nor that celebrated improvisatore vocalist, Mr. W. B. Harrison, nor an elegant collation at the St. James could persuade the public that Marion was either salubrious or free of mosquitoes.* There remained no alternative but for Giles, Wales & Co. and a few close associates to dig into their pockets to produce the necessary cash. Not that this practice was new. Over the years whenever the factory had needed funds,

---

*The Marion Building Company, *Land Auction Held at Marion — June 20th, 1872*, map with advertisement. As might be expected, Higginbotham complains about the damp and the mosquitoes.

AT A COST OF TWENTY-FIVE CENTS AND ONE HOUR'S TIME

Parties may examine the property at any time prior to the Sale.

Maps and all information furnished at St. James Hotel, on the premises.

INTERESTING STATISTICS.

In ten years New York has grown 14 per cent. and New Jersey 40 per cent. Of the 2,220,627 people living in and around New York within a radius of 40 miles,

6 per cent.	live in	Westchester County.
24	"	Long Island.
24	"	New Jersey.
41½	"	City of New York.

Over one-half as many persons in the New Jersey Suburbs as in the great Metropolis itself. The remaining 5½ per cent. live in Rockland, Orange, and Richmond Counties, and Fairfield County, Conn.

The cut which we present of the *largest and finest* Watch Manufactory in the world, is probably one of the best commentaries that could be made upon the almost miraculous growth of Jersey City, competing with the magical erection of Western cities, in rapidity of progress, and surpassing them in the extent and splendor of her improvements. Strange as it may seem, there are persons still living who remember when Jersey City, in 1802, comprised only a ferry house and one dwelling, and the entire population consisted of Major Hunt, John Murphy, and Joseph Bryant, and their families. In 1834 the population had reached 1,500; in 1840, 3,038; in 1850, 6,856; in 1856, 21,715; in 1870, over 100,000. The vast overflow of New York population into accessible economical suburbs, would naturally gravitate along the line of transit, and Jersey City is on the

GREAT AMERICAN HIGHWAY

of the world's travel, and being incorporated with the beautiful precincts of

BERGEN AND HUDSON CITY,

offers for all purposes of residence or profitable investment, beyond all doubt, superior advantages to any suburb of the great metropolis. It was in view of these facts, that the location was selected by Mr. Giles for the site of the Watch Manufactory, where over six hundred male and female artisans are employed in the delicate art of watch-making. Inasmuch as such an establishment requires accessibility, salubrious air, and healthy surroundings,

THE REASONS

that actuated him may be equally important to others.

ITS ACCESSIBILITY.

Situated directly on the Pennsylvania Railroad, adjoining the Marion Station, and immediately at the proposed site of the new splendid passenger depot now contemplated; 31 trains to and fro daily from Cortlandt and Desbrosses Street ferries; time, from ferries,

ONLY EIGHT MINUTES.

Single Fares, including Ferriage, 15 cts.

COMMUTATION, FIFTY DOLLARS,

Excursion Tickets, to and fro, 25 cts.

Business Men, to whom Time is Money,

THINK OF IT.

With a business anywhere below Canal Street you are practically nearer a home in Marion than if you lived at 23d Street, New York, and instead of hanging by a strap in filthy cars, subjected to greasy market baskets, the miasm of foul air, and the common dangers of drunken bullies and pickpockets, you cross the river in warm, comfortable boats (less liable to hindrance and delays, from ice and storm, than street cars), and are whisked, by splendid palace cars, in a few minutes, to your cozy home, at a cost of merely stage fare in New York. Beside these excellent facilities for steam transit, any who prefer it have

THREE LINES OF HORSE CARS.

One from *Jersey City Ferries*, Cortlandt and Desbrosses Streets; one from *Hoboken Ferries*, Barclay and Christopher Streets; and one from *Pavonia Ferry*, Chambers and 23d Streets, running every 15 minutes.

In addition to which

The Midland Hudson Connecting Railway,

Tapping every Trunk Line leading out of New York, will also have a station upon this property on Dales Avenue, with half-hour trains day and night.

An Extensive Water Front

ALONG THE

HACKENSACK RIVER,

Below the Bridge, with 90 feet water in the channel and 10 feet at the bluff bank, affords unusual facilities for

Transportation by Water.

It is also worthy of especial notice, that immediately opposite, on the west bank of the river, the Pennsylvania R. R. Co. have completed the purchase of 100 acres of Mr. S. N. Pike, upon which they will erect the most extensive, complete, and finest Machine Shops, Car Factories, and Locomotive Works probably in the world, thus bringing the material and speculative advantages of a

Huge Manufacturing City

to bear upon this property, while the local disadvantages are obviated by the interposition of the river.

The Extension of Broadway to the river,

and continuation of Pavonia Avenue from the Hudson River to the Hackensack, thence to Newark (which is in full view of this property), will render the facilities of

ABOVE AND LEFT. 90. Inner pages from Marion Building Company's June 20, 1872, land sale. Citing population growth of Jersey City from 1500 in 1834 to over 100,000 in 1870 and the many advantages living in Marion had to offer, Frederick Giles and his associates hoped this sale would provide them with the funds desperately needed to expand their product line. Contemporary copy writers might chuckle at the prose of the early 1870's, but it makes for interesting reading. Where else could one buy a home site ... "and instead of hanging by a strap in filthy cars, subjected to greasy market baskets, the miasm of foul air, and the common dangers of drunken bullies and pickpockets ... cross the river in warm, comfortable boats ... and be whisked, by splendid palace cars, ... to your cozy home." The prose did not overcome the fact that home sites in Marion were right on the Hackensack River salt marsh (Figure 88) with more mosquitoes than the "daily air currents drawn landward from Newark Bay" could blow away.

---

access from all directions complete. In fact, besides the land facilities, and surrounded by the Hackensack, Passaic and North Rivers and Newark Bay, its means of water transit are unsurpassed anywhere.

THE HEALTH

of a location is always of primary importance, and a few remarkable facts regarding the proverbial

Salubrity of Marion

are interesting. Situated on the western slope of the

BERGEN RIDGES,

and running to the river, the property is directly in the line of the

DAILY AIR CURRENTS

Drawn landward from Newark Bay.

And besides the cool breezes of summer and warm airs of winter, which are as regular as the tides, ministering to one's personal comfort in those trying seasons, they also serve most effectually to waft away Miasm, if any exists, and at the same time

Peremptorily drives off Mosquitoes,

For certainly neither of these banes of human comfort exist at Marion. So again

WE CONFIDENTLY ASSERT

that viewed in every aspect, this is the finest opportunity yet offered to secure valuable lots at speculative prices, and comfortable healthy homes within a few [mi]les of the business centre, at moderate rates.

IT IS A MATTER FOR

Simple Calculation and Plain Common Sense.

If property in Jersey City has advanced 1000 per cent. in ten years, and is

STILL, THE CHEAPEST IN THE VICINITY OF NEW YORK,

THE CONCLUSION IS EVIDENT THAT

JERSEY IS STILL THE PLACE TO BUY.

And it is equally clear to all who wisely interpret the history of the past, that JERSEY CITY will continue to expand until the whole of

HUDSON COUNTY

WILL BE ONE VAST CITY,

With Marion the centre.

Rivalling in wealth and prosperity, the great Metropolis itself, being on the mainland, webbed with railroads, permeated by rivers, and accessible by sea and land to the

COMMERCE OF THE WORLD.

In fact the tunnel through which must pour the products of the South and West into New York. And as this generation has witnessed its rise from such small beginnings, may not the next be able to record a consummation so very probable. The property now to be disposed of consists of

835 LOTS IN ALL.

And whatever are offered will positively be sold to the highest bidder. The streets and avenues are graded, the main ones paved and flagged and sewered, supplied with gas and water. The water mains being tapped before reaching the lower part of the city, the pressure is sufficient to force it 90 feet above the spire of the factory. In order to perpetuate the present agreeable surroundings, the lots, excepting the water fronts, being intended for residences, are sold subject to a nuisance clause.

they had been supplied by Giles, Wales & Co. Their return for this financing had been in the form of United States Watch Company stock.[7] The danger, of course, in this was that more and more the store's assets were in the form of the factory's paper. A further liability in this situation was that Giles, Wales & Co. were in fact the guarantors of the watch factory's bonds.[8] All of this was well and good, but should the day arrive that economic conditions placed severe strain on both the store and the factory at once there could be serious trouble.

While the amount of money raised was enough to finance the retooling, the rate at which it appeared was all too dependent on Giles, Wales & Co.'s ability to make profits. The critical importance of Fayette's resignation now became all too apparent; it was the loss of his share of the business that delayed everything. Fortunately, the store's profits despite his loss, remained high.[9] Nonetheless the lack of ready cash stretched out retooling for almost a year. It is believed that some details, such as the patent application for Frederick's reversing barrel, were delayed even longer.*

The effects of the strain caused by the slow down were noticeable in the actions of the front office. Most important, their anxiety led to a series of premature announcements about the new line of watches. As early as September, 1872, they had a price list printed and distributed in anticipation of being ready for the all-important Christmas trade. Alas, as November neared only the first examples of the three-quarter plate model were available for the American Institute Fair.[10] It was well into 1873 before the three-quarter plate and quarter plate reached the market in quantity. This final delay of the new models, although brief, meant that the opportunity to immediately recoup their tool costs during the Christmas season of 1872 was lost.

Unlike all of the company's earlier designs, the three-quarter plate and quarter plate and bridge designs were never patented. The lack of patents, of course, introduces an element of uncertainty into the identity or identities of the designers of these watches. It seems quite probable, however, from the elegant curves of its plate that the three-quarter plate was largely the work of Frederick Giles. On the other hand, the quarter plate and bridge model shows none of these characteristics. In fact if anything, it is quite reminiscent of Fayette's designs for the original 10 size watch. Were it not for the fact that this watch comes long after Fayette had severed his connection with the factory there would be every reason to ascribe its design to him. Although it cannot be proved it seems likely that someone else, probably Frederick, reworked Fayette's original 10 size design so that it would accept a stem-winding mechanism and would vaguely match the three-quarter plate in style. Regardless, both watches were well designed and neither demonstrates the faults of their predecessors. *The real tragedy of the three-quarter plate and the quarter plate and bridge models is that they came so late in the company's history.*

There is no certain way of telling exactly when Frederick and Julia had at last abandoned their separate ways. It seems likely, however, that she first came to New York after the death of Bertha's child to comfort her sister-in-law. Whatever, she was soon settled in Marion at the St. James. Her presence and deep concern were, unfortunately, not enough to relieve Frederick of the problems that had begun to prematurely age him. The elegant young twenty-four-year-old had by thirty-two become a bearded and gaunt old man.

Julia was not alone in her concern for Frederick. His brothers, despite the fact that the great Chicago fire had seriously disrupted their business, were also worried about his health.* Charlie, who planned to marry in mid-March of 1873, used the occasion to woo Frederick away from the grind. Since there had been some strain between the Giles brothers because of the Fayette affair, Frederick was at first reluctant. At last, assured that his younger brother would not attend, he consented.[11] This journey to Chicago was destined to be the only vacation that Frederick would ever take.

---

*Their business moved from 142 Lake Street to 83-5 State Street.

CHAPTER 10 REFERENCES

1. Higginbotham, p. 182.
2. Crossman, Adams Brown Reprint, pp. 37-50, and *Serial Numbers With Description of Waltham Watch Movements* (Waltham, 1954), p. 14.
3. Crossman, pp. 94-5, and *Elgin Watch Materials Catalog* (Elgin, N.D.), p. 1.
4. United States Watch Company, *Price List of Movements*, September 2, 1872.
5. The National Board of Health, *Sanitary and Topographical Map of Hudson County, N.J.* (Hoboken, 1880). At that time the edge of the salt marsh crossed Broadway just west of Dales Avenue.
6. *Laws of New Jersey* 1872, Chapter CLXXIV, p. 439.
7. Crossman, pp. 86-7.
8. Case No. 1197, List of Creditors.
9. Giles, Wales & Co. Sales and Profit —
   Sales: 1871 — $556,129.20
         1872 — $584,771.73
   Estimated Profit: 1871 — $85,759.28
                       1872 — $90,176.18.
10. *American Horological Journal*, VI:115.
11. Charles K. Giles to Julia Giles, February 6, 1873, and Charles K. Giles to Frederick A. Giles, February 24, 1873, Burns Collection.

---

*It is hard to find any other explanation for this device being advertised in the *Price List of 1872*, but its patent application not being made until December 1, 1873. (See Chapter 17 for additional data and some theories.)

Chapter 11
The Panic of '73

In the autumn of 1871 Chicago was devastated by a vast conflagration started, so we are told, by Mrs. O'Leary's cow. Tradition also tells us that on Thursday, September 18, 1873, the economy of the United States was struck by a severe panic which resulted when Jay Cooke & Co. was destroyed by its attempt to finance construction of the Northern Pacific Railroad. While both legends no doubt contain an element of truth, they ignore the combustibles for the spark. In the summer of '71 Chicago was a sea of tinder looking for a flame — any size would do. So also was the U. S. economic scene in 1873. The hard truth was there was very little that was right about the post Civil War U. S. economy. It was as Mark Twain described, the gilded age — the heyday of Daniel Drew, Jim Fiske, Jay Gould, and William Marcy Tweed. Every ill had been tolerated, if not enjoyed; gross speculation, corruption, over expansion and unsound banking practices were but a few of these abuses. Somehow they had all forgotten that their prosperity was dependent on foreign capital. When an era of war and hard times struck Europe it became impossible to attract further support for an American economy that had taken on the semblance of a circus. Perhaps Washington could have helped, but the Grant administration, when its time was not occupied by the latest scandal, pursued a policy more concerned with currency than people. In the end, the United States would have a sound dollar based on gold, but at the price of the longest depression in its history.[1]

As one might expect, the watch and jewelry industry was particularly hard hit by the panic. The only reason many firms did not immediately succumb to the stagnation was that the occasional small recessions that had taken place since 1862 had kept them lean enough to survive. Strength, however, was sometimes not enough and a number of companies were ultimately driven into bankruptcy. The watch manufacturing industry, always under capitalized and usually over expanded, found the going extremely difficult. Even the American Watch Company had to suspend work for four months.[2]

The panic had come at the worst possible time for the United States Watch Company. They had gambled heavily on the success of the new models. Then there had been the delay which had cost them the Christmas season of 1872. Finally, just a few months after the new watches had been introduced everything collapsed. Business was at a standstill. Some of the factory's backers, such as A. S. Hatch, whose banking house was forced to declare insolvency, were hard hit by the panic.[3] The luxury trade for which several important grades had been intended, particularly those based on the quarter plate and bridge design, would take years to recover.

91. "United States Watch Co." grade, SN 235,003, 16s, ¼ plate and bridge, 15j. The market for luxury trade, prestige products of USWC was especially damaged by the panic of 1873. Production delays on the ¼ plate well into 1873 explain the relative scarcity of these particular examples.

By early 1874 a complete change had taken place in the watch manufacturing industry. An era of sharp competition was signaled by heavy price cutting. The leadership of this movement was the lumberman president of the Elgin National Watch Company, Thomas M. Avery.[4] His intent was purely and simply to force the weaker firms in the market, such as the manufacturers of cheap Swiss watches, to the wall. While he temporarily succeeded in doing severe damage to the Swiss, Avery was much more effective in harming his American competitors. Not only was their labor more costly than the Swiss, but also they lacked the cost analysis studies that Elgin was now applying. The United States Watch Company in particular was not ready to meet this kind of pressure. Frederick had been cutting prices and costs, but he really could not match what Avery had in mind. Giles' earlier emphasis on distinctive but costly design features, and the large size of the factory with its resultant high operating costs now began to haunt him. The fact that he had not been able to arrest the decline in popularity of full plate United

States watches hurt even more. Even if he undersold Avery, he could not match Elgin's more popular designs.[5] The quarter and three-quarter plate watches might be superior to anything the man from Illinois had to offer, but they did not fit well into an American market which was growing increasingly conservative.

Even before September, 1873, Frederick Giles began to take certain financial actions which hinted that things were not going well for his empire. As early as January, Giles, Wales & Co. had begun to borrow small sums of money. It was not a lack of profits at the store that caused the need for these funds; the business was in late 1872 and early 1873 at its financial peak.* The factory, however, was fast becoming a pitiless monster which gobbled everything and still demanded more. Frederick, driven by an unshakable belief in the new models, had signed these early notes with the large wholesale jewelry house of Enos Richardson and Company. During the next nine months he borrowed over fifteen thousand dollars from this firm.[6] By late 1874, Giles, Wales & Co. owed large sums of money to Dodd, Hedges & Co., another wholesale jewelry house,[7] to Sylvester M. Beard, a spice merchant and a major stockholder in the factory,[8] and to Thomas G. Brown, a Newark manufacturing jeweler.[9] There was, of course, little hope that Giles, Wales could repay these debts at any time in the immediate future, and these creditors were well aware of this fact. It was a gamble, but all had a large stake in the factory and turning back could only bring about instant disaster.

Not everyone who was connected with the factory was in the position to risk his fortune in an enterprise that was evidently becoming hazardous. The key employees in particular began to cast about for positions which held better promise for the future. Among these early refugees the most important were George E. Hart, the factory's superintendent in charge of tooling, and John Logan, their hairspring maker.[10] Previous to Logan's appointment the factory had purchased all of its hairsprings from James Bottum. Mr. Bottum, never a person to stick to one thing or place for long, had not been entirely dependable. His problem seems to have been that he was torn between lathes, tools, watches, and springs, and when he tired of one he would take up one of the others. Further, an eccentric, he kept moving back and forth among three adjacent stores on Broadway.[11] Ultimately the peripatetic gentleman's habits wore out the company's patience. Somehow, springs must have a certain ability to impose their nature on their makers. John Logan was no more likely to stand still than James Bottum. He had only been with the factory a couple of years when he resigned. The immediate purpose for his leaving was to form with George Hart the watchtool building firm of Hart, Logan & Co.[12] This, however, did not last long. By May of 1874 Logan had moved on to other adventures at Walthum, Massachusetts, and Hart was forced to find another partner. This he found in the person of Isaac Denman. Unhappily, Hart's early partnerships were no more stable than the affairs of hairsprings — the firm of Hart & Denman lasted for less than a year.[13]

Thus 1874 saw the United States Watch Company down to their last superintendent, Henry J. Lowe. By midyear, however, a majority of the stockholders had had their fill and the number of superintendents became academic. With the factory closed Giles' position had become desperate; unless something was done quickly it would soon all be over. He and his close colleagues in short order hit upon the solution of forming a new company to operate the factory. This new firm, the Marion Watch Company, was incorporated in New Jersey, on July 20, 1874.[14] The State of New Jersey had changed its incorporation procedures since 1872 and it was no longer necessary to obtain legislative approval for this action.*

92. "Marion Watch Co." 18s single-sunk dial. This company was incorporated in New Jersey on July 20, 1874. The same basic movements were used after this date, but with progressively poorer finish and new dials signed with the Marion name.

The origins of the Marion Watch Company name were really quite simple. It had been used by locals for years and the United States Watch Company had used it as a title for a grade of watch almost as long. Likewise, the list of directors and major stockholders contains no real surprises. Frederick A. Giles, William A. Wales, George C. F. Wright, Andrew J. Wood, and

---

*Giles, Wales & Co. — Sales and Profit shows sales of $357,602.99 for 1873. The estimated profit for these sales is $55,145.06. It should be further noted that the word "crippled" appears next to the 1873 entry. This, combined with the fact that the sheet stops here, leads one to the conclusion that Frederick felt that the panic was his undoing.

*It was this change which also permitted holding companies that would make New Jersey infamous during the Trust era.

William S. Wyse are listed as serving on an equal shares basis. Wood, a former merchant in fats and oils, and Wyse, an employee of the Harper Bros. publishing house, had been large stockholders in Giles's enterprises from the earliest years.[15] Although the capital limit stated in the incorporation papers was eight hundred thousand dollars, a more significant figure was the minimum capitalization figure of twelve thousand five hundred dollars. Imagine, the seriously under-capitalized Marion Building Company was begun with twice as much only eight years earlier. The degree to which things had deteriorated can be measured as well by the fact that each of the directors had only to raise two thousand five hundred dollars to pay for his portion of twenty-five shares.[16]

Doubtlessly some readers will assume that the birth of the Marion Watch Company meant the death of the United States Watch Company. In truth, this was not the case; rather the old company entered a sort of limbo in which it owned, but did not use the factory, its site, the machinery, and the goods in process were the property of the old corporation.* Actual manufacturing, however, was the responsibility of the Marion Company. In this way the United States Watch Company and its stockholders were protected from further losses. Their only real concerns were now taxes, deterioration, old debt, and the possibility that the new company might not pay its way. As things turned out, the fears were well based; the Marion Watch Company made more watches than money (and more dials than watches). The immediate losses, however, were borne by Giles, Wales & Co. They were forced to continue stripping their accounts and using every bit of their credit to keep the factory running.

Frederick's apparent plan was to delay everything until after the Christmas season of 1874. While business had not returned to pre-1873 levels, he hoped it might be sufficient to stabilize his financial problems. Unfortunately, it was a goal to be neared, but not quite touched. In December of 1874 the patience of the wholesale jewelry houses of Felix and Schwitter, and Klinger, Rupp & Held ran out and they began suits against Giles, Wales & Co. in the New York Supreme Court. While the claims against the firm were small, the former for four hundred sixty-five dollars four cents and the latter three hundred ninety-seven dollars forty-four cents, they shook Giles' already weakened empire. On the 16th of December the Supreme Court issued writs of attachment against the property of Giles, Wales & Co. As required by the writs, the sheriff of the City of New York immediately seized goods equal in value to the judgements from the store. A minor matter in terms of the total stock at 13 Maiden Lane, but it had the effect of being totally disrupting. These circumstances forced those persons and firms having large or special interests in the New York store, the Marion Building Company or the United States Watch Company to action.[17] On December 21st Enos Richardson & Co., Marsh, Coe & Wallis, attorneys; S. M. Beard, Sons & Co.; Dodd, Hedges & Co.; Charles A. Crane, real estate agent; Thomas G. Brown and Isaac H. Wright filed an involuntary bankruptcy petition against Frederick A. Giles, William A. Wales and George C. F. Wright.[18]

Until comparatively recent times federal bankruptcy laws have had little popularity in the United States. The unprecedented long survival of the current 1898 act resulted largely from public acceptance of it as a necessary evil. Undoubtedly the income tax law, the mobility of contemporary commerce and the government's move into economic management have played important roles in the contemporary view of the matter. As for the demise of earlier laws, James Angell Mac-Lachlan suggests that "Probably the most effective single reason [for their repeal] was the elementary fact that financial failures present an unhappy situation where it is hard to satisfy the persons who must share the loss that the legal institutions invoked are not largely to blame. In hard times both creditors and debtors were dissatisfied. They could not get along without bankruptcy, and they could not get along with it, so statutes passed after much travail were shortly repealed."[19]

Three times before the passage of the 1898 act federal bankruptcy laws were tried in the United States and each time they were repealed. At the time of the Giles, Wales & Co. bankruptcy action the most successful of these ill-fated laws, that of 1867 or more precisely its June 24, 1874, revision, was in force. Seemingly this act had many commendable features and it was in some respects superior to the 1898 law. Some readers, however, may find its provisions a trifle archaic and even harsh. Insolvency, for instance, could be as little as a single creditor with a provable claim of two hundred fifty dollars which had gone fourteen days (forty after 1874) past due. Discharge from voluntary bankruptcy after 1874 required that a quarter of the creditors possessing one third of the valuation join in the petition for release. Consent, however, was not required in involuntary proceedings. There were, further, sixteen categories of grounds for denying discharge because of misconduct.[20] Critics of the time also claimed that the law favored intrastate over interstate creditors. One suspects, however, this complaint resulted more from problems of distance and communication than anything else.

"In the matter of Frederick A. Giles, William A. Wales and George C. F. Wright, case number 1124 under the 'Act to Establish a Uniform System of Bankruptcy Throughout the United States of March 2, 1876'" represents as typical a legal matter as it is possible to imagine. Or to put it very simply, since each was attempting to maximize his own interests they were bound to use every tactic short of fraud to gain advantage for themselves. Had Klinger, Rupp & Held, or Felix & Schwitter been willing to content themselves with their share of the assets of Giles, Wales & Co. they would have forced the firm into bankruptcy in the U.S. District Court for the Southern District of New York. But this is exactly what they did not want. They wanted one hundred cents on the dollar plus interest and costs. Thus as shrewd and aggressive creditors

---

*The Charter of the United States Watch Company was not dissolved until 1924 — New Jersey Secretary of State.

93 and 94. "United States Watch Co." grade, SN 206168, 18s, 15j with Marion Watch Co. dial. Very poor quality when compared to USWC earlier examples of this grade. Raggedy gold arbor cup in polished disc is the only gold feature in this item. Only things well done are the frosting, damaskeening and silver index ... all probably done in earlier days. Note lack of chamfering and finishing of plates. Flat hairspring, brass jewel settings, barrel has no provision for stopwork. Nickel movement has gilt pillars and barrel. Under dial, note lever setting mechanism and continuation of poor finish. Circa late 1874 or 1875.

they chose the state courts and treated the matter as merely a bill that was past due.[21]

The positions of Frederick and his major creditors are not so quickly explained. Simple logic might suggest that the easy out would have been to pay off the screamers and be done with it. Although there were some risks involved, this would have been the best route for Frederick. It would have immediately bought him the one thing he needed most — time. Unfortunately, he could not pay. He had the assets but they were in the form of goods, mostly watches, and their liquidity was problematic.[22]

Frederick's major creditors, Richardson, Beard and others, were in the main perfectly willing to grant him all of the time he needed. Since each was in one way or another deeply involved financially in one or more of his enterprises, they had every reason to aid him. To a man they still had faith in Giles' ability to succeed. The actions, however, of the group led by John G. Klinger, senior partner in Klinger, Rupp & Held, were something they felt they could not afford to let slip by. If Klinger got his way, every small creditor would follow and soon there would be nothing left but a well picked carcass. From the view of the major creditors the only hope was an arrangement. One which on one hand gave Giles, Wales & Co. time and on the other granted the fullest possible protection for all of the creditors.*

---

*In other words, the situation was quite orthodox to bankruptcy situations.

What the major creditors had in mind and what Frederick agreed to was bankruptcy, or at least going far enough into such a proceeding to establish a court backed compromise. Why they chose an involuntary over a voluntary proceeding is a matter open to debate. Most likely they were not totally certain that the court action could be stopped short of actual bankruptcy or that a sufficient percentage of the creditors would be willing to sign the compromise papers. Less debate, however, need center on why they elected to institute the action in the New Jersey District Court. John Klinger and his lawyers might later object that the New York federal court should have been used — but Klinger had had his chance. Richardson, Beard, and for that matter Frederick, had every reason to prefer New Jersey. There it would be far easier to gain sufficient delay to force a compromise.

The first step, filing the bankruptcy with William Muirheid, the U. S. Commissioner for the District of New Jersey, went well enough. There was, of course, an amusing moment or two — process had to be served on Frederick and his partners in New York City.[23] Further, most of the major creditors had to swear out their petitions before Henry Allen, a New York Registrar in Bankruptcy.[24] No matter, in short order John Linn, attorney for the major creditors was able to get an injunction that stopped Mr. Klinger and his associates.[25]

In this instance, the word stopped has to be treated in its relative sense. Klinger and his attorneys might be restrained by the injunction, but nothing in this world or the next would get them to cease their efforts to collect. As for signing the compromise papers, poor George Wright! He had been assigned the task of buttonholing creditors. This of course, included a visit to the dragon's lair at 15 John Street. It must have been a ghastly scene. The friendly, always bubbling George trying vainly to talk to angry, hard-nosed old man Klinger. Felix and Schwitter's recent defection to the ranks of the compromisers had done nothing to improve Klinger's disposition. After the old man's acid reply to the subject of the papers, George offered to pay him in goods. It was all to no avail; a debt was a debt, and to oppose him in court was a great wrong as far as John G. was concerned. He wanted his money plus interest and costs and nothing else would do.[26]

Fortunately George Wright's search for creditors who would agree to compromise met with success. More than half were willing to agree to an arrangement.[27] The risk had worked, the major creditors would get what they wanted and Frederick would get time to solve his problems. Even Klinger, Rupp & Held were able to get some satisfaction, although not the dire punishment John G. felt Giles, Wales and Wright deserved. The court had, when a settlement was assured, allowed them to proceed with their action in the New York court.[28] By this time Frederick had been able to raise enough money to pay them. And so on May 11, 1875, Hamilton Wallis, A. H.'s son and Frederick's attorney, went before Judge Nixon in Trenton and moved for a discontinuance. Two weeks later it was granted.[29]

CHAPTER 11 REFERENCES

1. Robert Sobel, *Panic on Wall Street* (London, 1960), pp. 154-196.
2. Moore, *Timing a Century*, p. 54.
3. Sobel, p. 181.
4. "Watch Factories of America" (revised), *American Jeweler*, April, 1905, pp. 182-83.
5. Abbott, *Watches and Men*, pp. 8-10. Also for the effects of price cutting and competition on the Waltham Company see: *Timing a Century*, pp. 69-70 and 73-74.
6. Case 1124, "Petition," December 22, 1874, and "Deposition by Frank H. Richardson," December 21, 1874.
7. Case 1124, "Petition" and "Deposition by David C. Dodd, Jr.," December 21, 1874.
8. Case 1124, "Petition" and "Deposition by Sylvester M. Beard," December 21, 1874.
9. Case 1124, "Petition" and "Deposition by Thomas G. Brown," December 21, 1874.
10. Crossman, p. 211.
11. Ibid., pp. 173-74. Mr. Bottum's changes of trade and moving habits were gleaned from the *New York City Directories, 1854-79*.
12. This is based on Crossman; the firm does not appear in the *Newark City Directory*.
13. Holbrook, *Newark, N.J. City Directory 1874-75*. The listing states that they are "manufacturers of watch machinery, small lathes and tools for jewelers and watchmakers, molds made and jobbing done on all kinds of light machinery."
14. "Certificate of Incorporation, The Marion Watch Company," filed July 30, 1874, and recorded in Book "D" of Incorporations, New Jersey Secretary of State.
15. Crossman, Adams Brown Reprint, p. 84, and *New York City Directories*.
16. "Certificate of Incorporation, The Marion Watch Company."
17. Case 1124, John G. Klinger, "Affidavit," May 4, 1874.
18. Case 1124, "Petition," December 22, 1874.
19. James Angell MacLachlan, *Handbook of the Law of Bankruptcy* (St. Paul, Minn., 1956), p. 11.
20. Ibid.
21. Case 1124, John G. Klinger, "Affidavit"
22. Ibid.
23. Case 1124, "Order to Show Cause," December 22, 1874, returned January 2, 1875.
24. Case 1124, "Depositions."
25. Case 1124, "Injunction," January 5, 1875.
26. Case 1124, John G. Klinger, "Affidavit."
27. Case 1124, "Consent for Discontinuance," filed May 10, 1875.
28. Case 1124, "Order Dissolving Injunction," filed May 4, 1875.
29. Case 1124, "Order for Discontinuance," signed May 25, 1875.

ABOVE. 95. SN 25508, 19j, Breguet hair spring, gold settings throughout. Best quality made, circa 1870 . . . "the most expensive watch in America."

96. SN 19923, 15j, flat hair spring, flush gold settings, good quality. First step down, introduced early 1873 before Panic hit.

DECLINING STANDARDS OF USWC PRODUCTS

BELOW. 97. SN 206214, 15j, flat hair spring, flush gold settings (some had brass). Second step down, introduced about time Marion Watch Co. formed in July, 1874. Poor quality, after Panic of 1873.

98. SN 202604, 15j, flat hair spring, flush brass settings. No butterfly cut out, as dies worn out. Third step down, circa 1875-76. Very poor quality.

Chapter 12
Giles, Wright & Co.

Until this point in our story there is little that would distinguish Frederick Giles from the thousands of businessmen whose careers came to flower during the war years. In an age when men of great power and wealth were often colorful personalities he was hardly noteworthy. Even the gradual erosion and ultimate collapse of his dreams during the seventies was not a unique experience. But from his earliest days Frederick had been driven by a sense of obligation. First there had been his orphaned brothers and sisters, then had come Julia and her brothers, later there had been the directors and the stockholders; even when one or the other failed to reciprocate his loyalty, he carried on. In the end, this sense of obligation became his undoing. It would have been far better if his creditors had refused to grant him one minute more. Frederick Giles, unfortunately was that breed of individual who did not know how or when to quit.

Had it been within the creditors' power to grant the Christmas season of 1872, or, for that matter, the season of 1873, then the gift would have been meaningful to Giles, Wales & Co. As it was, time without additional capital had become meaningless. In a nation in depression where capital was desperately short, the chances of a firm in difficulty finding new support were slim. While Frederick Giles and George Wright were willing to face these odds, William Wales was not. One brush with bankruptcy was enough; Wales did not even wait for matters to be resolved. On February 1, 1875, he threw in his hand and withdrew from the partnership.[1] Whether he was able to salvage anything from his years with the firm is open to question. While ordinarily the court would have seized his estate, Case 1124 seems to have never progressed to this point. Whatever, he soon found sufficient capital to open a small importing business at 12 John Street.[2] He must have found some success in this venture since he soon moved on to 1½ Maiden Lane, a far more desirable address.[3] Wales surprisingly enough had not had his fill of watch manufacturing ventures. In 1876 he joined the recently organized Auburndale Watch Co. at Auburndale, Massachusetts.*

As difficult as the situation of the new firm of Giles, Wright & Co. was, another firm in Frederick Giles' empire was in even worse condition. From the very beginning the Marion Building Company had been one of those business schemes that just never seemed to get anywhere. Frederick, always too busy to give it leadership, had delegated his authority to the board of directors.[4] There had never been enough capital to carry out the original plan of developing its lands into a vast industrial park. In fact, there had not even been sufficient money to reclaim the marshy acreage that lay nearest to the Hackensack River.[5] The secondary scheme of turning the area into a residential suburb had been a dismal failure. A few individual and row houses had been constructed but neither these nor the building lots had sold well.[6] Now that Giles, Wright & Co. and the watch factory could no longer afford to pay even the minimal subsidies that the building company once enjoyed, its situation had become hopeless. With foreclosure by its mortgage holders looming near, there was no way that receivership could be avoided.[7]

Although it had been the source of so many of Frederick's troubles, the factory was in better shape than one might expect. Still its condition could hardly be described as good. The company's growing shortage of funds had begun to seriously affect critical matters; machinery maintenance and replacement were now almost totally ignored. In earlier years the factory had produced some extremely low grade watches, such as the "G. A. Read" and "North Star," but their skimpy workmanship and lack of quality had been a matter of intent. *Now even the better grades began to show evidence of declining standards.* All too often plates bore ugly chatter marks where careless or inexperienced operators of the plate lathes had forced their tools too rapidly into the metal. Edges of the once elegant butterfly apertures now showed bending and tearing caused by dies and punches that had been driven once too often against the unrelenting nickel. Even the dials, a few short years ago the factory's pride and joy, had slipped into insipid mediocrity. Things were fast falling into a tragic circle of declining funds, wearing machinery and disappearing key employees.*

By June of 1875 the slipping fortunes of the watch factory had caused many of its principal employees to drift away. Ever since the resignations of George Hart and John Logan, the people on whom the company depended had one after another packed their bags and left. None of these departures, however, equals or leaves so many unresolved questions as those of Sylvanus Sawyer, a major stockholder, and Henry J. Lowe, the factory's superintendent. Sawyer, as it ought to be

---

*This date is open to conjecture. While Wales is listed in the 1877-78 directory, no place of business is given. Unfortunately, Crossman is of no help on this point since his dating in the latter part of the United States Watch Company article is wildly inaccurate.

*This information was gleaned from a large-scale microscopic inspection of United States watches conducted a number of years ago by Bill Muir.

remembered, was a noted inventor who had been successful both with his inventions and his commercial ventures. Too late, he discovered that watch manufacturing was far more difficult and much less remunerative than anything he had previously encountered. Elsewhere he had been a genius, but in the watch business he was just another sucker.

After sitting for years and watching his investment in the United States watch factory evaporate, Sawyer formed the notion that perhaps he could do better if he were to found his own factory. While there are no surviving records to explain the exact course of events, there is enough information from other sources to support a pretty good surmise. On one hand Frederick was desperate for cash and on the other the factory's machinery was under-utilized. What then could have been more reasonable than Sawyer approaching Giles with an offer to acquire some of the surplus equipment? Logical, yes. But the man from Fitchburg wanted the newest and best machines in the plant, the 16 size quarter plate and bridge tooling. The offer must have wrenched at Frederick's heart, but he was really in no position to argue. He needed the money more than he needed the machinery. And so the deal was consummated, but neither man was to profit. The money was far less than Frederick needed and Sawyer's Union Watch Company was not successful. Later scholars have attributed this failure to three factors: his inability to secure outside capital, the serious illness that forced Henry Lowe to retire, and Sawyer's belief that he was too old for the strain of managing a watch factory. This may be true, but it seems more likely that Sawyer's skills and interests were really in the area of tool building rather than watchmaking. Certainly the fact that he converted the machinery to this end rather than placing his redesigned version of the 16 size United States watch into production seems to support this view.[8]

In the latter half of 1875 Frederick's pursuit of money to save his faltering enterprise simply knew no bounds. Not only was the factory stripped of any equipment that might be considered surplus, but he also scoured the countryside in search of loans. Unfortunately, the burden of a debt of over three hundred and fifty thousand dollars which grew daily, family problems and his own weakening health had begun to weigh heavily on Frederick. His once keen sense of ethics became blurred by the enormity of his problems. On June 28th, using twelve thousand five hundred dollars worth of now practically worthless United States Watch Company first morgage bonds as collateral, he borrowed in the name of the Marion Watch Company ten thousand dollars from the Windham National Bank of Windham, Connecticut. Then less than two months later, on August 13, he borrowed nine thousand five hundred dollars from the same bank, this time in the name of the all but defunct Marion Building Company.[9] During the period of June, 1875, to January, 1876, Giles was able, by means of similar tactics, to make loans totalling over sixty-four thousand dollars from various banks in Connecticut and New Jersey.[10] The tackiest part of this matter was that almost half of this sum came from the First National Bank of Jersey City where Hamilton Wallis was a director.[11] In mitigation it ought to be said that Frederick sincerely believed that he could repay these obligations, but it was largely a matter of self-deception.[12]

Nor were the banks the only victims of these methods. Despite the loans the factory was often unable to meet its payrolls. Inasmuch as it had issued scrip for many years, all that was necessary to cover the shortages was to persuade the local merchants to accept the paper for goods but to delay redemption. One must suspect that George Wright had a great deal to do with smooth talking the various dry goods merchants, grocers and coal dealers into accepting such an arrangement. Smooth talk, however, could do little to cover the fact that by December 31, 1875, Giles and Wright's debts had risen to over four hundred seventy thousand dollars.[13]

99. The USWC conducted much of its local business in scrip. Its use was continued until a late date, even though it became increasingly more difficult to redeem.

The Windham Bank loans were the last instance when Frederick's brothers were willing to give him their total financial support. Despite the rivalry between the two older brothers, the various Giles family businesses had tended to work together through the years. Both E. A. Giles & Co. and Giles Bros. & Co. had in the past acted as outlets for United States Watch Co. products.* Further, they had been willing to assist Frederick in such matters as obtaining loans. But when it became increasingly obvious that the Marion and New York

---

*Giles, Wales & Co. advertisements usually refer to Giles Bros. & Co. as their Chicago agent. The fact that Edwin handled United States Watches is evident from his correspondence with Frederick.

City firms were in considerable financial difficulty, William began to give less and less support to his elder brother's operations. Whether it was his own experience with the ill-fated United States Clock and Brass Co. or an unwillingness to make a total commitment to an elder brother with whom he was often at odds is not known. Whichever, after his endorsements on the various Windham notes William totally withdrew.[14] Perhaps he was right. Edwin, out of loyalty to Frederick, stuck it out a little longer and came near to losing his business.[15]

It must have seemed to Frederick that the collapse of his fortunes stalked him so closely that it could no longer wait until he arrived at work. In February of 1874 he and Julia had had a second daughter, Grace. But tragedy had to intrude even here, and the baby died on September 23, 1875.[16]

Giles, Wright & Co.'s creditors finally began to lose faith in the firm's ability to make good its debts and on December 31, 1875, new bankruptcy proceedings were instituted. Once again the petition was filed with Judge Nixon at the New Jersey District Court. The plaintiffs, with the exception of Charles A. Crane, had not been involved in the old case and unlike their predecessors they were in earnest. Case 1197's new men were William M. Sutton, Thomas M. James and Samuel Clark of Jersey City and William S. Wyse of New York City.[17] On January 18, 1876, Frederick A. Giles and George C. F. Wright were declared to be bankrupt as of the beginning of the year.[18] Thus in accordance with the statute their total estate as of January 1, 1876, was seized by the court and placed in the hands of an assignee, William Muirheid.[19] Muirheid, a Jersey City attorney and a U. S. Commissioner for the district was chosen, as was the practice, by the firm's creditors.

Thus the state of affairs that emerged on January 18, 1876, was that Giles and Wright were bankrupt and their estates had been confiscated by the court. While the United States Watch Company, the Marion Watch Company and the Marion Building Company were viable and at least marginally solvent, their control had now passed from Giles and Wright to William Muirheid, the assignee.* This, of course, meant that the factory could not be operated since Giles, Wright & Co. could not provide working funds. Or could they?

---

*Just how viable the Marion Watch Company and the Marion Building Company were at this point is open to some question. The point is, however, that their obligations fell upon Giles and Wright rather than the corporations. Before someone suggests this was unjust, it should be pointed out that Frederick stated in his correspondence that this is what he wanted; see Frederick Giles to E. A. Giles, January 5, 1876, and January 16, 1876, Burns Collection.

CHAPTER 12 REFERENCES

1. "Copartnership Notices," February 1, 1875, source unknown, Burns Collection.
2. Trow, *New York City Directory 1875-76*.
3. Trow, *New York City Directory 1876-77*.
4. Letter of August 6, 1865.
5. *Sanitary and Topographical Map of Hudson County*.
6. *Land Auction Held at Marion;* also Hopkins, *Atlas of the County of Hudson and the State of New Jersey*.
7. Case 1197, "Notice of Motion to dissolve Injunction & Affidavit," February 12, 1877.
8. Frederick Mudge Selchow, "The Watch Company of Fitchburg, Massachusetts," NAWCC BULLETIN 13:914-922. Also, Abbott, pp. 91-92, and Crossman, pp. 158-159.
9. Case 1197, "Exhibits — Petition," May 16, 1876.
10. Case 1197, "List of Creditors."
11. Ibid., also George H. Farrier, ed., *Memorial of the Centennial Celebration of the Battle of Paulus Hook* (Jersey City, 1879), p. 175.
12. Frederick Giles to E. A. Giles, January 5, 1876, Burns Collection.
13. *New York Times*, January 11, 1876.
14. Case 1197, "Windham Bank Co. Petition & Exhibits," dated May 16, 1876.
15. Frederick Giles to Edwin Giles, January 16, 1876, Edwin Giles to Frederick Giles, May 11, 1879, Burns Collection.
16. Headstone, Locust Hill Cemetery, Montague, Massachusetts, G. Robert and Carolyn Burns, oral.
17. Case 1197, "Petition of Creditors for Adjudication of Bankruptcy," filed January 1, 1876.
18. Case 1197, "Adjudication of Bankruptcy," signed by Judge John T. Nixon, January 18, 1876.
19. Case 1197, "Memorandum" signed by Staats S. Morris, Registrar, February 16, 1876.

100. Ellis and William Elias began their New York "enterprises" with a dollar store business located in the basement of the Grand Central Hotel at Broadway and Bleeker Street. It was the store with striped awnings (left of horse carriage parked on street) in this vintage photograph.

Chapter 13
Ellis Elias

New York, Jan. 5, 1876

Mess. Giles Bro. & Co.
    Chicago, Ill.
Gent.

The interest on the notes held by the Windham Bank is somewhat in arrears, and they talk a little ugly. Don't hardly think they will do anything. If they do they will of course have to first sell the U. S. W. Co. 1st Wty bonds which they hold as collateral at 80 ct. on the $1. before they could go to the matter or endorse & then only for any deficiency there might be between the collateral and the amount of the notes. What is the law out there? If they should attempt anything couldn't you keep them out of it for a year or two? We have arrangements on foot which will probably enable us to pay the notes inside of a year & take up the bonds.
In haste yours truly

Giles & Wright[1]

New York, Jan. 5, 1876

E. A. Giles Esq.
    Dubuque, Iowa
Dear Ed.

We find that in order to get capital to work with, so that we can make money to pay up our old obligations that we must first relieve ourselves from any legal responsibility on them, so that a person putting in capital would not be in danger of having it gobbled up on the old matters. To do this some of our creditors are going to put us through bankruptcy which will free us legally. We can then get capital and with capital to handle the watch works, can make money fast, and go on & voluntarily pay 100 ct on the dollar with interest, and it is the only way we can do it. I suppose the people who hold your notes particularly the banks will go for you as soon as it is known that we are in bankruptcy, but it is better to do it even if we have to start you in business again soon as we get capital ahead.
In haste your aff. bro.

Fred[2]

Considering the state of Giles and Wright's business affairs the last thing one might expect is that they would become involved in still another scheme to operate the factory. Yet on January 5, 1876, they, William Wyse and Andrew Wood, incorporated the Empire City Watch Company.* Since each director member of this corporation was to put up two thousand five hundred dollars, somehow Frederick and George had to have acquired the money after the first of January.[3] It is apparent from Frederick's letters to his brothers that he had made some sort of arrangement, but with whom? The answer is to be found in Crossman's version of the story: "The factory was still kept going, however, as Mr. Giles still had hope for the future of the enterprise. He interested Mr. W. H. Elias of New York City, who furnished money for the payroll and agreed to take pay in movements."[4]

As usual Crossman's remarks are marred by an error; there was no W. H. Elias, but rather three brothers, Ellis H., William M. and Richard H. Elias.[5] The typo need not bother us since it is corrected by several other sources. His "W. H." Elias is none other than the illustrious Ellis H. You will no doubt remember that we encountered him once before in relation to a certain "Great Geneva Watch Company." Space did not permit a proper introduction then, but now the moment has arrived to take a look at this different kind of genius.

Ellis and his brothers first arrived in New York about 1869 after being driven from Cincinnati for swindling. By 1875 their activities had earned them this kind of critical acclaim in the *New York Times:* "There is a class of thieves who seldom fall within the clutch of the law, but who continually plunder the public. They are not the 'big bonanza brokers' who suffer equally with those they have induced to speculate. Their ventures are always successful as far as they are concerned, and in every case their victims are surely robbed. These men have so studied the power of the law and its technicalities that they have succeeded in bringing swindling to a science, and the criminal statutes framed with a view of protecting the innocent as well as punishing the guilty, fail to embrace their systematized swindling. They tread successfully the hairline which divides swindling which cannot be punished from open robbery that can. Thus when a swindler advertises that he will send a first class steel engraving of George Washington to any address for twenty-five cents and forwards in return a three-cent postage stamp, it is evidently a swindle, but yet one the law cannot reach. For many years the Police of this city have regarded the Elias brothers as the most dangerous representatives of this class of swindlers in this city."[6]

It was by no means easy to become an important, let alone the most dangerous, representative of any class of criminal in the 1870's. Competition for such distinctions in that era of civil corruption, corporate skulduggery, western badmen, in short, wall to wall crookedness, was pretty stiff. For their part, the Elias brothers were hardly wont to avoid competition or for that mat-

---

*Which is all the more fascinating since Wyse was one of the parties who petitioned for Giles and Wright's bankruptcy.

105

ter notoriety — their very success depended on an energetic visibility. Ellis, the leader and brains of the family, in particular made every effort to make his presence known to the public. The tall, pallid, bright-eyed young man with flowing moustache who loved fast horses somehow never seemed to find this task very difficult.[7]

Despite the fact that the press often referred to them in collective terms, the Elias brothers were actually very distinct individuals. Richard, for instance, claimed at every opportunity that he was in no way associated with his brothers. Further, he made it publicly quite clear that he was shocked by the thought that anyone would connect him with their schemes. No matter, even operating under the name Humphrey did not separate him in people's minds from his brothers.[8] On the other hand, William worked closely with Ellis and often their enterprises, especially the less questionable ones, bore his name. For his part, Ellis, while making himself quite conspicuous, never did anything — legal or ill — in his own name. After all visibility is one thing, culpability is quite another.*

Ellis and William first attracted attention when they opened their dollar store in the basement of the Grand Central Hotel at Broadway and Bleeker Street. The dollar store, an invention of Ellis', was an entirely legitimate business which bought up distressed over-inventoried, and bankrupt merchandise and resold it at a dollar per item.[9] Even Ellis' strongest critics spoke kindly of this phase of his career.[10] Unfortunately, he was by nature a restless man and the attractions of honest trade could not hold his interest for long. Some sources suggest that after a year or so in the dollar store enterprise he went broke and thus discovered that a 'bankrupt sale' could do wonders for business. Perhaps, but it seems more likely that Ellis just naturally discovered that the idea of getting something for nothing was a tremendous business lever. Whatever, the discovery prompted him to hold 'bankrupt sales' here, there and everywhere.[11]

After their adventures with dollar stores and bankruptcy sales Ellis and William ostensibly operated a fancy goods store at 667 Broadway.† Actually, this business served as a front for Ellis' various less-than-legal schemes. The legitimate store and the fact that each of his questionable enterprises was on the surface run by one of his clerks seemed to do wonders in protecting him from arrest.[12] Even so, on occasion he did run afoul of the law. He was never convicted and strangely enough these arrests seemed to come at times that were much to his advantage. It was not until long after his death that a possible reason for his immunity to inconvenient police actions became apparent. But that comes later in our story.

---

*The various city directories always listed family businesses in William's name, but the police, press, and Post Office were never deceived on this point.

†Trow's *New York City Directory* lists them at this address from 1868-1878. It should be noted that there is some conflict in the various sources and this may well have been the original dollar store.

Since Ellis was constantly hatching new schemes, discarding old ones and even operating several at once, it is a bit difficult to keep track of his activities.* The *New York Times*, however, made at least one attempt: "They . . . started a gift distribution lottery swindle, which was carried on in various forms until they started it on a grand scale in Broadway, between Houston and Spring Streets. In this place they made twenty thousand dollars in two days, when Captain Edward Walsh, then of the fourteenth precinct, raided on them, and found that the wheel contained nothing but blank tickets, two barrels full of which were seized, that the pianos offered as prizes were hired, and that the money gifts were bogus. They next started a swindle on Bond Street under the pretense of giving a concert in the Academy of Music for the benefit of the Cuban insurgents. They issued circulars signed 'Donna Emilia Huranez,' offering tickets for sale at one dollar each. The swindle was conducted by one of Elias' clerks named Constantine. This swindle was broken up by a *Times* police reporter, and Constantine was arrested by Detective Rielly. Then came the Swiss Watch Company swindle, Elias purchased a large number of worthless unmatched watch movements, put them in oriode cases, opened a large store on Broadway, and announced that he had purchased the stock of a Geneva Watch Company. This swindle failed, and was quickly followed by the Milton Gold jewelry swindle, which proved remunerative. After this the Elias Brothers seemed to have dropped out of notice."[13]

Interesting, but one has to question the reporter's notion that the Great Geneva Watch Company was unprofitable. It hardly seems logical that Howard & Co. would have gone to the trouble of bringing legal action against a business that was doing badly. As for Ellis dropping out of sight — absurd. He would not have known how! In truth, he was just busy working the "sawdust swindle" circuit for a while.

Legend has it that Ellis invented the sawdust swindle. If so, it was by far the most vicious of all of his schemes. Usually an Elias plot was one of simple misrepresentation. That is, if he said he had a watch for sale, such was the case — it was just that the value was a trifle inflated. Sawdust was different, however, since it contained both the promise to sell counterfeit money and the element of blackmail. The idea was to lure the intended victim into purchasing a quantity of bogus money. For his real money the sucker would receive a box which supposedly contained the counterfeit, but which was filled with sawdust. To insure their escape before the ruse was discovered, the bunko men always made sure that the box was not opened until after they had fled the scene. Their main protection, of course, was that even after discovering the deception there was little the victim could do since it was his intended dishonesty that made theirs possible.[14]

Ellis may have invented the sawdust swindle, but it really was not his style. It was far too intimate a game;

---

*Comstock devotes three chapters as well as a liberal sprinkling of references to Ellis in his book *Frauds Exposed*.

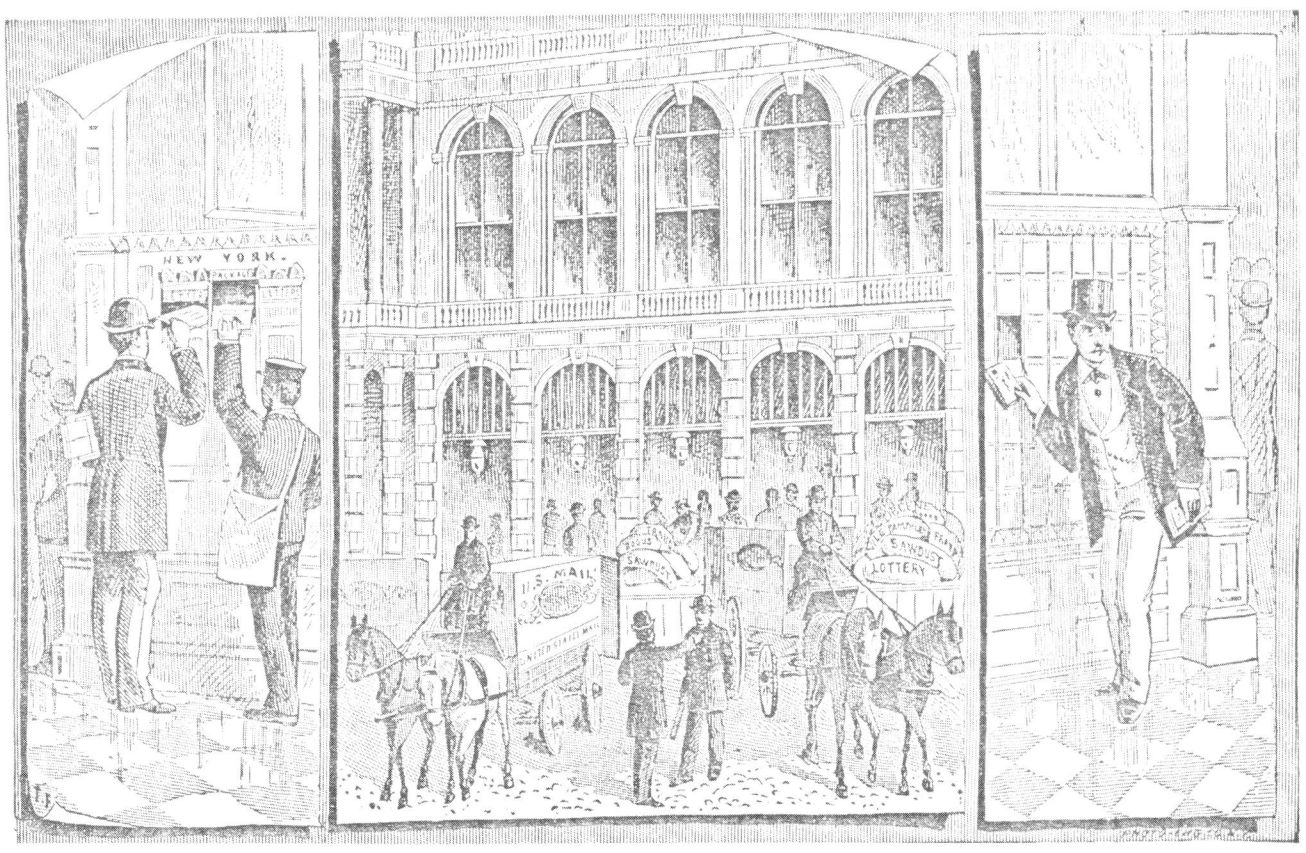

101. Frontispiece from "Frauds Exposed" Anthony Comstock, 1880. Ellis Elias was one of the "main characters" in this book, with three chapters as well as a number of references devoted to him. There is some speculation that the gentleman portrayed "receiving his mail" in this engraving is our Mr. Elias.

he preferred working to a mass audience. Thus it is hardly surprising that after about a year he invented still another scheme. When a creative soul responds to frustration his solution is likely to be rather interesting. In September, 1875, Ellis Elias hatched what was perhaps his wildest merchandising scheme. At that time sixteen thousand of the following circulars were placed in the mail:

"A Chance for Everybody
William M. Elias & Co.'s
General Average Sale
of
$4,000,000 worth of goods annually
Send your orders and trust to luck.
We guarantee to return at least double the amount invested in valuable and useful goods.
Many of our sealed boxes contain
Silk Dress Patterns, Camel Hair Shawls, Diamond Sets of Jewelry, Gold and silver watches and many other useful and valuable goods.

Our New System
We have arranged with a large number of manufacturers, importers and jobbers in this country and Europe to dispose of $4,000,000 worth of their surplus stock of goods annually on the general average plan.

This new system is founded on a true financial basis which enables manufacturers, importers and jobbers to convert into money their surplus stock with comparatively little loss. As it is impossible for manufacturers and jobbers to know a year ahead what demand there will be for their particular line of goods, the result is that larger amounts are in many cases manufactured and imported than can be disposed of through ordinary channels of trade, therefore this new system known as the general average sale was inaugurated by us for the purpose of disposing of all classes of fancy goods.

As there will be many ask how we can afford to give at least double the value of the amount invested and still make money — we give the following explanation, in the first place manufacturers, importers and jobbers make at least 100 per cent profit on their goods which are sold to the retail dealer, who makes from 75 to 150 percent profit when he disposes of his goods, then in many cases, the parties who furnish us with goods lose at least one-half what the goods cost to produce — the result is that an article that cost to manufacture say $10 would be sold to the retailer for $20, and he would ask about $35 for it. In some cases those who buy goods at retail pay even larger profit than this on original cost of the goods they purchased, of course on some lines of goods the profits are not so great now, when we say we guarantee to give each purchaser at least double the amount of his investment in goods, we figure the goods at a retail price, which is the only way they can be had unless bought in large quantities.

How a General Average is Made of the Goods.

When the manufacturers, importers and jobbers send their goods to our depository to be realized on, we make an estimate of their retail value then they are divided into lots which are packed in boxes and sealed. The boxes are then put into six different departments which are marked $5, $10, $15, $20, $25 and $50. When orders are received for boxes we simply go to the department they are ordered from and put the address on them, then they are sent by express, so we do not know what the boxes contain when they are shipped.

Thus you can see it depends on your luck in reference to getting the most valuable articles; of course, you are more apt to get the most valuable articles in the high priced boxes, still you have a proportionate chance in even the $5 boxes. The beauty of the general average system is that parties with a small amount of money have an opportunity of securing the most valuable goods.

While we offer diamonds, gold and silver watches, camel hair shawls, silk dress patterns, and other valuable goods in our general average boxes as inducements, we don't wish our customers to understand that we are engaged in any game of chance or lottery, but on the contrary, we claim our sales to be strictly legitimate, by which we can furnish more goods for the money than can be bought on any other system.

We have had many inquiries in reference to exchanging goods — we state in answer that it is utterly impossible under our system of doing the general average business to make any exchanges whatever. Parties ordering goods will please state whether they wish a $5, $10, $15, $20, $25 or $50 box. You can send the price of the box by draft, P. O. money orders or in registered letter; or if you prefer, we will send the box by express, C.O.D. Always state by which express you wish your goods sent.

Address all orders and communications to
William M. Elias & Co.
no 667 Broadway, New York"[15]

One not familiar with the nature of the human animal might assume that Ellis' general average scheme was doomed from the start, but nothing could be further from the truth. After all, claw machines seldom sit idle at penny arcades, Klondike is by far the most played game of solitaire, and if you remain unconvinced, then pause for a moment on the habits of the average driver. Of course the general average scheme was a success — a huge one! It did a business that made the dollar store with its honest bargains look ridiculous. On the other hand it was not exactly the sawdust game, something of value had to go into those packages. Most of the merchandise was easy to acquire, but watches were a sticky proposition.

The acquisition of watches was one of the most difficult problems which faced Ellis Elias, and for that matter any firm outside normal channels. It must be understood that honor and reputation had nothing to do with the matter; legitimate trade meant jobbers, jewelers and watchmakers. Things had not always been that way. At one time watches were relatively easy to obtain, but as the American industry matured it tended to make every effort to curtail free markets. Watch manufacturers lobbied strenuously for tariffs to protect themselves from foreign imports. By the mid-1870's these efforts and the efficiency of the American factories had driven their only real competition, the Swiss, from the market place. For their part, the jobbers, watchmakers and jewelers exerted great pressure upon the manufacturers to eliminate sales outside of the "legitimate trade." Factories that sold directly to consumers or allowed their watches to be used for premiums often found themselves in great difficulty.* The situation clearly smacked of trust and monopoly, but government action lay several decades ahead. As for Ellis Elias he did what every dealer not in the legitimate trade dreamed of; he decided to acquire his own watch factory.

---

*Crossman and Abbott are filled with the sad stories of companies that tried. In fact, only one man, John C. Dueber, had the guts and power to buck the legitimate trade and get away with it!

CHAPTER 13 REFERENCES

1. Giles and Wright (Frederick Giles) to Mess. Giles, Bro. & Co., January 5, 1876, Burns Collection.
2. Frederick Giles to Edwin Giles, January 5, 1876, Burns Collection.
3. "Certificate of Incorporation, Empire City Watch Company," filed January 10, 1876, Book "D" of Incorporations, p. 275, New Jersey Secretary of State.
4. Crossman, Adams Brown Reprint, p. 89.
5. *New York Times*, September 17, 1875, and April 8, 1879; Crossman's dates are, of course, totally inaccurate.
6. *New York Times*, September 17, 1875.
7. *The Sun*, June 24, 1881, p. 116.
8. *New York Times*, April 8, 1879.
9. *New York Times*, June 24, 1881, p. 5:3; *The Sun*, June 24, 1881.
10. Anthony Comstock, *Frauds Exposed* (1880) p. 163.
11. *New York Times*, June 24, 1881.
12. Comstock, p. 163.
13. *New York Times*, September 17, 1875.
14. Comstock, pp. 196-201.
15. *New York Times*, September 17, 1875.

Chapter 14
Royal Gold and Empire City Schemes

The arrangement between Frederick Giles and Ellis Elias antedated the Giles and Wright bankruptcy by at least several weeks. Unfortunately, if there were ever any records of this meeting they have never surfaced. All that is known for certain is that Ellis had agreed that once the bankruptcy had been consummated he would assume the cost of operating the factory and accept repayment in the form of watch movements.* Whether he and Frederick had any understanding about the nature and quality of these watches is not known. One thing is certain, however. Giles had no intention of committing his more valuable creations to Ellis' dubious schemes. No — Elias would have to make do with the obsolete and the less than perfect.

Of course, Ellis Elias may have had no objection to the idea of receiving the factory's unsold movements and seconds. The notion of perfect high grade Marion-built watches being used in General Average packages or some lottery scheme was ludicrous. Elias' primary interest was, after all, in obtaining the most, not the best, for his money. Merchandise of higher intrinsic value would preclude volume and would be simply too overpriced for his needs. Even if Ellis had wanted the more valuable timepieces there was hardly a sufficient number of these rare inventory items on hand to fill his needs.† The only way they could be had was to resume their production. And it was now common knowledge that when Marion produced quality its actions were at best ponderous.

Frederick Giles, having been a promoter throughout most of his career, in all probability found the line between his efforts and Ellis' razor thin. After all, a watch merchant who kept three chronometers was in no position to be squeamish. And certainly the difference betwen a blurb by the Marion Building Company and the literary efforts of William M. Elias and Company was extremely small. Had Frederick ever been aware of the ethical niceties in question, the desperation of his situation had long ago dulled these sensitivities. It no longer mattered to what use Ellis might place their watches; the essentials would be maintained — the factory would run, Frederick would be president and George Wright treasurer.

Matters had not, however, reached the point of total indiscretion; Ellis Elias' visibility in the affairs of the watch factory was kept to the lowest possible level.

---
*This is, of course, what Crossman and Giles' letters were referring to.
†Higher grade Marion-built watches were never a common item.

Whether this was due to Frederick's wishes or resulted from Elias' usual habit of avoiding direct involvement is a moot point. The elegant bunko artist was simply the tall thin man who was nowhere to be found. If he was a partner in the Empire City Watch Company, the firm's records do not reflect it. Neither did the factory name a watch after its new benefactor. In fact, all of the dealings between the two men seem to have taken place within a company which was also nowhere to be found.

Royal Gold American Watch was neither a corporation nor a company.* It never had a home although its watch movements claimed New York as their place of origin. Even the name has a curious ring to it. Yet, despite its lack of tangibility this arrangement, Royal Gold, lasted almost a year and a half and about four thousand watches appeared under its label.

To fully comprehend the nature of the Royal Gold arrangement it is necessary once again to point out the problem of string saving. Every American watch factory produced a fair number of watches that could not be marketed because of various defects. Marion, in par-

---
*Neither the New Jersey nor the New York Secretaries of State have been able to locate Royal Gold in their records. It should also be noted that "company" does not appear in the name.

102. *Typical upper dial layout for "Royal Gold American Watch." Note no "Co." designation.* **Dials generally have plain or single sunk seconds.**

103 and 104. 18s full plate, SN 38918, and 14s ¾ plate, SN 253092, "Royal Gold American Watch" products. The "extra jeweled" designation on both examples was probably much to the liking of Ellis Elias. The 18s has empty top escape and pallet caps. Because these products were surplus stock that had been reworked, their serial numbers can be found throughout the entire USWC number range. Both of these examples are gilt, stem and key wind.

ticular, had been plagued through the years by a surfeit of imperfect movements. During the early years of the factory untrained personnel and flaws in the machinery had created problem after problem. Later, the scarcity of funds had prevented proper maintenance of the equipment. In this context, microscopic analysis of late Marion movements tell an appalling tale of worn and crumbling dies. Still, all of this would have meant little to this segment of our story but for management's habit of never throwing anything out. By the mid-1870's it is estimated, based on subsequent events, that the Marion factory had in excess of five thousand imperfect movements in storage.

It was these five thousand plus flawed timepieces that were the centerpiece of the arrangement between Frederick and Ellis Elias. The plan was quite simply to overhaul and repair these seconds. Once this was accomplished, the watches would then receive Royal Gold dials and engravings. As for casing, there is no way of knowing where this task was accomplished, but it seems most likely that Giles' people would have had to do this task. Thus, the Rejected Giles And Wright watches were made ready for marketing by Ellis H. Elias as Royal Gold American Watches.[1]

One might well ask where were William Muirheid, the assignee, and the major stockholders while all of this was going on? Did not one of these men worry about the damage that associating with the Eliases might do to his reputation? The answer seems to be that to a man they were perfectly willing to acquiesce to any plan that might salvage Giles, Wright & Co. and the factory. (The probable reason for this is that people in Jersey City had a good deal to lose financially if the factory went under.) Anyway, Elias' role in things was for the moment pretty well concealed and his money had stabilized things sufficiently for the Empire City Watch Company to get started.

It should have been evident from the beginning that Ellis Elias would have to buy an awesome number of Royal Gold watches if the Empire City scheme was to be implemented. Frederick's plans for his new company were as overambitious as any of his earlier dreams. Even before the Empire City Watch Company had produced a single movement it rushed to press an elaborate price list. This sheet contained no less than twelve different model names, many of which were to be offered in full, three-quarter plate and quarter plate versions. Further, not only were the usual run of sizes to be produced, but the bulk were to be manufactured in nickel. How Frederick intended to accomplish this feat without the quarter plate machinery and with the full plate machinery in desperate disrepair is not so problematic as it might seem. Aside from the five thousand second rate watches that were to be used in the Royal Gold scheme, the factory also had on hand parts for over ten thousand satisfactory movements.*

---

*This figure was arrived at by adding the estimated number of Empire City watches to the estimated number of watches later completed by the Howard brothers under the Independent label. (Because of the random serial numbers, these estimates are difficult to confirm.)

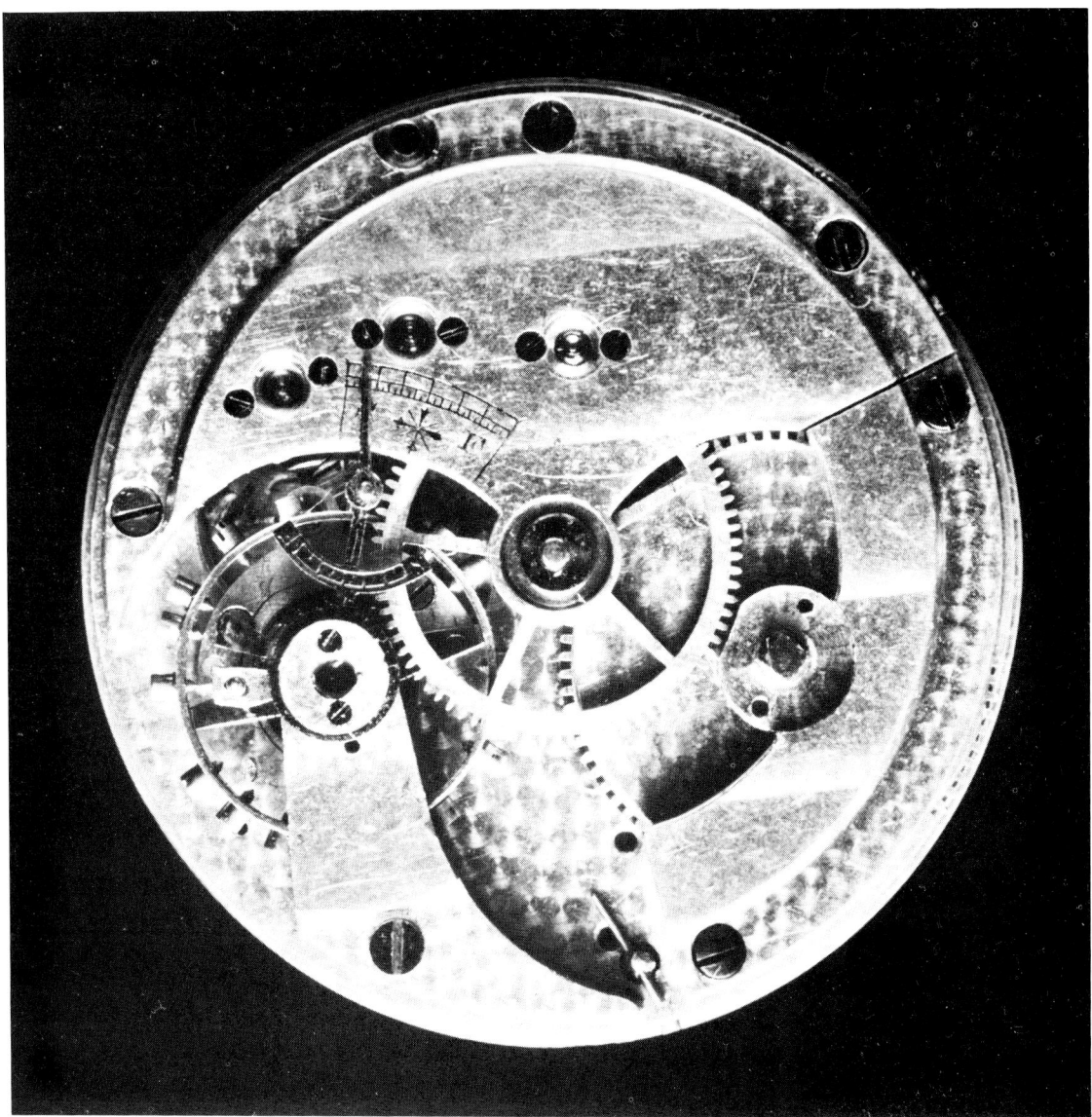

105. 10s, ¼ plate and bridge, nickel unfinished USWC movement. No serial number or grade on top plate; no patent dates on balance cock or barrel cover. SN 248239 on pillar plate under dial. Some USWC unsold, satisfactory movement stock like this surviving example were completed with appropriate dials and engraving as Empire City Watch Company products. The stock of reject, or imperfect movements and parts were the source for Royal Gold products.

But everything depended on Ellis and his willingness to continue financing the factory. For the moment his enthusiasm for his new toy was unbounded, but the future was a different question. New forces had begun to stir in New York. A spirit of reform was in the air and Ellis was one of the prime targets of the reformers. Until now he had been fairly well protected from problems. Although often in court, his preference for remaining behind the scenes and his ironclad arrangements with the police had secured him from difficult situations.

The primary source of Ellis' immunity from the law was Captain Alexander S. Williams of the Twenty-ninth Precinct. Williams, a tough Nova Scotian, had early earned the nickname "Clubber" for his over-generous use of his locust night stick. For years honest citizens were awed by the absolute respect the criminal element seemed to have for him.[2] What they did not realize was that Williams had long ago achieved an alliance with the local underworld. They either did it his way or he and his strong arm squad would break their heads. It was Williams who said upon his appointment to the Twenty-ninth, "I've had nothing but chuck steak for a long time, and now I'm going to get a little of the tenderloin." The label stuck, and the precinct which ran from Fourteenth to Forty-second Street and from Fourth to Seventh Avenue became New York's infamous Tenderloin district.[3]

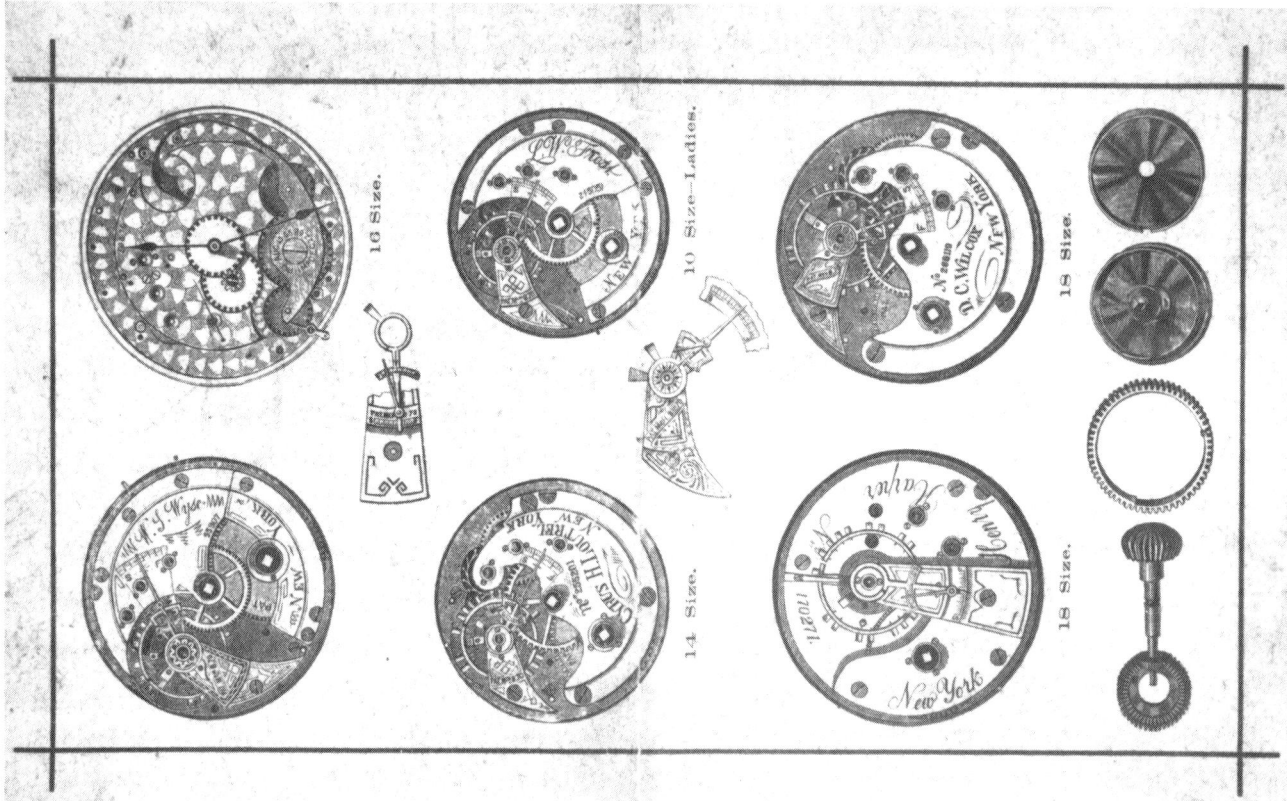

FOR THE TRADE ONLY.

EMPIRE CITY WATCH COMPANY

MANUFACTURERS OF

KEY AND PENDANT WINDING

WATCHES

No. 13 Maiden Lane,

NEW YORK.

FACTORY,

JERSEY CITY, N. J.

straint, but also dealing with the assignee who represented the claims of Giles and Wright's creditors, the demands of various lienholders, and the confusion caused by three watch companies that had never quite died. Had any of these companies had the good fortune to have fallen into bankruptcy things would have been simpler, but Frederick's methods of operation had left the United States and Marion firms derelicts and the Empire City Company an unabandoned wreck. Nothing was simple; the bondholders, for instance, found their relationship with Muirheid, the assignee, to be one of total ambiguity. On one hand, since Giles, Wright & Co. were the guarantors of the factory's bonds they were the firm's creditors.* Yet, on the other, events had forced the bondholders to prejudice the claims of all other creditors by laying claim to Giles and Wright's primary asset — the factory. This ambiguity or dualism of role extended to other things; Frederick, for instance, was executor of his elder sister's late husband's estate. Thus, by virtue of the deceased brother-in-law's holdings in the factory, he possessed the right of vote at any meeting of the creditors. Nor was this a minor point; Frederick had made skillfull use of this power in the previous bankruptcy.[14] So it can be seen that instead of a simple matter of selling off the factory's assets to recover what they could, the bondholders were facing serious problems.[15] In truth it would take a year of complex moves and the outlay of considerable cash before they could salvage one penny of their original investment.

One of the immediate results of this delay was that Frederick and the Empire City Watch Company were given a reprieve of sorts. There was no question of reopening the factory; it was closed and as far as the bond and mortgage holders were concerned it was going to remain so. There were, however, no immediate plans for the hotel and few objected to Frederick's using it as a headquarters for selling off the watches that had been completed.[16] After all, what could have been better for the bond holders? At no further expense to themselves they stood a chance of recovering at least a small part of their losses.

Unfortunately there was no possibility that Frederick and his close associates could recover their losses. Whatever their sins had been they were now to pay a price that was totally out of proportion with events. In particular, Frederick, George Wright and Alexander Wallis, upon whom the worst was to fall, had given everything they could to make the factory a success. Perhaps they had misled investors with overly optimistic promises, but their motives had been above reproach. Each man believed implicitly in the factory's prospects and had spent nearly every cent he possessed to further this faith. Even when they had plunged to the depths of bad judgment, as in the Windham Bank and Ellis Elias affairs, they had done so trusting that future events would justify their actions. Now that it was obvious even to them that the situation was hopeless and that it would be impossible for them to redeem their obligations, they were crushed emotionally. Each privately cursed the fates that had destroyed the dream.

George Wright had clung to the dream longer than anyone might have expected. Despite his hearty and bubbling manner he was hardly a lion, yet he had managed to outlast both Fayette and Billy Wales. Only his brother Henry, now reduced to the title of clerk at the St. James, demonstrated a stronger will to stick it out.[17] It was to be expected that Henry, no stranger to adversity, would carry on, but George's will to hold out rose from different sources. An eternal optimist, he kept believing that somehow things would turn out all right. Certainly there was a way of looking at this situation that could have led him to see a bright side. For years he had been merely the "Co.", but now despite bankruptcy he was a real partner with his name above the store and the title Secretary-Treasurer of the Empire City Watch Company.[18] All of this prestige permitted him to do something that he had not been willing to consider for many years — he decided to marry.

A familiar pattern emerged with George's marriage to Annie Curll Hayes at New Haven on May 17, 1876; almost at once there was a marked change in his relationship with his sister and brother-in-law.[19] It is difficult to know what role Annie might have played in this alteration. The truth of the matter is that almost nothing is known about her. The press once referred to her as an invalid, but since there were no details given it is hard to say whether this was an important factor.[20] Whatever, something at last gave George the insight to question the course that Frederick was following. The fact that he had been very close to his sister, Julia, and felt that he owed Frederick a great personal debt must have made this process extremely difficult.[21] Family obligations, however, did not, by early 1877, stop him from reaching the conclusion that he must take a stand somewhere.

There is no exact information to tell us what at last drove George to desert Marion; circumstantial evidence, however, leads to a single conclusion. Even so, considering all that he had been through, the event that caused him to leave seems almost trivial. On June 2, 1877, Captain Williams, who had begun to smart under Anthony Comstock's gathering attacks, had swooped down on Elias' General Average Store and arrested Ellis on a charge of conducting a lottery.[22] Hurriedly bailed out by his bemused employees, the elegant bunko artist found little humor in the situation.[23] After all, it was his clerks who were supposed to take the risks and the ridiculous part was that the "Centre" was far closer to a legitimate business than anything he had done in years. George Wright's reaction, however, was much different. Even though the Royal Gold arrangement had been dead for months, he was terrified by the storm of publicity that broke with Ellis' arrest. In a matter of days Wright had severed his connection with Marion and moved to the city. Although he would continue at the store, he swore that he would never again have anything to do with watch manufacturing.*

---

*It should be noted that Giles, Wright & Co. consists of Frederick Giles and George Wright at this point because Wales had left the firm.

*The exact date of George's leaving the Empire City Watch Company is uncertain, however at one point in the 1877 corporate report Frederick lists him as "G. C. Fright."

109. Frederick Giles was already fast losing his health when he finally left the St. James Hotel in mid-1878 and returned to Montague, Massachusetts. There, on June 18, 1879, he died of tuberculosis. He was only 44 years old. The effect of USWC on Frederick is readily apparent by contrasting this photograph to the one shown in the frontispiece.

Frederick had tried to convince George that they had little to fear from the scandal, however he just could not seem to muster his old skills at persuasion. At another time he might have been able to react more forcefully, but recently he had not felt up to it. He did not seem to have any energy these days and recently he had developed a rather persistent cough that he could not shake. Julia kept begging him to go away for a rest, but Frederick was afraid that if he left Marion even for a brief period the bond holders might bar his return. In private moments of reflection, however, he had to admit to himself that there really was not much left to fight for. Even after the Empire City Watch Company died in late 1877 he could not bring himself to move out of the St. James. And so rather than move his residence to the city each day he would use a bit more of his ebbing strength for the journey to the store on Maiden Lane. Not that it mattered, he really did not have the will to be of much assistance to George.* By mid-1878 his illness had progressed to the point that he at last gave in to Julia's entreaties and wearily boarded the cars for Montague.

Once back in the hills of western Massachusetts he regained some of his old strength and even began to consider plans for the future. Despite their former dif-

---

*George's resignation from the watch factory had not also included his resignation from Giles, Wright & Co. In fact one view of his leaving the factory suggests that he did it to protect the store from criticism.

ferences, Fayette had been begging him to come to Switzerland. There it was hoped that Frederick would have a better chance to regain his health. Further, Fayette wanted his elder brother to join him in his watch exporting business. Although Frederick had reservations, he at last consented to go.[24] Unfortunately, everything was delayed by the still hanging Giles and Wright bankruptcy matter. The case had dragged on interminably. One creditor, Frederick Schroeder, had only recently gone so far as to file an affidavit with the court accusing Giles and Wright of using the bankruptcy to conceal vast sums of money and land.[25] In truth, George in New York was struggling to keep the store alive, while Frederick and his family in Montague were nearly destitute. Even so the affidavit delayed Judge Nixon's release of Giles and Wright from bankruptcy until April 1, 1879.[26]

The winter of 1878-9 had proven to be a difficult time for Frederick. He could no longer pretend that his illness was a protracted case of bronchitis or a weakness of the lungs — the evidence of tuberculosis was all too apparent. By spring he was grimly aware that his situation was hopeless, nor could he hide this fact from Julia. Even so, they made a brave attempt to keep the terminal nature of his illness from Frederick's brothers and sisters.[27] Frederick, however, did not take the same position with his neighbors. To all who would listen he would repeat the litany of disasters that had befallen his attempt to found a watch factory. His claim that had not illness interrupted his career he would have acquired another fortune was, alas, never put to the test.[28] Frederick Asa Giles died at Montague on June 18, 1879, of tuberculosis.[29]

CHAPTER 15 REFERENCES

1. Frederick Giles to E. A. Giles, January 5, 1876, Burns Collection.
2. Crossman, Adams Brown Reprint, p. 211. U. S. Patent 142,924, Logan and Hart, September 16, 1873. U. S. Patent 151,121, Hart and Logan, May 19, 1874.
3. Holbrook, *Newark City Directory 1875-6*.
4. Holbrook, *Newark City Directories 1876-7* and *1877-8*.
5. Ibid.
6. Edwin A. Battison, "The Auburndale Watch Company: First American Attempt Toward the Dollar Watch." Contributions from the Museum of History and Technology, Paper 4, in *United States National Museum Bulletin 218*, pp. 49-68. Washington, DC: Smithsonian Institution, 1959.
7. "The Death of George E. Hart," *Jewelers' Circular Weekly*, January 3, 1906, pp. 42-3; "George Edwin Hart," *American Jeweler*, January, 1906, p. 6.
8. Crossman, p. 148; Abbott, p. 95.
9. Ambrose Webster to Charles S. Crossman, January 12, 1888.
10. William B. Fowle, *Account Book*, quoted in letter of January 5, 1968, Edwin A. Battison to William Muir, copy in author's collection. The complete entries are as follows:

1876 Sept. 28	Adv. Geo. E. Hart & Co. bill		480.80
Oct. 20	G. E. Hart & Co.		350.00
Nov. 22	Geo. E. Hart & Co.		1365.18
Nov. 24	Geo. E. Hart & Co. (three minor entries)		52.38
Nov. 25	Geo. E. Hart & Co. a/c contract		919.13
Dec. 14	Geo. E. Hart & Co. a/c contract		246.72
Dec. 14	Geo. E. Hart & Co. extra bills		222.85
1877 Jan. 15	Geo. E. Hart & Co. contract		1141.66
Jan. 15	Geo. E. Hart & Co. extras		211.03
Mar. 9	Geo. E. Hart & Co. contract & extras		979.17
1878 Apr. 19	Advanced to Hart & Co. on a/c machinery		1000.00
June 22	Pd. Geo. E. Hart & Co. bill of 19 June		565.99
Aug. 10	Geo. E. Hart & Co. bill (etc.)		1812.13
Aug. 31	Geo. E. Hart & Co. bill		1886.63

11. Case 1197, "Notice of motion to dissolve Injunction & Affidavit," February 13, 1877.
12. Ibid.
13. G. Robert and Carolyn Burns, oral.
14. Case 1124, "Consent for Discontinuance."
15. Case 1197, "Notice of motion to dissolve Injunction & Affidavit."
16. Empire City Watch Co. "Statement by Corporation transacting Business in the State of New Jersey," filed June 30, 1877. Also see Giles & Wright entries, Boyd, *Jersey City Directory 1877-78*.
17. James Gopsill, ed., *Jersey City Directory 1877-78*.
18. Empire City Watch Co. "Statement . . ."; Boyd, *Jersey City Directories 1876-77* and *1877-78*.
19. Wright Family Bible.
20. "Trade Gossip," *Jewelers' Circular and Horological Review*, March, 1884, XV:2:62.
21. Christabel Burns to Freeman H. McMillan, October 11, 1860.
22. District Attorney (Manhattan), *Register of Felonies, December 1876 - July 1877*, New York Municipal Archives.
23. *New York Times*, June 4, 1877, p. 8:2.
24. Bertha Giles to Frederick and Julia Giles, June 16, 1879; Fayette Giles to Frederick and Julia Giles, June 22, 1879, Burns Collection. It is obvious from a postcard which Fayette wrote to Frederick on April 15, 1878, that the brothers had met previously in New York and resolved some of their differences.
25. Case 1197, "Specifications Against Discharge," filed July 19, 1878.
26. Case 1197, "Final Discharge of Bankrupt — Frederick A. Giles," "Final Discharge of Bankrupt — George C. F. Wright," April 1, 1879.
27. E. A. Giles to F. A. Giles, May 11, 1879; W. A. Giles to Julia Giles, May 25, 1879; Bertha Giles to Frederick and Julia Giles, June 16, 1879; Fayette Giles to Frederick and Julia Giles, June 22, 1879, Burns Collection. Fayette's letter of June 22nd at last acknowledged the seriousness of the situation.
28. "Frederick A. Giles — Obituary," clipping — source unknown, probably a Greenfield, Mass., newspaper, Burns Collection.
29. Ibid. Copy of Record of Death — "Frederick A. Giles," The Commonwealth of Massachusetts, June 18, 1879.

Chapter 16
Aftermath

Even before Frederick's death the United States Watch factory had come to the end of its existence. For many months after its closing in 1877 it had stood silent while those who claimed possession bargained over the details of its disposal. At last, in March of 1878, the following advertisement appeared in the trade papers:

"For sale — The building and machinery of the late United States Watch Co. at Marion (Jersey City), N.J. The property will be sold upon very favorable terms, and if desired, the moveable machinery will be sold separately from the real property. Apply to Jas. A. Alexander, 2 Cortland St., New York City."[1]

The ad must have provoked more than a few chuckles among the denizens of Maiden Lane. It was going to be a long time before anyone would be crazy enough to open another watch factory in Jersey City. But James Alexander was a persistent man; if he could not find a single buyer, several would do. After a few months his patience at last paid off; he succeeded in locating three groups who among them would purchase everything. Not surprisingly, the moveable machinery had attracted the most interest. A harried Alexander found himself having to iron out disputes between two firms that often wanted the same items. On one hand there were E. D. and C. M. Howard of Fredonia, New York, C.O.D. watch merchants and manufacturers of patent medicine. They, much like Ellis Elias, had been forced to watch the legitimate trade dry up their sources of supply and now to save their business they had to acquire a factory. Pitted against them were Hart & Sloan of Newark. George Hart was once more gambling that his intimate knowledge of the factory would prove profitable.

It was once stated in an article about the Peoria Watch Company that the Howards got 75% of the machinery and Hart, Sloan & Co. the remainder. This now seems doubtful.[2] It is much safer to say that the Newark firm acquired closer to half of the machines. And one can assume that George made sure that it was the better half. As for the Howards, they gathered up their equipment as well as what remained of Marion's horde of unfinished watches and shipped the lot to Fredonia. The tale of the headaches and heartaches that befell the watch manufacturing ventures of these intrepid patent medicine men has been told several times and thus need not be repeated here. Suffice it to say that they produced many fine watches, but that the trade had little appreciation for their efforts. After three different attempts — the Independent, Fredonia and Peoria Watch Companies — and a brief stint at sub-contracting for the Non-Magnetic Watch Company what remained of the old United States machinery that the Howards had purchased and the last factory that they had an interest in, Peoria, became the Parsons Horological Institute. Later a part of the machinery was destroyed when the school burned, but before the fire many items had been sold. Unfortunately, the name of the buyer has yet to come to light.

Even before he had conceded that the United States Watch factory would have to be sold piecemeal, James Alexander had turned the idea of approaching George Hart over in his mind. But since Hart was still in the midst of rebuilding machinery for Auburndale, Alexander had at last dismissed the thought. After all, everyone knew that Hart, Sloan & Co.'s resources were spread to their limit. No one, however, had reckoned with George's determination — his firm was going to be the force in the biggest used machinery buy of the age or else! His immediate problem was completing the Auburndale contract before other buyers for the Marion machinery arrived on the scene. He almost made it, too! By the end of August, 1878, he had sent off the last of the machinery to Mr. Fowle and was able to turn his attention to the Marion matter. Unfortunately, despite the profitability of the Auburndale deal Hart, Sloan & Co. had nowhere near enough money for anything on the scale of what was contemplated. Before they could proceed they would have to seek help. It took the better part of a month to secure the necessary promises of loans and in the interim the Howards had appeared on the scene.[3] At first, Hart assumed that his carefully laid plans were destroyed, however it soon became obvious that the Howards might actually prove to be a benefit. Driven by their ignorance and the needs of their planned factory they sought out the very items that Hart wanted least. There were, of course, times when both parties wanted the same machine, but if anything the Howards acted as a check on Hart's acquiring instincts.

George Hart was most certainly capable of acquiring things merely to fulfill a certain pride of possession. One could hardly expect a mechanic reared in the world of Frederick Giles to be otherwise. Yet George generally knew the difference between a good and a poor buy. Further, he was driven by different motives; he had learned that the one sure way of making money in the watch manufacturing field was to build the machinery

OPPOSITE PAGE. 110. The Maiden Lane Jewelry Center, looking east from Broadway circa 1885, shortly after the demise of George C. F. Wright & Co. at the old 13 Maiden Lane address. At the time of this engraving, some 308 firms occupied a space of less than two blocks.

and let others struggle with the problems of making watches. Thus, while the Howards bought pain and grief with their share, he was acquiring sufficient equipment to upgrade his firm. Let the boys from Fredonia search out production lines, he would select only those machines that could be counted to produce a profit. In particular, Hart sought out the testing and machine shop equipment before he turned to pinion cutters and their ilk. The wisdom of his thinking was first evident in the fact that while, in May of 1879, the Howards were struggling to get their factory organized, he could advertise in the trade papers:

"Hart & Sloan
Builders of
Watch Machinery
and Makers of
Intricate Mechanical Instruments
363 & 365 Market Street
Newark, N. J.

Having purchased a large lot of watch machinery, which we have fitted up and have now at work, are prepared to take orders for all kinds of small work. Gauges, parts of watches and fine instruments of every description.

We have also for sale a lot of small lathes and special tools for watchmakers and amateurs suitable for repairing Watches & Clocks. We can also furnish all kinds of new Watch Machinery, or special tools for clocks or other fine work; and small screws of every description from 220 to 40 threads to the inch diameter to correspond."[4]

Whether or not Hart and Sloan were successful in the small parts aspect of their venture went unrecorded, but it must be noted that soon the firm turned exclusively to the manufacture of watch making machinery. In that field they were to prosper for many decades. Crossman records that they were responsible for supplying Ezra F. Bowman with the machinery for his watchmaking venture. Later, they were involved with the tooling of the famous Waterbury Watch factory. This contract ended George Hart's connection with the company that he had founded. From 1883 onward he served with the Waterbury Co. and then with its successor, the New England Watch Company. During his twenty-three years with these firms Hart rose from their mechanical expert to general superintendent, the post which he held at the time of his death in December, 1905.[5] As for the old Hart, Sloan & Co., it became Sloan and Chace and was still in the machinery trade until recent times.

It was generally only those individuals, such as Hart, Wales and Fayette Giles, who had left the United States Watch factory before its tragic final plunge, that lived for many years after. Alexander Hamilton Wallis, for instance, who had lent so much of his reputation in aiding the company to find backers and who had later fought on in the desperate attempts to save it, survived Frederick Giles by only four days, although he was only sixty-one years of age. The fact that the First National Bank of Jersey City, of which his son, Hamilton, was a director, had lent the factory vast sums of money destroyed his spirit. In the end he succumbed to intermittent fever followed by typhoid.[6]

The next individual whose untimely end can be traced to the Marion disaster may come as a surprise to those who believe that evil grants a certain degree of immunity to those who pursue it. Ellis H. Elias' venture into watch manufacturing had swallowed up a good part of his ill-gotten fortune. By late 1878, he and his brother William were declared bankrupt.[7] Not being inclined to accept defeat, Ellis attempted to fraudulently dispose of some of his property. The effort did not succeed and the court declared the entire bankruptcy to have been "conceived and consummated in fraud by men who live by fraud."[8] This combined with the fact that Anthony Comstock made Ellis a prominent subject in his 1880 book, *Frauds Exposed*, left the once wealthy bunko artist in straitened circumstances. In December of 1880 he suffered an illness which at first appeared to be a cold; however it lingered on, gradually worsening. By late May the disease finally forced his confinement to his room at the St. Cloud Hotel, where he died a month later at the age of forty-two.[9]

George Channing Fuller Wright was, of course, by agreement of all those who knew him, a nice guy. But in the end he fared no better than his antithesis, Ellis Elias. Wright, who had fled Marion rather than await the final closing of the factory's doors, had later struggled to salvage what he could of the once prosperous firm of Giles, Wales & Co. After Frederick's death he had taken in another partner, thus for a year the store at 13 Maiden Lane was known as Wright & Kent. Later he had carried on alone under the name of George C. F. Wright & Co. Despite the fact that he had restored the

111. 16s, 19j ¼ plate and bridge, SN 55904, stemwind, lever set, nickel frosted and damaskeened marked "Geo. C. F. Wright, New York." This is believed to be one of the last USWC products sold under Wright's final business name at 13 Maiden Lane, circa 1883. It could have also been his personal watch, or a model for a grade never put into production.

store to some of its former prosperity, his health already weakened by the Marion tragedy could not take the strain of the struggle. On February 8, 1884, he died at age forty-two, the victim of a stroke.[10]

Isaac Henry Wright had loyally stood by Frederick to the very end. Even after George had deserted Marion he had continued to act as a clerk at the St. James. Later, when Frederick had gone home to Montague, Henry and his wife followed. Alas, any hopes that he might have had of taking part in his brother-in-law's future business plans were dashed by Fredericks' death. Although he might have returned to New York or Boston to pick up the threads of his career, Henry chose to settle down to the peace of a small town. Until his death on December 30, 1891, he ran a small notions store and the local post office at Montague.[11]

It has already been stated that shortly after leaving the partnership William Wales became associated with the Auburndale Watch Company. While in the employ of that company he received a number of patents for stop watches, a pocket thermometer and a game counter. Although these items had some commercial success, the Auburndale Company did not. His hopes dashed for a second time, Wales never again became involved in watch manufacturing. In 1884 he returned to New York City where he opened a jewelry business at 16 Maiden Lane. After about five years he entered semi-retirement, a status which he maintained until his death on February 22, 1896.[12]

112. *William A. Wales died in New York City on the 22nd of February, 1896.*

After severing his connection with the Blooming Grove Park Association Fayette and his wife, Bertha, had returned to Switzerland. There he settled at Geneva where he set up a shop as agent for Jules Jurgensen selling the famous watchmaker's watches and chronometers.[13] For the next eighteen years he spent most of his time in his adopted country with only an occasional trip to the United States. Strangely enough, while he and Bertha corresponded with Frederick's widow, Julia, and had the opporunity to visit her while on their visits to the U.S., they could not bring themselves to make the trip to Montague. Their excuse was that the press of affairs and Bertha's indifferent health, particularly the fact that she suffered from chronic bronchitis, would not permit the journey.[14] In 1890 Bertha died and their daughter, Emma, married a Mr. Jay White of Lapeer, Michigan. It was at this point that Fayette's lifestyle underwent an almost total transformation. He left the jewelry business and became a reformer and an author of somewhat utopian works on socio-economic subjects. While his most important works, *A Century Onward* and *The Industrial Army*, are now forgotten, their advocacy of society operating as an immense cooperative was well received by the critics of the time. In a more practical vein Fayette played a role in the reformation of French university degrees. He also remarried, his second wife being an Emma Smalley of Boston. Alas, time permitted no more, and on December 5, 1897, he died of a heart attack while on a hunting trip to North Carolina.[15]

Despite their differences of personality the later careers of William, Charles and Edwin Giles were to become closely intertwined. While originally there had been two businesses, E. A. Giles & Co. of Dubuque, Iowa, and Giles Bros. & Co. of Chicago, Illinois, the strain placed on Edwin's firm by the collapse of Frederick's empire ultimately forced their consolidation in 1880.* At first William Giles headed the enlarged firm, however his retirement in 1882 raised Charles to this role.[16] During the next few years the new team of Charles and Edwin Giles was an outstanding success. In particular, Charles' abilities as an inventor gave them a position of leadership in the anti-magnetic watch protector field.† Alas, their good fortunes led to an unwise overexpansion of their firm. Just as Giles Bros. took over the entire fourth floor of the Chicago Masonic Building the nation's economy took a downturn. As a result, in May of 1893 they were forced into receivership.[17] For several weeks Charles and the firm's creditors struggled to reach a compromise, but neither side was willing to give sufficiently. Finally the Gileses were forced out and the company was taken over by a Mr. H. D. Spaulding of Boston. William Giles, however, who had remained on the sidelines during this affair had no intention of allowing the firm he had founded to

---

*There is some evidence that Fayette's business was also connected to Giles Bros. about this time. Unfortunately, there is no way of telling the degree to which it was interlocked.

†According to the September, 1885, *Jeweler's Circular and Horological Review*, William A. Wales had the New York agency for Giles Bros. anti-magnetic line.

disappear in this fashion. Two years later in 1895 with his backing, Charles, Edwin and W. G. Curtis were able to begin anew.[18] Beyond this point the exact details of the Giles brothers' story is not known. All that can be said is that William seems to have outlasted everyone. He died on December 17, 1913, in Phoenix, Arizona.[19] In the end he paid back the debt he owed Frederick by educating his daughter and granddaughter.[20]

The most difficult item James A. Alexander had had to dispose of was the United States Watch factory itself. The elegant building had acquired the reputation of being a white elephant and few business firms wanted it. Had not northern New Jersey been undergoing an economic revolution it might have sat empty for years. Fortunately, during the last quarter of the 19th Century the textile industry, particularly the silk business, lured by low labor costs had begun to settle in the area. The large well-lit structure seemed a natural for this use and soon an enterprising entrepreneur had acquired it and some of the surrounding land for conversion into a silk mill and workers' housing.[21] This prosperity did not last and soon the aging building began the dreary process of being passed from owner to owner. Each converted and altered it as their varying businesses demanded. At one time it was a factory for the manufacture of smoking pipes and at another its business was school furniture. Finally, buried in the grime and additions of the passing years it was dismantled in 1925 — leaving only a vague imprint on the ground.

Watch manufacturing, however, had not come to a permanent end in Jersey City. Further, there was an element of connection between the old United States Watch Company and the later firm, the New York Standard Watch Company. You will no doubt remember that Andrew J. Wood was a major stockholder in the United States Watch Company and a director of both the Empire City and Marion Watch companies. During the years these companies were in operation Wood served in an undistinguished way, but among those who stuck it out to the end it turned out that he possessed a unique attitude. It would seem that Mr. Wood was neither chagrined nor broken by the disaster. Quite the contrary! In the watch game he found an excitement that his old business, fats and oils, lacked. So when R. J. Clay and F. P. Markham announced in 1883 that they would form a watch company at Williamstown, Massachusetts, he had more than mere money to contribute. Thanks to Mr. Clay the new company held the patents on an absolutely unique worm gear escapement, but they lacked a stem-winding design. This, however, Mr. Wood could provide. He remembered Fayette's old internal ring gear plan, U.S. Patent #65,208 of May 28, 1867. Although it might be costly to make, its patent had just about run out and it was, most importantly, an effective design. Needless to say, it was soon adopted. Alas, the Williamstown Watch Company was one of those ill-starred affairs that never got off the ground; however Mr. Clay's watch, now complete with its Giles winding mechanism, was to get a second chance.*

We do not know if anyone who was connected with the old United States Watch venture ever took part in Jersey City's second factory, the New York Standard Watch Company. It must be recorded, however, that when the new factory got going over on Communipaw Avenue its first fruits were Mr. Clay's worm drive watch complete with Fayette's winding mechanism. Later, when the worm gear was discarded the company continued to use the Giles mechanism for its second model. But there the relationship ends, for the New York Standard Company was to enjoy the success that eluded the United States Watch Company.

---

*Crossman, pp. 195-197. The information about the internal gear wind mechanism is based on New York Standard and one suspected Williamstown watch.

CHAPTER 16 REFERENCES

1. *Jewelers' Circular and Horological Review*, "Special Notices," March and April, 1878, IX:2:XXXV.
2. William Muir, "Peoria and Non-Magnetic or Who Got the Scrap Iron?" NAWCC BULLETIN 17:489.
3. Dun and Barlow, *Mercantile Agency Reference Book*, 1879.
4. *Jewelers' Circular and Horological Review*, May, 1879, 10:5 XXII.
5. 'The Death of George E. Hart," *Jewelers' Circular Weekly*; "George Edwin Hart," *American Jeweler*.
6. *New York Times*, June 23, 1879.
7. "Trade Gossip," *Jewelers' Circular and Horological Review*, December, 1878, IX:10:237; *New York Times*, January 5, 1879.
8. "Trade Gossip," *Jewelers' Circular and Horological Review*, December, 1879, X:10:237.
9. *New York Times*, June 24, 1881, 5:3; *The Sun*, June 24, 1881, 1:6; *New York Herald*, June 24, 1881, 9:6.
10. "George Channing Fuller Wright," *Certificate of Death*, February 7, 1884, New York Municipal Archives; "Trade Gossip," *Jewelers' Circular and Horological Review*, March, 1884, p. 62; *The Sun*, February 9, 1884, 1:4.
11. From an obituary, original source unknown, information supplied by Carolyn Burns.
12. "Death of William A. Wales," *Jewelers' Circular*, February 26, 1896, p. 14.
13. Fayette Giles to Frederick and Julia Giles, June 22, 1879.
14. Bertha Giles to Julia Giles Root and Hattie Giles, March 21, 1887, Burns Collection.
15. *New York Times*, December 7, 1897, and "F. S. Giles is at Rest,"
16. Thomas W. Herringshaw, *National Library of American Biography* (Chicago, 1902), Vol. 2.
17. *Jewelers' Circular and Horological Review*, May 17, 1893.
18. *Jewelers' Circular and Horological Review*, May 24, 1893; May 8, 1895.
19. Estate of William A. Giles, Petition by Elizabeth H. Giles and William F. Giles, dated December 26, 1913, Probate Court of Cook County, Illinois.
20. Oral statement by Carolyn Burns.
21. "Trade Gossip," *Jewelers' Circular and Horological Review*, July, 1880, 11:6:120.

PRODUCTION

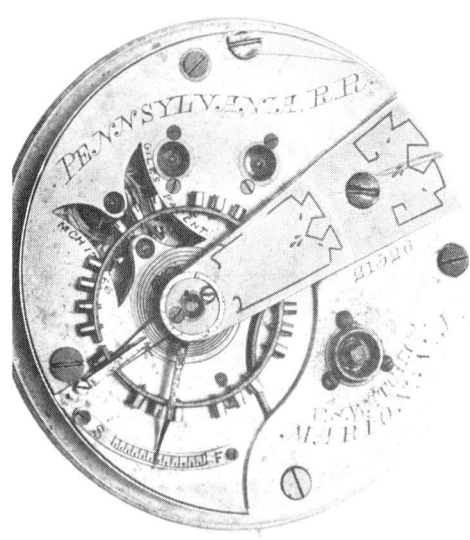

113. Full Plate — Butterfly, 18 size only.

114. Full Plate — No Butterfly, 18 size only (Elson regulator missing).

115. Three-Quarter Plate, 14 and 18 size (18 size illustrated).

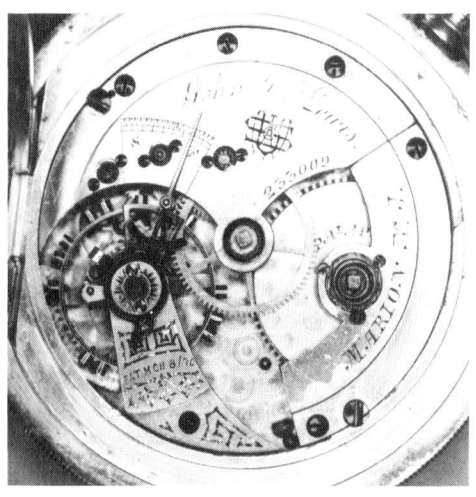

116. Quarter Plate and Bridge, 10 and 16 size (16 size illustrated).

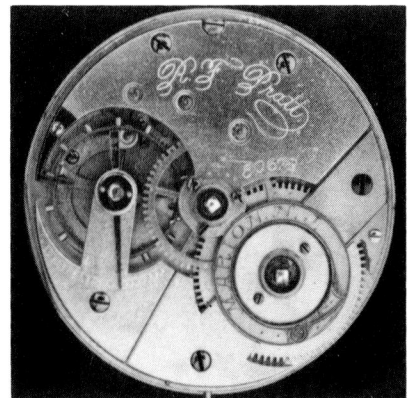

117. Quater Plate, Cock and Bridge (Swiss), 10 size only.

113 thru 117. Basic movement styles and sizes used by the USWC. These illustrations are shown as close to 1.5 times actual size as possible. (Other illustrations in this book are greater enlargements to show detail.) It is significant to note that the USWC 16s actually measures a good 17s, and this has been a source of confusion over the years.

Chapter 17
Products and Patents

Grades, Styles and Sizes

The end products are listed by grade name and approximate sequence of production below:

1867-69   Frederic Atherton & Co.
            Fayette Stratton (18 size only)
            George Channing
            Edwin Rollo

1869-72   United States Watch Co.
            Marion Watch Co.
            A. H. Wallis
            Wm. Alexander
            S. M. Beard
            Henry Randel
            John W. Lewis
            Asa Fuller
            R. F. Pratt (10 size only)
            Chas. G. Knapp (10 size only)
            Young America (14 size only)
            J. W. Deacon
            G. A. Read (18 size only)
            North Star

1872-74   I. H. Wright
            A. J. Wood (10 size only)

Other products, such as those produced for independent jewelers, wholesale houses, and others, will be treated separately.

Products of the USWC were manufactured in five (5) basic styles and four (4) sizes as illustrated in the following table and Figures 113 through 117:

Table II
USWC Movement Styles and Sizes

Style	10 Size	14 Size	16 Size	18 Size
Full Plate — Butterfly				*
Full Plate — No Butterfly				*
Three-Quarter Plate		*		*
Quarter Plate and Bridge	*		*	
Quarter Plate, Cock and Bridge (Swiss)	*			

These twenty grades, five movement styles, and four sizes, were available between 1870 and 1873 in over 200 variations.

Patents

Frederick A. Giles and his younger brothers were all mechanically inclined and were inventors of record, with registered watch-related patents to their credit. Perhaps the most distinctive feature of 18 size, full plate movements was the so-called "butterfly opening" in the plate which permitted examination of the escapement. This plate opening was registered as a "Design" patent and was not in the regular "Letters" series.

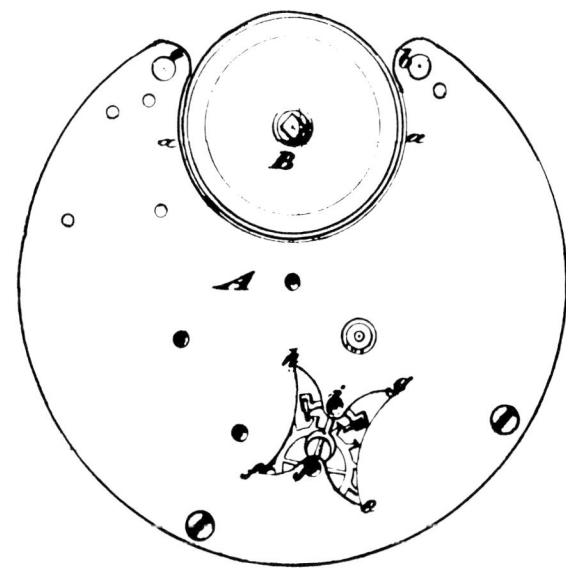

118. F. A. Giles Design Patent 2281 Drawing, illustrating cut out in the plate to facilitate escapement examination. It is interesting to note that Giles did not call the cut out by the well known term, "butterfly." A portion of the patent description states . . . "the sides, e g and f h, of the said opening have salient curvatures, and these meeting in points with the re-entering curvatures in the ends, give the opening a form having some resemblance to that of two crescents placed back to back." This patent was issued on March 13, 1866.

A listing of the Giles patents follows:

Pat. No.	Date	Details
43,490	July 12, 1864	Fayette S. Giles, New York, N.Y., "Improvement in Watch-Cases"

47,412	Apr. 25, 1865	Frederick A. Giles, 13 Maiden Lane, New York, N.Y., "Improvement in Winding and Setting Watches"
D2,055	May 9, 1865	Frederick A. Giles, "Design for the Top Plate and Balance Cock of a Watch"
49,397	Aug. 15, 1865	Frederick A. Giles, 13 Maiden Lane, New York, N.Y., "Improvement in Winding and Setting Watches"
D2,266	Feb. 27, 1866	Frederick A. Giles, "Design for the Top Plate of a Watch"
D2,281	Mar. 13, 1866	Frederick A. Giles, "Design for Watch Plates" Butterfly opening
D2,291	Apr. 3, 1866	William A. Giles, Chicago, Ill., "Clock Cases"
57,495	Aug. 28, 1866	Edwin A. Giles, New York, N.Y., "Improvement in Stem Winding Watches"
D2,525-6	Dec. 11, 1866	William A. Giles, Chicago, Ill., "Clock Cases"
65,208	May 28, 1867	Fayette S. Giles, New York, N.Y., "Improvement in Stem Winding Watches"
69,561	Oct. 8, 1867	Charles K. Giles, Chicago, Ill., "Cuckoo Clock"
D3,885-6	Mar. 8, 1870	Fayette S. Giles, New York, N.Y., "Design for the Top Plate, Barrel-Bridge and Cock for Watches"
145,939	Dec. 30, 1873	Frederick A. Giles, Jersey City, N.J., "Improvement in Watch-Barrels"
289,642	Dec. 4, 1883	Charles K. Giles, Chicago, Ill., "Anti-Magnetic Shield for Watches"

Note: Charles K. Giles received 22 additional patents on watch cases and related components from 1884 to 1889. His patents gave the Chicago based Giles Bros. & Co. a leadership position in the anti-magnetic watch protector field during the 1880's.

In addition to the patents of the Giles brothers at least one other inventor, Julius Elson, contributed to the design of USWC watches. Elson first appears as a watchmaker in Philadelphia in the late 1840s. By the early 1860s he had moved to Boston, where his watchmaking shop was located at 4 School Street. Besides being a first class watchmaker, Elson is also credited with four patents relating to watchmaking. Only one was used by the USWC, his double index regulator, letters patent 100,511 dated March 8, 1870. This device was designed to assist in rate adjustment by means of a compound or double index system. It first appeared on the company's movements in late 1872 and was used to the end of the factory's production. It was standard on most quarter and three-quarter plate movements and could be ordered special on full plate watches at an additional charge. Elson's regulator was unique and certainly in keeping with Frederick's demand for elegance.

Aside from Elson's patent regulator, there has been through the years some speculation about a possible connection between Arthur Wadsworth's patent spring barrel, letters patent 74,457 dated February 11, 1868, and Frederick Giles' reversible barrel letters patent 145,939 dated December 30, 1873. *Based on the data that now exists,* there are strong and compelling reasons for doubting these views:

(1) Wadsworth's barrel patent was assigned to himself and Robert Schell.

(2) Patent laws at that time required marking with the patent issue date. Elson's patent date was clearly shown on the balance cock, so if the Wadsworth patent was used why wasn't the patent date on the barrel?

(3) From the design standpoint there is very little connection between the Wadsworth and Giles barrels — they are different concepts.

(4) The Giles "patent applied for" reversible barrels introduced late in 1872 are virtually identical to the later barrels with patent date of "Dec. 30, '73."

All of this, of course, begs two important questions — why the USWC price list dated September 2, 1872, cited a "patent reversible barrel" and why did Giles mark the early barrels "patent applied for" when he did not apply for his patent until December 1, 1873. We do not know the exact answers to these questions except that Frederick, like the cannoneers in a certain old bawdy poem, was not stopped by "trifles." There is no evidence of any caveats or earlier patent applications, continuations or amendments on the patent which was granted within the month on December 30. False use of the term "patent" in advertising or "patent applied for" on any article is subject to a $500 fine for every such offense under 1985 patent law. In 1873, however, the fine was $100 and was limited to false marking of an article with the term "patent." Also in 1873, a patent could still be obtained on an article if application was made within two years from the date it was first sold. Giles was well within the two-year time frame limitation and did not violate the 1873 patent laws, so he had little at risk with premature patent publicity and patent application markings. (While the authors have no intention of supplying our readers with the poem, those who are interested in more details on any of these patents should write either the Patent Office or the NAWCC, Inc., Museum and Research Center, 514 Poplar Street, Columbia, PA 17512. Both charge a nominal fee for copies.)

Movements

In an advertisement of December 9, 1869, the USWC commented on their movements: "Manufacturers of all the grades of American watches, pendant winders and key winders, both nickel and frosted movements.

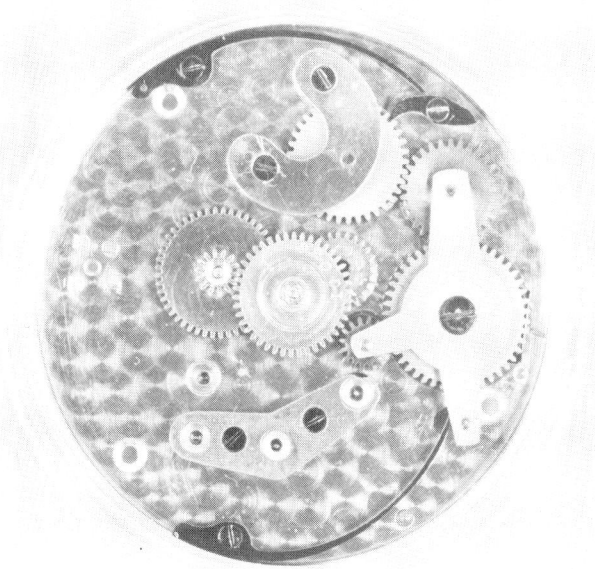

119 and 120. Movement and under dial views, "Fayette Stratton" grade, SN 10033, 18s, 15j, early example (circa 1868) of gilt damaskeening. Gold jewel settings located at the balance, escape and pallet arbor pivots. Particularly well finished balance with gold screws. Note circular gilt damaskeening under dial; repeated under pillar plate and on the barrel lid. Giles 1865 stem winding patent and 1868 improvement. Push button set.

For our late improvements in Stem Winding mechanism we claim a Strength, Simplicity, and Smoothness hitherto unattained in any other manufacture, at home or abroad."

A visitor to the wholesale showrooms in New York City had this to say: "In examining some of the finer movements made by this company, we find them made entirely of nickel; their beauty of finish excited our special admiration, and it is on this class of work that the 'damaskeening' is shown in all its beauty. The work finished in this style has the appearance of watered silk, and far surpasses anything of the kind ever before attempted."

Damaskeening was applied to both nickel and gilt movements, but was used primarily with nickel. Very few gilt damaskeened movements have been observed in surviving USWC watches, even though it was offered in the 1870, 1872, and 1873 price schedules as an additional cost item. It is significant to note that USWC pioneered the use of damaskeening on American watches and were responsible for many improvements in the application techniques. The earliest known example of gilt damaskeening at this time is a Fayette Stratton, SN 10033, illustrated in Figures 119 and 120.

The USWC was also unique in their use of frosting on the higher quality nickel movements. In some cases the complete movement was frosted, while in others only portions were frosted to achieve an attractive two-tone effect. Engraving and lettering on the finer nickel grades were richly enameled in black and steel work was gilded. This fine engraving, enameling, and frosting are illustrated nicely in Figure 121.

121. "A. H. Wallis" grade, SN 19061, 18s, 19j, full plate. A very early example (circa 1869) of nickel damaskeening and frosting with fine engraving nicely enameled. The balance cocks on USWC finer grades like this example seem to be "one of a kind" engravings. The USWC was one of few American companies to use this unique engraving and/or frosting of balance cocks.

Table III
USWC Grades and Movements
1st and 2nd Generation

	1867-1872		2nd Generation, 1872-1874				
	Full Plate	¼ Plate	Full Plate	¾ Plate		¼ Plate and Bridge	
Frederic Atherton & Co.	18		18	18	14	16	10
Fayette Stratton	18						
George Channing	18		18	18	14	16	10
Edwin Rollo	18		18	18	14	16	10
United States Watch Co.	18		18	18	14	16	10
Marion Watch Co.	18		18	18	14	16	10
A. H. Wallis	18		18	18	14	16	10
Wm. Alexander	18		18	18	14	16	10
S. M. Beard	18		18	18	14	16	10
Henry Randel	18		18	18	14	16	10
John W. Lewis	18		18	18	14	16	10
Asa Fuller			18	18	14	16	10
R. F. Pratt		10					
Chas. G. Knapp		10					
Young America					14		
J. W. Deacon			18	18	14		10
G. A. Read			18	18			
I. H. Wright			18	18	14	16	10
A. J. Wood							10
North Star	18		18	18	14		
Total Different Grades	20						
Total Different Movements	85	12	2	15	15	15	12	14

As can be seen in the chart in Table III, a total of twenty different name grades were produced by the USWC, encompassing some eighty-five possible movements. In addition, many other models were produced by this company which bear no identifiable marks relating to either USWC or "Marion." These will be covered in later chapters.

A major overhaul in USWC production started late in 1871 with plans to produce the 18 size in both three-quarter and full plate styles. As part of this production change over, other sizes and styles were added to the line, including a 14 size three-quarter plate and a unique quarter plate and bridge in 10 and 16 sizes. Two new inventions were added to most quarter and three-quarter plate models, the Giles reversible barrel and the Elson double index regulator. Production problems delayed the introduction of these second generation models until late in 1872.

Reason for these dramatic changes was undoubtedly the press of competition from the American, Elgin, and Howard Watch Companies, plus the fact that the USWC did not have a good, cheap watch to offer to the public. During this period, the lowest price of an 18 size full plate movement dropped from $20.00 ("Edwin Rollo") to $10.50 ("G. A. Read") with the latter movement having several improvements not found on the $20.00 movement.

Referring once again to the chart, one can see that the majority of the production involved ten of the original grades which were produced in Full Plate, 18 size on two different occasions, and in Three Quarter Plate in 14 and 18 size, plus Quarter Plate in 10 and 16 size, thus accounting for sixty different movements.

USWC movements were offered in an almost countless number of finish and style variations. Most USWC historians agree that these many variations contributed to a significant inventory control problem and associated high costs.

Only the "Fayette Stratton" grade, the second to be produced by the USWC, was dropped after its initial production runs and was not involved in the spin-off into other size movements.

Chapter 18
Prices of Movements and Cases

The relative grading of USWC models by the Company itself has been recorded in several USWC Catalogs and Trade Lists. Quite naturally, the prestige items lead the list, followed by the balance of the nickel movements, to the lesser items in gilt finish.

The following prices represent the basic catalogue price of a standard Full Plate, 18 size movement, without extras. Other size movements are specified when 18 size was not produced as part of a model series. This listing is presented to show the most expensive grade, down to the least expensive.

United States Watch Co.	$360.00
A. H. Wallis	115.00
Henry Randel	103.00
Frederic Atherton & Co.	90.00
Wm. Alexander	78.00
Marion Watch Co.	78.00
S. M. Beard	70.00
I. H. Wright	60.00
Fayette Stratton	57.00
John W. Lewis	57.00
George Channing	54.00
Edwin Rollo	41.00
Chas. G. Knapp (10 size)	40.00
R. F. Pratt (10 size)	38.00
Asa Fuller	31.00
A. J. Wood (10 size)	31.00
J. W. Deacon	25.00
Young America (14 size)	19.00
G. A. Read	18.00
North Star	12.00

Most grades were offered in several models which varied in price because of differences in movement style, size, jewel count, finish, and type of winding. Table IV (pages 134-136) illustrates a complete listing of all models known to exist between 1870 and 1873 with corresponding prices.

All USWC movements were guaranteed, with the exception of the "G. A. Read" grade. The exception reads: "All except the "G. A. Read" are warranted by special certificate." This is understandable in view of the fact that the "Read" grade was the least expensive movement of any regularly produced USWC model. This grade is also one of the most easily obtained USWC products due to the large numbers that were produced.

Documentation is lacking on the "North Star" grade, but its price can be estimated in the $10.00 to $12.00 range based on comparison to the "G. A. Read." The "North Star," like the Read, was probably not warranted by special certificate.

To the catalogue prices of the movements would be added the cost of a case and whatever extras were wanted. The cases ranged from the very cheapest up to Coin Silver, Sterling Silver and 14 and 18 Karat Gold. "Every variety of Gold and Silver, Diamond Set and Magic Cases constantly on hand and made to order. Also Stemwinding Minute Repeaters, 1/4-1/5 Split and Fly-back Seconds, for timing Horses, Artillerymen, etc."

From the beginning of production in 1867 up through the apex of their operations in 1869-1871, some movements were sold cased in fine 18K, 14K, or sterling silver cases distinctively marked with the USWC monogram. None of the surviving data gives any indication that the company ever manufactured their own cases. It is probable that they relied on case supply sources previously established through Giles, Wales & Co. A circa 1868 heavily chased, USWC monogramed 18K hunting case is illustrated in Figures 123 and 124. Prices of the cases varied with weight, engraving, decorations, and extras. For example, 1870 prices for a cased "United States" model ranged as follows in stemwind:

18K	$438 to $588
14K	$412 to $515
Sterling	$352 to $381

Some of the very early cases have been noted that are marked with an enamel inlaid USWC monogram on the dust cover. Later cases are marked with the

122. Very early (circa 1867) 18k hunting case marked with a blue, black and white enamel inlaid USWC monogram on dust cover and G. W. & Co. on inside covers. Note button set attachment on case rim and that pendant is a key wind type with stem winding crown. This case houses SN 1034, "Frederic Atherton & Co."

ABOVE. 123 and 124. Front and reverse views, 18k, heavily chased USWC marked hunting case (circa 1868). Note lock on push button to prevent accidental setting when lid is closed. Chain guards on bow, found on many surviving USWC cases. BELOW. 125 and 126. Inner case USWC monogram marks on two different 18k hunting cases. Some cases survive that are marked G. W. & Co. for Giles, Wales & Co. A few examples have also surfaced with matching case and movement numbers. Some original USWC cases will be marked with only the case maker's initials, such as J.L. for Jacques Laurent who made many cases for Giles, Wales & Co.

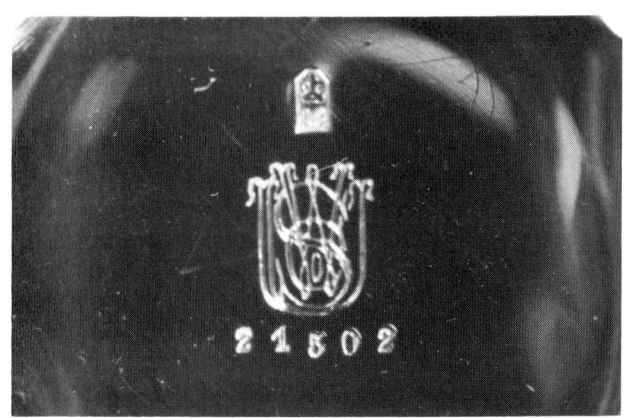 

127. *Front view, 18k, nicely engraved hunting case (circa 1870), marked G. B. & Co. for Giles, Brothers & Co. in Chicago. This drum style case houses a "Wm. Alexander" movement, SN 20293. The case serial number is 9684. 14k, sterling, and coin silver USWC marked cases have also been recorded.*

USWC monogram on the inside covers. It is significant to note that all original cases are not marked with the USWC monogram. In keeping with the style and practices of the times, movements would be matched to a case that met with the buyers' particular tastes. Some original USWC cases will therefore be marked with only the case maker's initials or his own distinctive mark, much like early E. Howard & Co. cases.

Information on one of the most probable manufacturers of USWC cases is provided by Warren H. Niebling in his 1971 book on American Watch Cases:

"Jacques Laurent started his case business in New York City in 1863. *He at first made cases for only one company, Giles, Wales & Co.* He made a large number of cases and was very successful. His brother, Emil, joined him and the company then was J. Laurent & Co. They were in New York City until 1878...."

The link to Jacques Laurent as a USWC case maker has been firmly established by the discovery of a finely engraved, drum style 18K hunting case marked "JL" housing an 18s full plate, button set USWC movement. Some USWC cases have been noted that are marked "G. W. & Co." for Giles, Wales & Co., and "G. B. & Co." for Giles, Brothers & Co. Other USWC case serial numbers have been noted that match the movement serial number; one matching numbers "George Channing" has been noted with a Montreal case maker's or jeweler's mark, "John Street" — another matching numbers "George Channing" is marked "C & A P" for C. & A. Pequignot, case makers in Philadelphia. It is unfortunate that relatively few original cased USWC products, particularly those in 18K, have survived the passage of time. The high price for gold cases coupled with high priced USWC movements found a very small market in the late 1860's and 1870's.

Table IV
Summary of USWC Models and Prices
1870-1873

Grade	Model	Reference Date	Movement Style	Movement Sizes	Jewels	Variations	Finish*	Original Prices KW	SW
United States Watch Co.	21	1870	Full Plate	18s	19	1	N, PD, RGJS, E, GIP, BHS, HCIP, BS, PB	—	$332
	22	1870	Full Plate	18s	19	1	N, PD, RGJS, E, GIP, BHS, HCIP, BS, PB	$307	—
	23	1870	Full Plate	18s	19	1	G, GJS, GIP, BHS, HCIP, BS, PB	—	294
	24	1870	Full Plate	18s	19	1	G, GJS, GIP, BHS, HCIP, BS, PB	269	—
	17	1872	¼ Plate & Bridge	10s & 16s	19	4	N, DF, RGJS, E, GIP, BHS, HCIP, GS, PRB, PDIR	420	450
	18	1872	¾ Plate	14s & 18s	19	4	N, DF, RGJS, E, GIP, BHS, HCIP, GS, PRB, PDIR	370	400
	19	1872	Full Plate	18s	19	2	N, DF, RGJS, E, GIP, BHS, HCIP, GS, PB	330	360
	21	1872	Full Plate	18s	19	2	Same as 19 except PD, BS	270	300
	21½	1872	¼ Plate & Bridge	10s & 16s	19	4	Same as 21 except PRB, PDIR	360	390
	22½	1872	¾ Plate	14s & 18s	19	4	Same as 21 except PRB, PDIR	310	340
	23	1872	Full Plate	18s	19	2	G, GJS, GIP, BHS, HCIP, BS, PB	240	270
	23½	1872	¼ Plate & Bridge	10s & 16s	19	4	Same as 23 except PRB, PDIR	300	330
	24½	1872	¾ Plate	14s & 18s	19	4	Same as 23 except PRB, PDIR	260	290
United States Watch Co.	10	1873	¼ Plate & Bridge	10s & 16s	15	4	N, PD, GJS, E, FHS, HCP, GS, PRB, PDIR	295	325
	11	1873	¾ Plate	14s & 18s	15	4	N, PD, GJS, E, FHS, HCP, GS, PRB, PDIR	285	315
	12	1873	Full Plate	18s	15	2	Same as 11 except PB	270	300
	14	1873	¼ Plate & Bridge	10s & 16s	15	4	G, GJS, FHS, HCP, GS, PRB, PDIR	175	205
	15	1873	¾ Plate	14s & 18s	15	4	G, GJS, FHS, HCP, GS, PRB, PDIR	165	195
	16	1873	Full Plate	18s	15	2	Same as 15 except PB	150	180
A. H. Wallis	50	1870	Full Plate	18s	19	1	N, PD, GJS, E, FHS, HCIP, BS, PB	—	199
	51	1870	Full Plate	18s	19	1	N, PD, GJS, E, FHS, HCIP, BS, PB	174	—
	52	1870	Full Plate	18s	19	1	Same as 50 except only adjusted HC	—	139
	53	1870	Full Plate	18s	19	1	Same as 51 except only adjusted HC	114	—
	52	1872	Full Plate	18s	17	2	N, PD, GJS, FHS, HC, BS, PB	100	115
	52½	1872	¼ Plate & Bridge	10s & 16s	17	4	Same as 52 except PRB, PDIR	135	150
	53½	1872	¾ Plate	14s & 18s	17	4	Same as 52 except PRB, PDIR	110	125
Henry Randel	54	1870	Full Plate	18s	15	1	N, PD, GJS, FHS, HC, BS, PB	—	120
	55	1870	Full Plate	18s	15	1	N, PD, GJS, FHS, HC, BS, PB	103	—
	54	1872	Full Plate	18s	15	2	N, PD, GJS, FHS, HC, BS, PB	87.50	102.50
	54½	1872	¼ Plate & Bridge	10s & 16s	15	4	Same as 54 except PRB, PDIR	120	135
	55½	1872	¾ Plate	14s & 18s	15	4	Same as 54 except PRB, PDIR	95	110
Wm. Alexander	56	1870	Full Plate	18s	15	1	N, PD, GJS, FHS, BS, PB	—	90
	57	1870	Full Plate	18s	15	1	N, PD, GJS, FHS, BS, PB	73	—
	56	1872	Full Plate	18s	15	2	N, PD, GJS, FHS, BS, PB	63	78
	56½	1872	¼ Plate & Bridge	10s & 16s	15	4	Same as 56 except PRB, PDIR	87.50	102.50
	57½	1872	¾ Plate	14s & 18s	15	4	Same as 56 except PRB, PDIR	70	85

Name	#	Type	Size			Features		
S. M. Beard	58	Full Plate	18s	15	1	N, PD, GJS, FHS, BS, PB	$ —	$ 79
	59	Full Plate	18s	15	1	N, PD, GJS, FHS, BS, PB	61	—
	60	Full Plate	18s	15	2	N, PD, GJS, FHS, BS, PB	55	70
	58½	¼ Plate & Bridge	10s & 16s	15	4	Same as 58 except PRB, PDIR	70	85
	59½	¾ Plate	14s & 18s	15	4	Same as 58 except PRB, PDIR	60	75
John W. Lewis	60	Full Plate	18s	15	1	N, PD, GJS, FHS, BS, PB	—	64
	61	Full Plate	18s	15	1	N, PD, GJS, FHS, BS, PB	46	—
	60	Full Plate	18s	15	2	N, PD, GJS, FHS, BS, PB	42.50	57.50
	60½	¼ Plate & Bridge	10s & 16s	15	4	N, PD, GJS, FHS, BS, PB	55	70
	61½	¾ Plate	14s & 18s	15	4	Same as 60 except PRB, PDIR	48	63
Frederic Atherton & Co.	25	Full Plate	18s	19	1	G, GJS, FHS, HCIP, BS, PB	—	161
	26	Full Plate	18s	19	1	G, GJS, FHS, HCIP, BS, PB	144	—
	27	Full Plate	18s	19	1	Same as 25 except only adjusted HC	—	101
	28	Full Plate	18s	19	1	Same as 26 except only adjusted HC	84	—
	27	Full Plate	18s	17	2	G, GJS, FHS, HC, BS, PB	75	90
	27½	¼ Plate & Bridge	10s & 16s	17	4	Same as 27 except PRB, PDIR	115	130
	28½	¾ Plate	14s & 18s	17	4	Same as 27 except PRB, PDIR	82.50	97.50
Fayette Stratton	31	Full Plate	18s	15	1	G, GJS, FHS, HC, BS, PB	—	114
	32	Full Plate	18s	15	1	G, GJS, FHS, HC, BS, PB	96	—
	33	Full Plate	18s	15	1	Same as 31 except not adjusted	—	64
	34	Full Plate	18s	15	1	Same as 32 except not adjusted	46	—
Marion Watch Co.	29	Full Plate	18s	15	1	G, GJS, FHS, HC, BS, PB	—	86
	30	Full Plate	18s	15	1	G, GJS, FHS, HC, BS, PB	69	—
	29	Full Plate	18s	15	2	Same as 29 except not adjusted	63	78
	29½	¼ Plate & Bridge	10s & 16s	15	4	Same as 29 except not adjusted, PRB, PDIR	95	110
	30½	¾ Plate	14s & 18s	15	4	Same as 29 except not adjusted, PRB, PDIR	70	85
I. H. Wright	33½	Full Plate	18s	11	2	G, FHS, BS, PB	45	60
	34½	¼ Plate & Bridge	10s & 16s	11	4	Same as 33½ except PRB, PDIR	60	75
	34¾	¾ Plate	14s & 18s	11	4	Same as 33½ except PRB, PDIR	52.50	67.50
George Channing	37	Full Plate	18s	15	2	G, FHS, BS, PB	35	53
	37¼	Full Plate	18s	15	2	Same as 37 except N Top Plate, Barrel Bridge & Cock	38.50	53.50
	37¾	¼ Plate & Bridge	10s & 16s	15	4	Same as 37¼ except PRB, PDIR	50	65
	38¼	¾ Plate	14s & 18s	15	4	Same as 37¼ except PRB, PDIR	42.50	57.50
Edwin Rollo	40	Full Plate	18s	15	2	G, FHS, BS, PB	20	38
	43	Full Plate	18s	11	2	G, FHS, BS, PB	25.50	40.50
	43¼	¼ Plate & Bridge	10s & 16s	11	4	Same as 43 except PRB, PDIR	38.50	53.50
	43¾	¾ Plate	14s & 18s	11	4	Same as 43 except PRB, PDIR	31.50	46.50
R. F. Pratt	38	¼ Plate, Cock, & Bridge	10s	15	1	G, FHS, BS, (Swiss Made)	30	—
Chas. G. Knapp	39	¼ Plate, Cock, & Bridge	10s	15	1	G, FHS, BS, (Swiss Made)	35	—
	39½	¼ Plate, Cock, & Bridge	10s	15	1	G, FHS, BS, (Swiss Made), except with Damaskeening	50	—

*Note: Finish codes on page 136.

Table IV (Continued)

Grade	Model	Reference Date	Movement Style	Movement Sizes	Jewels	Variations	Finish	Original Prices KW	Original Prices SW
Asa Fuller	43½	1872	Full Plate	18s	11	2	G with N Barrel Bridge, FHS, BS, PB	$ 19.50	$ 31.50
	44¼	1872	¼ Plate & Bridge	10s & 16s	15	4	Same as 43½ except PRB, PDIR	31.50	45.50
	44½	1872	¾ Plate	14s & 18s	11	4	Same as 43½ except PRB, PDIR	25.50	37.50
J. W. Deacon	45	1872	Full Plate	18s	11	2	G, FHS, BS, PB	15.50	25.00
	45⅛	1872	¼ Plate & Bridge	10s	15	2	Same as 45 except PRB, PDIR	26.50	37.50
	45¼	1872	¾ Plate	14s & 18s	11	4	Same as 45 except PRB, PDIR	21.50	31.00
Young America	45⅓	1872	¾ Plate	14s	7	2	G, FHS, BS, PRB	11.75	19.75
	45¾	1872	¾ Plate	14s	11	2	G, FHS, BS, PRB	13.25	21.25
A. J. Wood	47	1872	¼ Plate & Bridge	10s	15	2	G, FHS, BS, PRB	20.00	31.50
G. A. Read	46	1872	Full Plate	18s	7	2	G, FHS, BS, PB	10.50	18.50
	46½	1872	¾ Plate	18s	7	2	Same as 46 except PRB	11.75	19.75

Variation Summary:

Grade	Quantity
United States Watch Co.	54
A. H. Wallis	14
Henry Randel	12
Wm. Alexander	12
S. M. Beard	12
John W. Lewis	12
Frederic Atherton & Co.	14
Fayette Stratton	4
Marion Watch Co.	12
I. H. Wright	10
George Channing	12
Edwin Rollo	12
R. F. Pratt	1
Chas. G. Knapp	2
Asa Fuller	10
J. W. Deacon	8
Young America	4
A. J. Wood	2
G. A. Read	4
Grand Total	211

Notes:
1. Extras were available on many models at extra cost, such as gilt damaskeening, enameling, frosting, adjustments, etc. See Grade Descriptions for this detail.
2. Reference date applies to available price schedules. Variations have been noted in both jewel count and finish.
3. Original prices apply to movement only.
4. Grades with lower jewel count & poorer finish are usually 1874 and later vintage.
5. No movement prices available on North Star grade, but it is known to be cheapest USWC product with an uncompensated, flat balance wheel.
6. Variations apply to count of size and winding differences only.

Finish Codes:
- N — Nickel
- G — Gilt
- PD — Plain Damaskeen
- DF — Damaskeen & Frosted
- RGJS — Raised Gold Jewel Settings
- GJS — Gold Jewel Settings
- E — Enameled
- GIP — Gold Index Plate
- BHS — Breguet Hairspring
- HCIP — Adjusted Heat, Cold, Isochronism & Positions
- HCP — Adjusted Heat, Cold, & Positions
- FHS — Flat Hairspring
- HC — Adjusted Heat & Cold
- GS — Gilded Steel Work
- BS — Blued Steel Work
- PB — Patent Butterfly
- PDIR — Patent Double Index Regulator
- PRB — Patent Reversible Barrel

Chapter 19
USWC Grade Descriptions

"Frederic Atherton & Co."

This grade was the first to be produced by the USWC and was put on sale to the public in June 1867. The original model was 18 size, full plate with butterfly opening and six pair extra jewels and two extra conical pivots, cap jeweled, in gold settings. It was the first stemwind, button set, 19-jewel American-made watch. Serial numbers started with 1001. The second generation of this grade (1871) had five pair extra jewels and did not include the extra conical pivots, cap jeweled, in gold settings. This grade was also produced in 10, 14 and 16 size, as described below.

The majority of the Atherton models that have been recorded have been 18 size and were referred to as "stem-winders." Some key-wind models have been noted.

The "Frederic Atherton & Co." grade was supposedly named after Giles and Wales, "Frederic" being the first given name of Giles, and "Atherton" the second given name of Wales. If this is so, it is not known why the "k" in "Frederick" was deleted when referring to this model. Giles, in signing patent applications for example, always wrote his name "Frederick," with a "k."

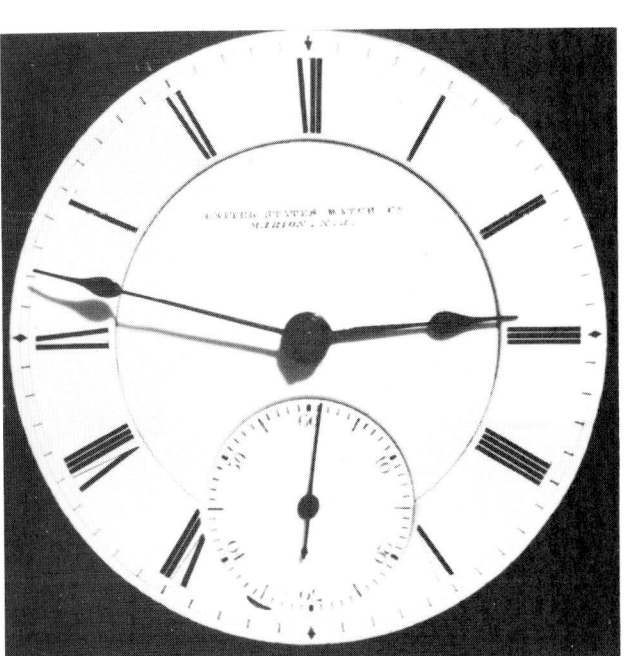

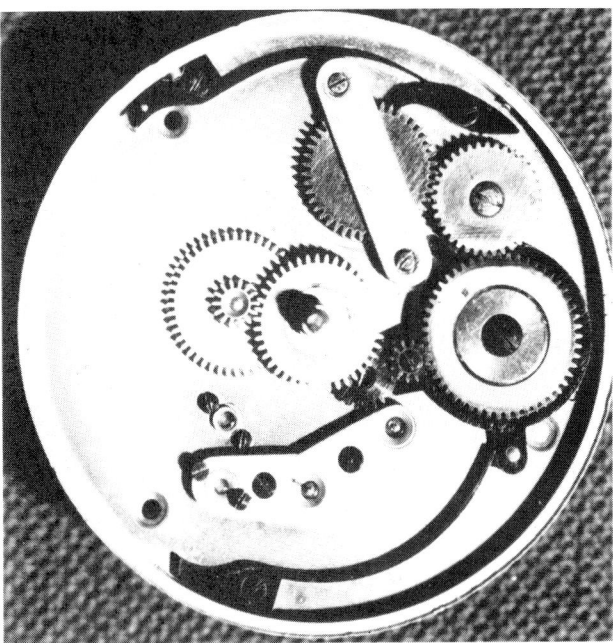

128, 129 and 130. Three views, "Frederic Atherton & Co.," SN 1023, 18s, 19j, full plate. Earliest example of this grade yet to turn up, it is pure stem wind with the early 1865 patent. Double recessed dial with both USWC and Marion, N.J. Typical flat hairspring. This was probably produced in mid-year, 1867.

131. "Frederic Atherton & Co.," SN 1532, 18s, 19j, full plate. Stem wind, typical button set, circa 1868. First generation, Model 25. Movement price in 1868 was $161.

132. "Frederic Atherton & Co.," SN 9119, 18s, 17j, full plate. Stem wind with 1868 improvement. Circa 1872, after this grade had its jewel count reduced from 19j. Movement price in 1872 was $90.

One speculation is that by dropping the "k" better balance was provided with "Atherton" . . . both then had eight letters and the engraving height of the letters "c" and "n" with the same. The shorter name version also gave additional space for the engraving.

The choice of name for this grade may have been inspired by the Waltham "Appleton Tracy & Co." grade.

Factory descriptions follow:

No. 25 (1867)	18 size, Full Plate Lever Movement, Straight Line Escapement, Hardened and Tempered Hair Springs, Exposed pallet jewels, six pair extra jewels, and two pair extra conical pivots, cap jeweled, in gold settings, Sprung over Expansion Balance, adjusted to heat and cold, Isochronism and position. Sunk Seconds, Stem wind and set.
No. 26 (1867)	Same as No. 25 but Key wind and set.
No. 27 (1867)	Same as No. 25 but not adjusted to Isochronism and position. Stem wind and set.
No. 28 (1867)	Same as No. 27 but Key wind and set.
No. 27 (1871)	18 size, Full Plate Lever Movement, Straight Line Escapement, Hardened and Tempered Hair Springs, Exposed pallet jewels, five pair extra jewels, Sprung over Expansion Balance, adjusted to heat and cold. Patent Dust Bands, Sunk Seconds. Both Key wind and Stem wind movements.
No. 27½ (1871)	10 size and 16 size, Quarter Plate and Bridge, Same grade as No. 27 (1871). Patent Reversible Barrel and Patent Double Index Regulator. Both Key wind and Stem wind movements.
No. 28½ (1871)	14 size and 18 size, Three-Quarter Plate. Same grade as No. 27 (1871). Patent Reversible Barrel and Patent Double Index Regulator. Both Key wind and Stem wind movements.

Extras offered on the three 1871 models were:

If adjusted to positions	$50.00
If "Damaskeen" finish	13.50
Reversible Barrel on Full Plate model	3.00
Double Index Regulator on Full Plate model	6.00

"Fayette Stratton"

This grade was also initially produced in 1867. It was the only grade to be produced solely in 18 size, full plate, and thus does not appear in the 10, 14 and 16 size versions. No explanation is available why this grade was dropped from production schedules in 1868/1869 other than Frederick's unhappiness with Fayette when he left the company to pursue other interests. Evidence is available that "Fayette Stratton" movements were cased with "Marion Watch Co." dials (ca. 1874) suggesting that surplus "Fayette Stratton" movements were still in stock.

Both Crossman and Abbott incorrectly report the "Fayette Stratton" as an all nickel movement — it was always made in gilt.

The "Fayette Stratton" grade was named after Fayette Stratton Giles, a younger brother of Frederick. Fayette Giles was active in both the Giles/Wales organization and the USWC. He is also credited with four watch patents.

Factory descriptions follow:

No. 31 (1867)   18 size, Full Plate Lever Movement, Straight Line Escapement, Hardened and Tempered Hair Springs, Exposed pallet jewels, four pair extra jewels, Sprung over Expansion Balance, adjusted to heat and cold. Sunk Seconds. Stem wind and set.

No. 32 (1867)   Same as No. 31 but Key wind and set.

No. 33 (1867)   Same as No. 31 but not adjusted to heat and cold.

No. 34 (1867)   Same as No. 33 but Key wind and set.

"Damaskeen" finish on the above watches was offered at $15.00 extra.

barrel bridge, and cock. Some of the surplus movements have been noted with "Marion Watch Co." dials.

The movement was named after George Channing Fuller Wright, a brother-in-law of Frederick Giles. Wright was a junior member of the Giles/Wales organization and an official of the USWC.

133. "Fayette Stratton," SN 2729, 18s, 15j, key wind and set (Model 32), adjusted heat and cold. Second grade produced by USWC, it was the only one discontinued from production in the second generation models.

"George Channing"

This grade was initially produced in the 1867-1868 period. Only one version of this model was produced during this period, an 18 size, full plate lever movement in gilt. The second generation of this grade was made in 10, 14, 16 and 18 size with nickel top plate,

134 and 135. "George Channing" grades in first (SN 3907) and second (SN 15707) generation models. Both 18s, 15j. SN 3907 is all gilt, while SN 15707 has nickel damaskeened top plate, barrel bridge and cock. Both movements nicely finished. Note difference in the regulator types between the two.

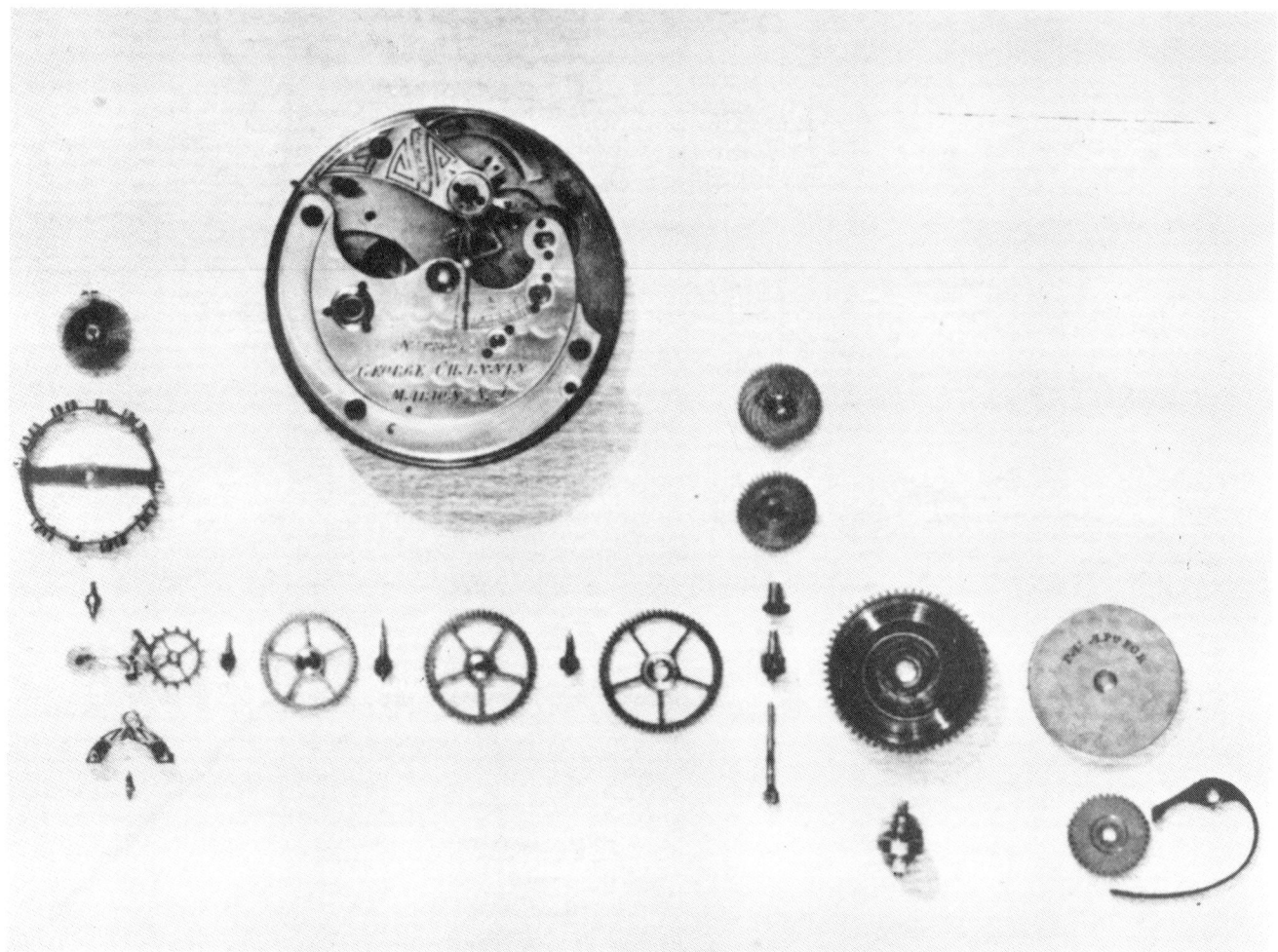

136. "George Channing" grade, 18s, 15j, second generation model in ¾ plate with nickel damaskeened finish top plate. SN 286043 with patent double index regulator and patent applied for reversible barrel. Since the Giles patent was granted on his reversible barrel on December 30, 1873, this example might indicate that the SN 286043 was reached by USWC prior to that date.

Factory descriptions follow:

No. 37 (1867) 18 size, Full Plate Lever Movement, Straight Line Escapement, Hardened and Tempered Hair Springs, Exposed pallet jewels, Sprung over Expansion Balance, four pair extra jewels. Sunk Seconds. Both Key wind and Stem wind movements.

Extras offered on No. 37 were:
If "Damaskeen" finish $15.00
If adjusted to heat and cold 10.00

No. 37¼ (1871) 18 size, Full Plate, NICKEL Top Plate, Barrel Bridge and Cock, Lever Movement, Straight Line Escapement, Hardened and Tempered Hair Springs, Exposed pallet jewels, four pair extra jewels, Sprung over Expansion Balance, Patent Dust Band. Sunk Seconds. Both Key wind and Stem wind movements.

No. 37¾ (1871) 10 size and 16 size, Quarter Plate and Bridge. Same grade as No. 37¼. Patent Reversible Barrel and Patent Double Index Regulator. Both Key wind and Stem wind movements.

No. 38¼ (1871) 14 size and 18 size. Three-Quarter Plate. Same grade as No. 37¼. Patent Reversible Barrel and Patent Double Index Regulator. Both Key wind and Stem wind.

Extras offered on the three 1871 models were:
If adjusted to heat and cold $6.00
Reversible Barrel on Full Plate model 3.00
Double Index Regulator on Full Plate model .. 6.00

The standard 18s, full plate models are shown in Figures 134 and 135, while a dis-assembled three-quarter plate model is shown in Figure 136.

"Edwin Rollo"

This grade was initially produced in the 1867-1868 period; an 18 size, full plate lever movement with butterfly movement opening. It was subsequently released in the 10, 14, 16 and 18 sizes as noted below. These models were in continuous production during the period 1867-1874 and are probably among the easiest to acquire for the collector. They are also one of the more available and interesting models to collect, due mainly to the fact that they were in continuous production, thus affording an opportunity to see changes and improvements as they were incorporated into "Edwin Rollo" movements in succeeding production runs.

The grade was named after Edwin Rollo Pratt, a salesman for both the Giles/Wales organization and the USWC and was introduced as a competitive model to Waltham's "P. S. Bartlett" grade.

Figure 137 shows the standard first model in a full plate style that is stem wind (Model 40) with 15j; Figure 138 shows the second model (Model 43) with 11j; and Figure 139 illustrates second model in ¼ plate and bridge style in 10s (Model 43½).

Factory descriptions follow:

No. 40 (1867)   18 size, Full Plate Lever Movement, Straight Line Escapement, Hardened and Tempered Hair Springs, Exposed pallet jewels, Sprung over Expansion Balance, Seven and one half pair jewels. Both Key wind and Stem wind movements.

Extras offered on No. 40 were:
If "Damaskeen" finish $15.00
If adjusted to heat and cold 10.00

No. 43 (1871)   18 size, Full Plate Lever Movement,

137. "Edwin Rollo," SN 6534, 18s, 15j, full plate, stem wind, Model 40, first generation.

138. "Edwin Rollo," SN 142836, 18s, 11j, full plate, key wind, Model 43, second generation in the lower jewel count. It is significant to note that some second generation "Edwin Rollos" have been reported with a 15j count.

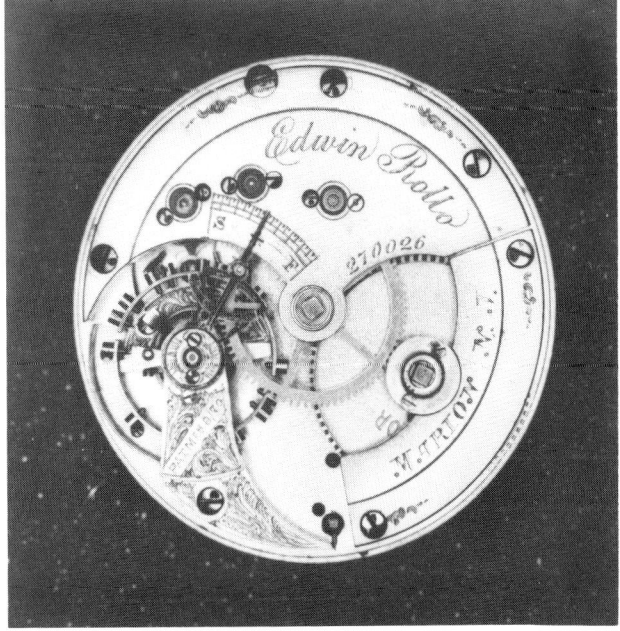

139. "Edwin Rollo," SN 270026, 10s, 11j, ¼ plate and bridge, key wind, Model 43¼, with Elson patent double index regulator (March 8, 1870) and patent applied for reversible barrel. Apparently produced prior to December 30, 1873, when patent was granted. Nicely finished example.

Frosted, Straight Line Escapement, Hardened and Tempered H a i r Springs, Exposed p a l l e t jewels, Sprung over Chronometer Balance, Eleven jewels, Patent Dust Band. Both Key wind and Stem wind movements.

No. 43¼ (1871) 10 size and 16 size, Quarter Plate and Bridge. Same grade as No. 43, Sunk Seconds. Patent Reversible Barrel and Patent Double Index Regulator. Both Key wind and Stem wind movements.

No. 43¾ (1871) 14 size and 18 size, Three Quarter Plate. Same grade as No. 43, Sunk Seconds. Patent Reversible Barrel and Patent Double Index Regulator. Both Key wind and Stem wind movements.

Extras offered on the three 1871 models were:
If "Damaskeen" finish $10.00
If adjusted to heat and cold 6.00
If with Sunk Seconds Dial, No. 43 only 1.00

"United States Watch Co."

The "United States Watch Co.," the grade with the most publicity from 1869 to 1874, is today one of the most elusive to find and comment upon. When one considers that this grade, first produced in 1869, was the most expensive movement made in the United States at that time, listing up to $450.00, it follows that they very probably had short production runs, and as a result, very few may exist at this time. Unfortunately, much confusion surrounds this movement.

Both Crossman and Abbott incorrectly reported this 1869 movement as a "nickel, ¾ plate, 16 size gold train movement. . . ." Dr. Stephens, in his article in the June 1950 NAWCC BULLETIN, reported: "I have two movements which fit this description except that they are 18 size and have brass trains instead of gold."

Research of available information indicates that the original "United States" movement was produced in February, 1869, and was a gilt, 18 size, full plate lever movement with butterfly opening, with fine rubies in gold settings but without gold train. The serial numbers were in the 12,001-12,025 range. The earliest nickel, 19j "United States" models to surface in the survey were in the 24,001-24,100 range and production started late in 1869. In fact it appears that all USWC movements produced from 1867 through late 1870 were 18 size and full plate with the exception of the 10 size Swiss-made ladies' watch which was ¼ plate. It was not until 1872 that the ¾ plate movements were produced, and then in 14 and 18 sizes. The Crossman/Abbott referenced 16 size also appeared about this time, but with ¼ plate and bridge. No "United States" models have turned up with a gold train . . . another reference error from Messrs. Crossman and Abbott.

Neither surviving factory price schedules nor any parts lists give any indication at all that the "United States Watch Co." grades were ever produced with a gold train. Frederick Giles probably believed that the additional effort put into finish, frosting, engraving with enameling, damaskeening, and various gold and gilded feature trim was sufficient for his finer grades. Giles probably contributed to some of the confusion with complex price schedules, his use of a ¼ plate and bridge which is easily called a ¾, and the fact that his 16s models actually measure a good 17s.

The confusion is somewhat understandable. The "United States" grade was advertised as being made in four different sizes (10, 14, 16 and 18) with the 18 size being produced in three basic versions. Thus, we have:

1869 — 18 size, full plate, rubies in gold settings
1871
to
1873 — 18 size, full plate, rubies in gold settings
18 size, ¾ plate with double index regulator
16 size, ¼ plate and bridge
14 size, ¾ plate with double index regulator
10 size, ¼ plate and bridge

The confusion is further compounded by the fact that by early 1873 the "United States" grade was offered in over fifty variations (Table IV, page 134).

Referring back to Dr. Stephens' article, the photographs indicate that his 18 size movement is the ¾ plate with double index regulator bearing a patent date of 1870, and thus, is not the original full plate model.

Firm evidence for the release date of the original 1869 model was found in a testimonial dated May 8, 1869. It was a USWC brochure listing testimonials pertaining to various models, one of which was the "United States" grade, serial number 12,006. The testimonial read in part: "has been carried by me three months" which would move the purchase date back to at least February 8, 1869, which is earlier than the generally accepted release date of this model.

The theory has been advanced that the "United States" grade was produced by the USWC to compete against the high grade Howards and "American Watch Company" models. Giles insisted that his top prestige grade be available in a variety of sizes and models; they were given costly features and finishing touches with prices higher than any American watch on the market. However, in their efforts to achieve excellence in so many variations, the USWC was losing valuable ground to their competitors who were beginning to produce a good American watch at fairly reasonable prices.

Some "United States Watch Co." movements have been recorded with "Giles, Wales and Co." on the dial. Most "United States" models have double recessed dials and some have fancy script Old English USWC monograms or other extra details.

It is *significant* to note that all early (1869-72) factory descriptions show the "United States" model to be finely finished with 19 jewels. Early in 1873 USWC introduced 15-jewel versions of their "United States" grades at prices from $300 to $325, probably to more effectively compete with the lower priced "American Watch Co." grades of Waltham. These maintained a good finish with gold jewel settings. Until the fall 1873 Panic, all USWC premium products were maintained at higher prices than Waltham equivalent items. In subse-

 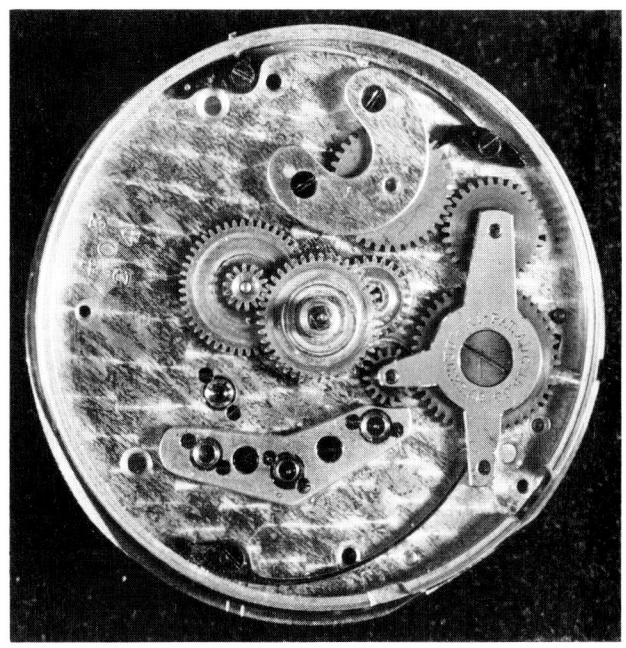

140 and 141. Movement and under dial views, "United States Watch Co." grade, SN 24043, 18s, 19j, interesting variation in frosted finish without damaskeening not listed as standard model (compare to SN 24042, adjacent serial number in Figure 68, page 78). Gilded steel work. Raised gold jewel settings for 3rd, 4th and balance wheel. Escape wheel and pallet arbor cup settings have nicely chamfered edges. Plate and case screws are flat and chamfered. Arbor plate polished and chamfered with gold arbor cup. Gold regulator index. All 19j models apparently have polished brass escape wheels, visible through cut out in plate. Winding square on barrel arbor is nicely polished and chamfered on four edges. Under dial view shows winding mechanism with yoke that is frosted. The yoke also has polished and chamfered edges which required additional machining operations. Screwed in gold jewel settings under dial. Lower cap jewel setting for the balance is also gold. The feature of gold settings throughout was a first in American mass produced watches. Later Model 72 "American Watch Co." grades and Elgin Model 72 and 91 convertible 21j watches would also have this feature.

quent financially insecure years, after the 1873 Panic, some were "traded down" with very poor finish, no gold settings, in 15 or less jewels. Even though prices were dropped significantly, the post Panic years produced a scant market for "United States" models in any variation. Trading down the finish and jewel count was also employed on other models in the final troubled years of USWC.

Factory descriptions follow:

No. 21 (1869)   18 size, Nickel, Best Quality, Full Plate Lever Movement, Straight Line Escapement, Exposed ruby pallets, Nineteen fine ruby jewels, three pair conical pivots, cap jeweled, with fine rubies in gold settings, Breguet hair spring, hardened and tempered, Compensation balance, accurately adjusted to heat and cold, Isochronism and position. Stem wind and set.

No. 22 (1869)   Same as No. 21 but Key wind and set.

No. 23 (1869)   18 size, Best Quality, Frosted plates, Full Plate Lever Movement, Straight Line Escapement, Exposed ruby pallets, Nineteen fine ruby jewels, three pair conical pivots, cap jeweled, with fine rubies in gold settings, Breguet hair spring, hardened and tempered, Compensation balance, accurately adjusted to heat and cold, Isochronism and position. Stem wind and set.

No. 24 (1869)   Same as No. 23 but Key wind and set.

"Damaskeen" finish on Nos. 23 and 24 was offered at $20.00 extra.

No. 17 (1871)   10 size and 16 size, Nickel, Quarter Plate and Bridge, Damaskeen and Frosted, Enameled, Gold I n d e x Plate, Gold Jewel Settings throughout, Gilded Steel Work. Best quality, Lever Movement, Straight Line Escapement, Exposed ruby pallets, Nineteen fine ruby jewels, three pair conical pivots, cap jeweled,

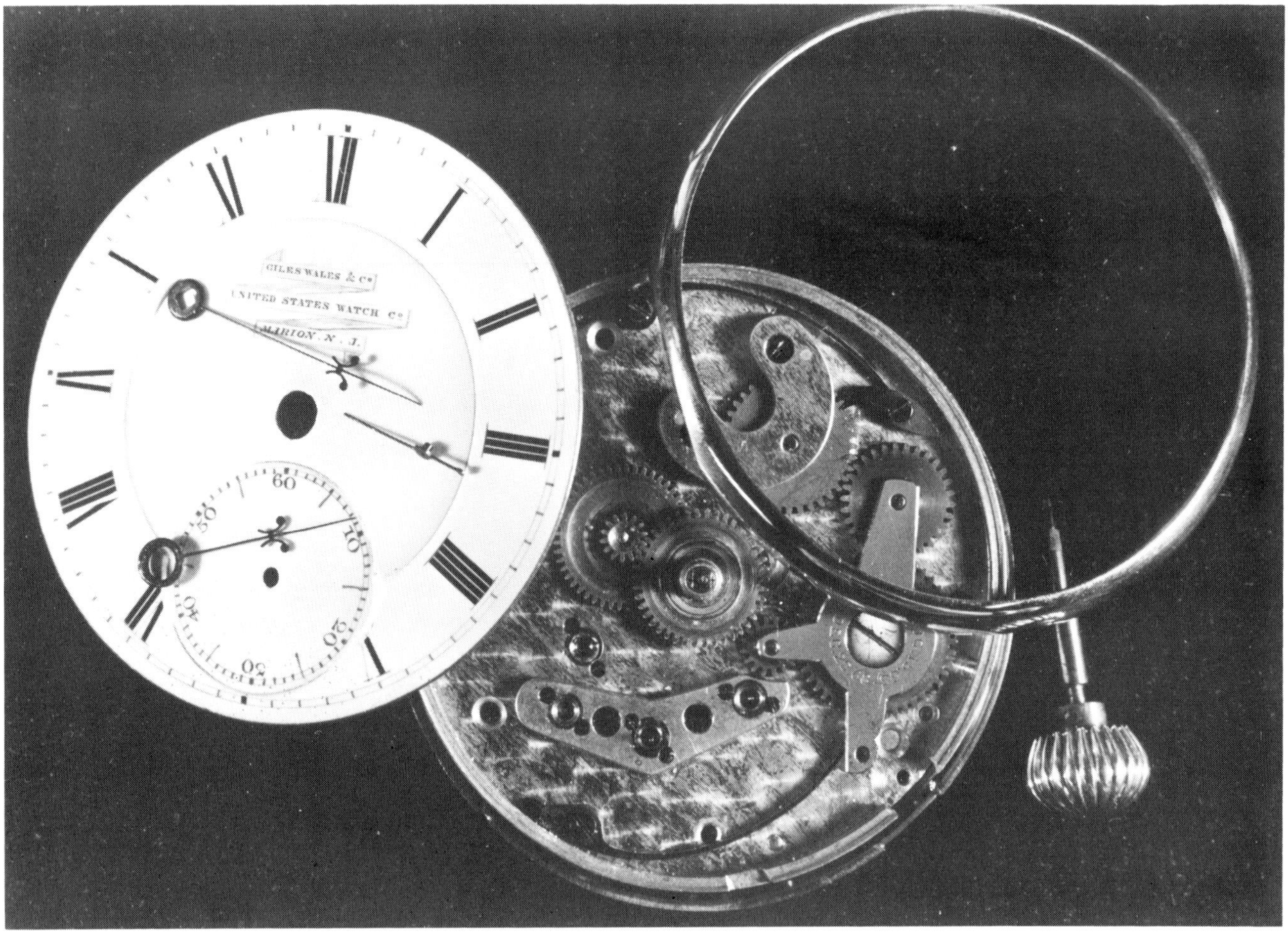

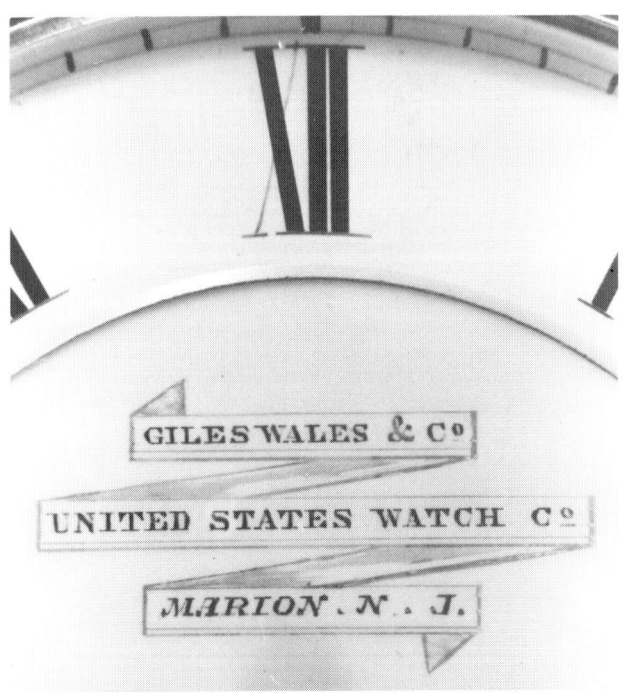
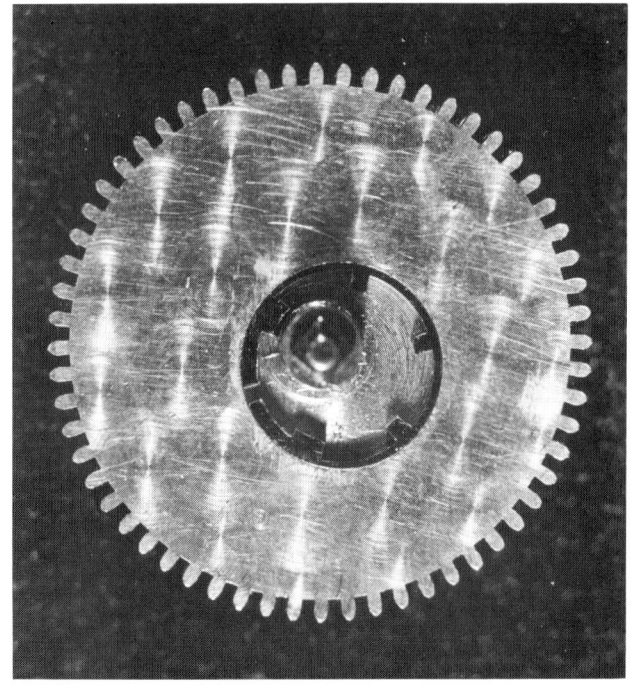

OPPOSITE PAGE. 142, 143, and 144. SN 24043 miscellaneous components, dial and stopwork enlargements to show detail. "United States" grades typically have fine double recessed dials with USWC logos or banner dials with both United States and Giles, Wales & Co. names. Color use in this dial is particularly noteworthy with black lettering in a red banner. The chapter bit with black numerals is the standard white and the recessed center bit is a light pastel blue. The stopwork view illustrates the almost unique system used in all first generation full plate watches even to Lewis and Rollo grades. The female outer ring will turn in its bore as it is induced to do so by the male piece on the arbor. This barrel is shown in the run down condition. The arbor will wind anti-clockwise 4¾ times before being fully wound up. This will give the full plate watches a going period of approximately 28½ hours on a single wind. This was important as USWC watches in full plate had no reversible barrel until the second generation models. The stopwork will prevent overstressing of the mainspring when the spring is suddenly very tightly wound. A typical mainspring might wind as many as 7 turns in the barrel before being tightly wound. This also permits using only the middle turns of the spring for more equal torque applied to the train during running.

145. "United States Watch Co.," SN 25508, 18s, 19j with same basic features as SN 24043. This particular model is pure key wind and key set, and has damaskeening in addition to frosting. The damaskeening as well as the balance cock engraving styles are both unique to each example in these finer grades.

146. SN 25508, under dial view illustrating no provisions for stem winding had ever been applied to this example. This is unusual as most models turned up in this grade have been stem wind with key winding capability. Note circular damaskeening pattern.

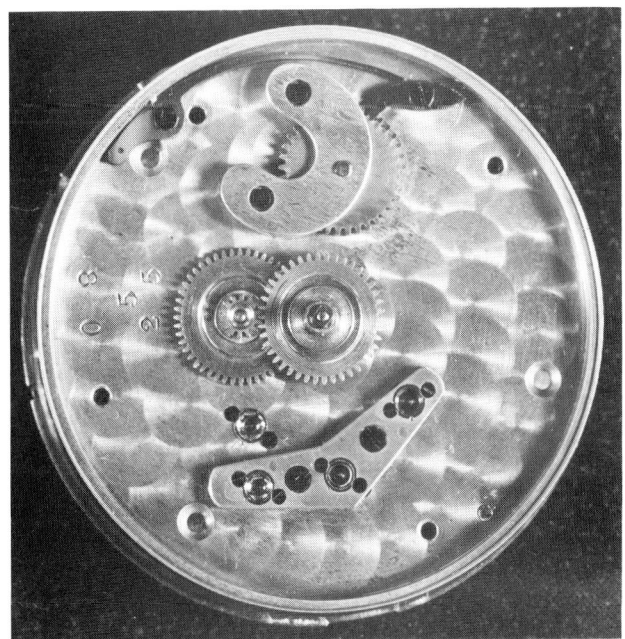

	with fine rubies, Breguet hair spring, hardened and tempered, Compensation Balance, accurately adjusted to heat and cold, Isochronism and position, Patent Reversible Barrel and Patent Double Index Regulator. Both Key wind and Stem wind movements.
No. 18 (1871)	14 size and 18 size, Nickel, Three-Quarter Plate, same quality as No. 17, Patent Reversible Barrel and Patent Double Index Regulator. Both Key wind and Stem wind movements.
No. 19 (1871)	18 size, Nickel, Full Plate, same quality as No. 17. Both Key wind and Stem wind movements.
No. 21 (1871)	18 size, Nickel, Full Plate, same as No. 19 except plain Damaskeen and blued steel work. Both Key wind and Stem wind movements.
No. 21½ (1871)	10 size and 16 size, Nickel, Quarter Plate and Bridge, same quality as No. 21 (1871). Patent Reversible Barrel and Patent Double Index Regulator. Both Key wind and Stem wind movements.
No. 22½ (1871)	14 size and 18 size, Nickel, Three-Quarter Plate, same quality as No. 21 (1871). Patent Reversible Barrel and Patent Double Index Regulator. Both Key wind and Stem wind movements.
No. 23 (1871)	18 size, Full Plate, Best Quality, Frosted Movement, Straight Line Escapement, Exposed ruby pallets, Nineteen fine ruby jewels, three

	pair conical pivots, cap jeweled, with fine rubies in gold settings, Breguet hair spring, hardened and tempered, Compensation Balance, accurately adjusted to heat and cold, Isochronism and position. Both Key wind and Stem wind movements.
No. 23½ (1871)	10 size and 16 size, Quarter Plate and Bridge, same grade as No. 23 (1871). Patent Reversible Barrel and Patent Double Index Regulator. Both Key wind and Stem wind movements.
No. 24½ (1871)	14 size and 18 size, Three-Quarter Plate, same grade as No. 23 (1871). Patent Reversible Barrel and Patent Double Index Regulator. Both Key wind and Stem wind movements.

Extras offered on the 1871 movements were:
Patent Reversible Barrel added to
 full plate movement .... $ 3.00
Patent Double Index Regulator added to
 full plate movement 6.00
If damaskeen finish on No. 23, 23½, 24½ ... 18.00
Patent Dust Band on full and three-quarter
 plate .... GRATIS

No. 10 (1873)	10 size and 16 size, Nickel, Quarter Plate and Bridge, exposed ruby pallets, fifteen ruby jewels, gold jewel settings, plain damaskeen and blued steel work, hardened and tempered hairspring, adjusted to heat, cold and positions. Patent reversible barrel and Patent double index regulator. Both Key wind and Stem wind movements.
No. 11 (1873)	14 size and 18 size, Nickel, Three-Quarter Plate, same quality as No. 10, Patent reversible barrel and Patent double index regulator. Both Key wind and Stem wind movements.
No. 12 (1873)	18 size, Nickel, Full Plate, same quality as No. 10. Both Key wind and Stem wind movements.
No. 14 (1873)	10 size and 16 size, Frosted Movement, Quarter Plate and Bridge, same quality as No. 10. Both Key wind and Stem wind movements.
No. 15 (1873)	14 size and 18 size, Frosted Movement, Three-Quarter Plate, same quality as No. 10, Patent reversible barrel and Patent double index regulator. Both Key wind and Stem wind movements.
No. 16 (1873)	18 size, Frosted Movement, Full Plate, same quality as No. 10. Both Key wind and Stem wind movements.

Extras offered on the 1873 movements were:
If enameled and frosted on No. 10, 11, 12 ... $15.00
If damaskeen finish on No. 14, 15, 16 .... 16.00

147. "United States Watch Co.," SN 53528, 18s, ¾ plate, 15j, very early Model 11, circa 1873. Stem wind with Elson double index regulator and patent applied for reversible barrel. Nicely damaskeened and frosted. Lever set. Note balance guard, typical of ¾ and ¼ plate and bridge movements, and dial feet screws.

148. "United States Watch Co.," SN 76502, 18s, ¾ plate, 15j, late Model 15 with Marion Watch Co. dial, circa late 1874. Stem wind, lever set. Note no patent date on cock, even though this model has the Elson regulator. Gilt (frosted) model.

149. "United States Watch Co.," SN 225301, 18s, ¾ plate, 15j. Nickel damaskeened and frosted with gilded steel work. Would appear to be a Model 11, but this variation has a Breguet hair spring. Standard Model 11 examples have a flat hair spring, while Breguet springs were used for 19j versions.

150. "United States Watch Co.," SN 235003, 16s, ¼ plate and bridge, 15j. Nickel damaskeened and frosted. Model 10, circa early 1874. Patent reversible barrel and double index regulator. Production delays on this model are responsible for their relative scarcity.

If gilded steel work on No. 10, 11, 12	5.00
Patent reversible barrel added to full plate movement	3.00
Patent double index regulator added to full plate movement	6.00
Patent dust band on full and three-quarter plate	GRATIS

"Marion Watch Co."

The name "Marion Watch Co." was used as a grade name prior to 1870, with the name engraved on the back plate in normal fashion. In all probability, it was the USWC response to the American Watch Company "Waltham Watch Co." grade. It is not to be confused with those movements having "Marion Watch Co." on the dial.

It is significant to note that "Marion Watch Co." dials did not come into existence until mid-1874 when that name was formally incorporated. In many cases these new dials were applied to existing USWC movements that had been previously produced. Prices at that time were lowered and the watches were completed with much poorer finish and, in some instances, a reduced jewel count.

Factory descriptions follow:

No. 29 (1869)   18 size, Full Plate Lever Movement, Straight Line Escapement, Hardened and Tempered Hair Springs, Exposed pallet jewels, four pair extra jewels, Sprung over Expansion Balance, adjusted to heat and cold. Sunk Seconds. Stem wind and set.

151. "Marion Watch Co.," SN 7646, 18s, full plate, 15j, gilt. Key wind and set, Model 30. This grade was probably named in response to the "Waltham Watch Co." grade of the American Watch Company. This particular example is circa 1869.

No. 30 (1869)   Same as No. 29 but Key wind and set.

Extra on the 1869 movements:
If adjusted to position $50.00

No. 29 (1871)   18 size, Frosted, Full Plate Lever Movement, Straight Line Escapement, Hardened and Tempered Hair Springs, Exposed pallet jewels, four pair extra jewels, Sprung over Expansion Balance. Sunk Seconds. Both Key wind and Stem wind movements.

No. 29½ (1871)   10 size and 16 size, Quarter Plate and Bridge. Same grade as No. 29 (1871). Patent Reversible Barrel and Patent Double Index Regulator. Both Key wind and Stem wind movements.

No. 30½ (1871)   14 size and 18 size, Three-Quarter Plate. Same grade as No. 29 (1871). Patent Reversible Barrel and Patent Double Index Regulator. Both Key wind and Stem wind movements.

Extras offered on the 1871 movements:
If adjusted to positions $42.50
If Damaskeen finish 12.50
Patent Reversible Barrel added to full plate movement 3.00
Patent Double Index Regulator added to full plate movement 6.00
Patent Dust Band on full and three-quarter plate .... GRATIS

152. "Marion Watch Co.," SN 76530, 18s, ¾ plate, 15j, stem wind, Model 30½, gilt. Regulator arm missing, patent December 30, '73, reversible barrel, circa 1874. Note USWC logo on movement of this example.

"A. H. Wallis"

The "A. H. Wallis" movement was initially produced in the 1869-1870 period. Costwise, it was the second most expensive movement produced by the USWC and, in many cases, was similar to the "United States Watch Co." movement. At least some of the movements have been noted with "Marion Watch Co." dials.

The movement was named after A. H. Wallis, a New York City attorney, who was also a stockholder in the USWC. Wallis was influential in many important decisions concerning the affairs of USWC, including selection of the factory location in Jersey City, New Jersey. His photo is shown in Figure 19 on page 43.

The fact that Frederick Giles selected this grade name for one of his best prestige items indicates a very high regard for A. H. Wallis. These grades had a flat hair spring. The first generation models were 19j; this was lowered to 17j in the second generation and later examples have even fewer jewels.

Factory descriptions follow:

No. 50 (1869)   18 size, Nickel, Full Plate Lever Movement, Straight Line Escapement, Hardened and Tempered Hair Springs, Exposed pallet jewels, six pair extra jewels, and two pair extra conical pivots, cap jeweled, in gold settings. Sprung over Expansion Balance, adjusted to heat and cold, Isochronism and position. Sunk Seconds. Stem wind and set.

No. 51 (1869)   Same as No. 50 but Key wind and set.

No. 52 (1869)   18 size, Nickel, Full Plate Lever Movement, Straight Line Escapement, Hardened and Tempered Hair Springs, Exposed pallet jewels, six pair extra jewels, and two pair extra conical pivots, cap jeweled in gold settings, Sprung over Expansion Balance, adjusted to heat and cold. Sunk Seconds. Stem wind and set.

No. 53 (1869)   Same as No. 52 but Key wind and set.

No. 52 (1871)   18 size, Nickel, Full Plate Lever Movement, Straight Line Escapement, Hardened and Tempered Hair Springs, Exposed pallet jewels, five pair extra jewels, Sprung over Expansion Balance, adjusted to heat and cold. Sunk Seconds. Both Key wind and Stem wind movements.

No. 52½ (1871)   10 size and 16 size, Nickel, Same quality as No. 52 (1871). Quarter Plate and Bridge. Patent Reversible Barrel and Patent Double Index Regulator. Both Key wind and Stem wind movements.

No. 53½ (1871)   14 size and 18 size, Nickel, Three-Quarter Plate. Same grade as No. 52 (1871). Patent Reversible Barrel and Patent Double Index Regulator.

ABOVE. 153 and 154. "A. H. Wallis" grades, SN 19034 .and SN 19061, both very early examples, 18s, full plate, 19j, stem wind, Model 50. Note the damaskeening, frosting and unique balance cock engravings. When contrasted to SN 19238 below and SN 19210 in Figure 69, it becomes apparent that the damaskeening patterns and balance cock engravings were a distinctive feature of this grade. The Wallis grades were the second most expensive movement produced by the USWC. BELOW. 155 and 156. Movement and under dial views, "A. H. Wallis." SN 19238, Model 50. All Wallis grades have flat hair spring, flush gold jewel settings, and gold arbor cups in machined finish disc. Blued screws, no chamfering. Improved winding mechanism with unchamfered yoke. The Model 50 movement sold for $199 in 1870. The second generation models had their jewel count lowered to 17j. After the panic of 1873, jewel count and finish standards were lowered further.

157. "A. H. Wallis" grade, SN 243009, 10s, ¼ plate and bridge, 17j, key wind, Model 52½. This second generation product sold for $135 in 1873. Patent double index regulator and patent applied for reversible barrel. Circa 1873. A finely finished ladies' watch, this example is in its original silver case.

158 and 159. Movement views, "Wm. Alexander" grade, SN 20296, 18s, 15j, stem wind and button set. Model 56. Damaskeened, frosted and enameled. Arbor cup has no disc and escape wheel is not polished. Side view illustrates gilt barrel with nickel snap over lid. Note hole in rim of pillar for tool insertion to let mainspring down.

Both Key wind and Stem wind movements.

Extras offered on the 1871 movements were:

If adjusted to positions	$50.00
If enameled and frosted	12.50
If enameled and frosted on cock only	2.50
Patent Reversible Barrel added to full plate movement	3.00
Patent Double Index Regulator added to full plate movement	6.00
Patent Dust Band on full and three quarter plate	GRATIS

"Wm. Alexander"

This movement was originally produced prior to 1870. It was named after William Alexander Giles, one of Frederick's younger brothers. It should be noted that a James Alexander was also a stockholder in USWC. . . . In this instance, Frederick may have been taking care of "two birds with one stone."

Factory descriptions follow:

No. 56 (1869) 18 size, Nickel, Full Plate Lever Movement, Straight Line Escapement, Hardened and Tempered Hair Springs, Exposed pallet jewels, four pair extra jewels, Sprung over Expansion Balance. Sunk Seconds. Stem wind and set.

No. 57 (1869) Same as No. 56 but Key wind and set.

Extras offered on the 1869 movements were:

If adjusted to heat and cold	$10.00
If adjusted to position	50.00

No. 56 (1871) 18 size, Nickel, Full Plate Lever Movement, Straight Line Escapement, Hardened and Tempered Hair Springs, Exposed pallet jewels, four pair extra jewels, Sprung over Expansion Balance. Sunk Seconds. Both Key wind and Stem wind movements.

No. 56½ (1871) 10 size and 16 size, Nickel, Quarter Plate and Bridge, same grade as No. 56 (1871). Patent Reversible Barrel and Patent Double Index Regulator. Both Key wind and Stem wind movements.

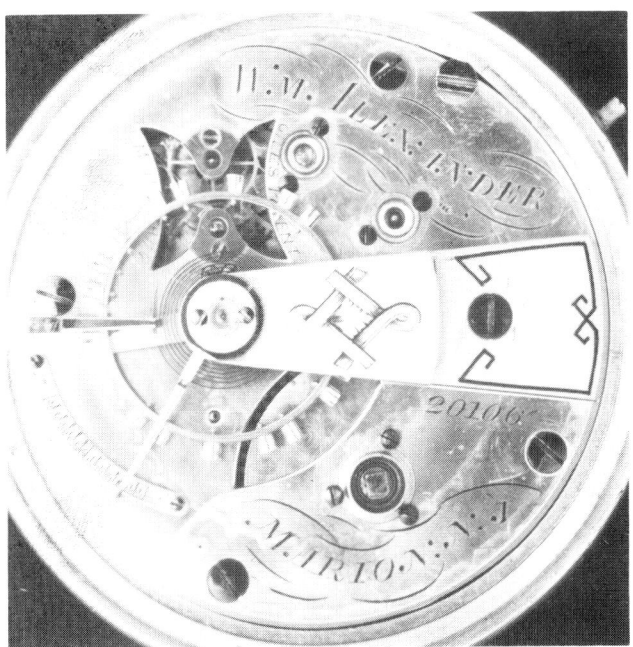

160. "Wm. Alexander" grade, SN 20106, 18s, 15j, stem wind and button set, Model 56. Interesting two-tone effect achieved through the combination of nickel damaskeening on movement plates with a frosted balance cock. This frosted balance cock was made available as a $2.50 option in the 1872 price list.

No. 57½ (1871) 14 size and 18 size, Nickel, Three-Quarter Plate, same grade as No. 56 (1871). Patent Reversible Barrel and Patent Double Index Regulator. Both Key wind and Stem wind movements.

Extras offered on the 1871 movements:
If adjusted to heat and cold	$ 9.00
If adjusted to positions	42.50
If enameled and frosted	9.00
Patent Reversible Barrel added to full plate movement	3.00
Patent Double Index Regulator added to full plate movement	6.00
Patent Dust Band on full and three-quarter plate	GRATIS

"S. M. Beard"

This movement was also initially produced in the 1869-1870 period. It was named after S. M. Beard, a tea merchant and senior member of the firm of Beard and Cummings of New York City. He was also an original stockholder and Director of the USWC.

Factory descriptions follow:

No. 58 (1869) 18 size, Nickel, Full Plate Lever Movement, Straight Line Escapement, Hardened and Tempered Hair Springs, Exposed pallet jewels, four pair extra jewels, Sprung over Expansion Balance. Sunk Seconds. Stem wind and set.

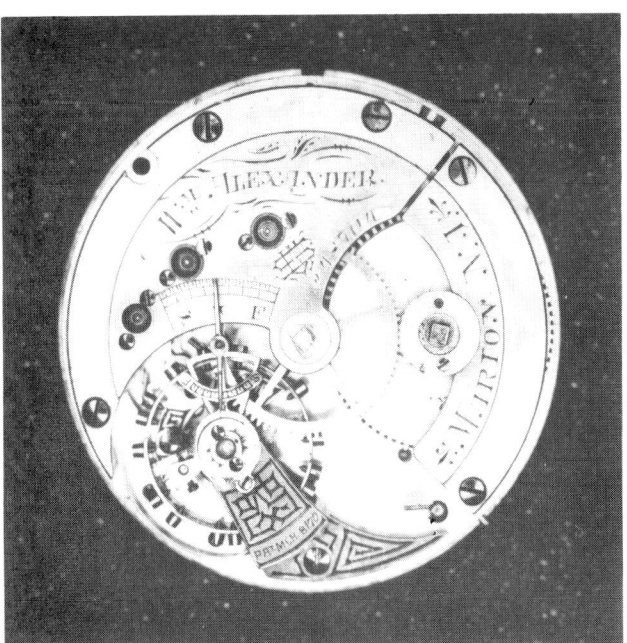

161. "Wm. Alexander" grade, SN 248704, 10s, ¼ plate and bridge, key wind and set, Model 56½. Patent applied for reversible barrel and patent double index regulator. Circa 1873. Note USWC logo on movement, a feature apparently used only for a short time.

162. "S. M. Beard" grade, SN 21293, 18s, 15j, stem wind and button set, Model 58. Frosted with a circular damaskeening pattern. Circa late 1869 or early 1870. Note no gold settings on escape wheel and pallet arbor jewels were used on this grade.

No. 59 (1869) Same as No. 58 but Key wind and set.

Extras offered on the 1869 movements were:
If adjusted to heat and cold $10.00
If adjusted to position 50.00

No. 58 (1871) 18 size, Nickel, Full Plate Lever Movement, Straight Line Escapement, Hardened and Tempered Hair Springs, Exposed pallet jewels, four pair extra jewels, Sprung over Expansion Balance. Sunk Seconds. Both Key wind and Stem wind movements.

No. 58½ (1871) 10 size and 16 size, Nickel, Quarter Plate and Bridge, same grade as No. 58 (1871). Patent Reversible Barrel and Patent Double Index Regulator. Both Key wind and Stem wind movements.

No. 59½ (1871) 14 size and 18 size, Nickel, Three-Quarter Plate, same grade as No. 58 (1871). Patent Reversible Barrel and Patent Double Index Regulator. Both Key wind and Stem wind movements.

Extras offered on the 1871 movements:
If adjusted to heat and cold .... $ 9.00
If adjusted to positions 42.50
If enameled and frosted 9.00
Patent Reversible Barrel added to full plate movement 3.00
Patent Double Index Regulator added to full plate movement 6.00
Patent Dust Band on full and three-quarter plate .... GRATIS

"Henry Randel"

This movement was initially produced in the 1869-1870 period. Several watches with this movement have been recorded with "Marion Watch Co." dials.

The movement was named after Henry Randel, one of the original stockholders in the USWC. He was a diamond merchant and was senior member of the firm of Randel, Baremore & Billings of New York City.

Factory descriptions follow:

No. 54 (1869) 18 size, Nickel, Full Plate Lever Movement, Straight Line Escapement, Hardened and Tempered Hair Springs, Exposed pallet jewels, four pair extra jewels, Sprung over Expansion Balance, adjusted to heat and cold. Sunk Seconds. Stem wind and set.

No. 55 (1869) Same as No. 54 but Key wind and set.

Extras offered on the above two movements were:
If adjusted to position $50.00

No. 54 (1871) 18 size, Nickel, Full Plate Lever Movement, Straight Line Escapement, Hardened and Tempered Hair Springs, Exposed pallet jewels, four pair extra jewels, Sprung over Expansion Balance, adjusted to heat

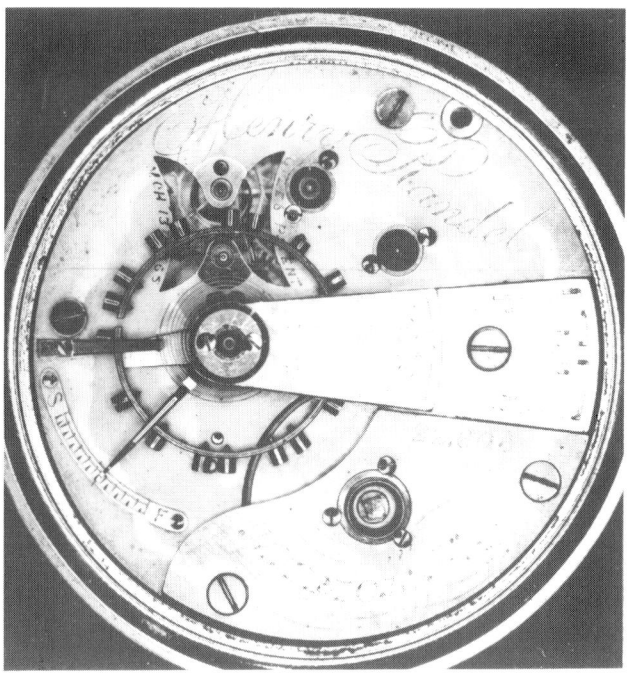

163. "Henry Randel" grade, SN 22606, 18s, 15j, stem wind and button set, Model 54. Frosted balance cock. This example illustrates a common problem with USWC products surviving today. Because the engraving is not very deep, the enameling over the years tends to wash out and must be restored. This first generation example is circa late 1869 or early 1870.

164. "Henry Randel," SN 205162, second generation of the same model, also 15j. Frosted and damaskeened. Interesting balance cock engraving. Circa 1874.

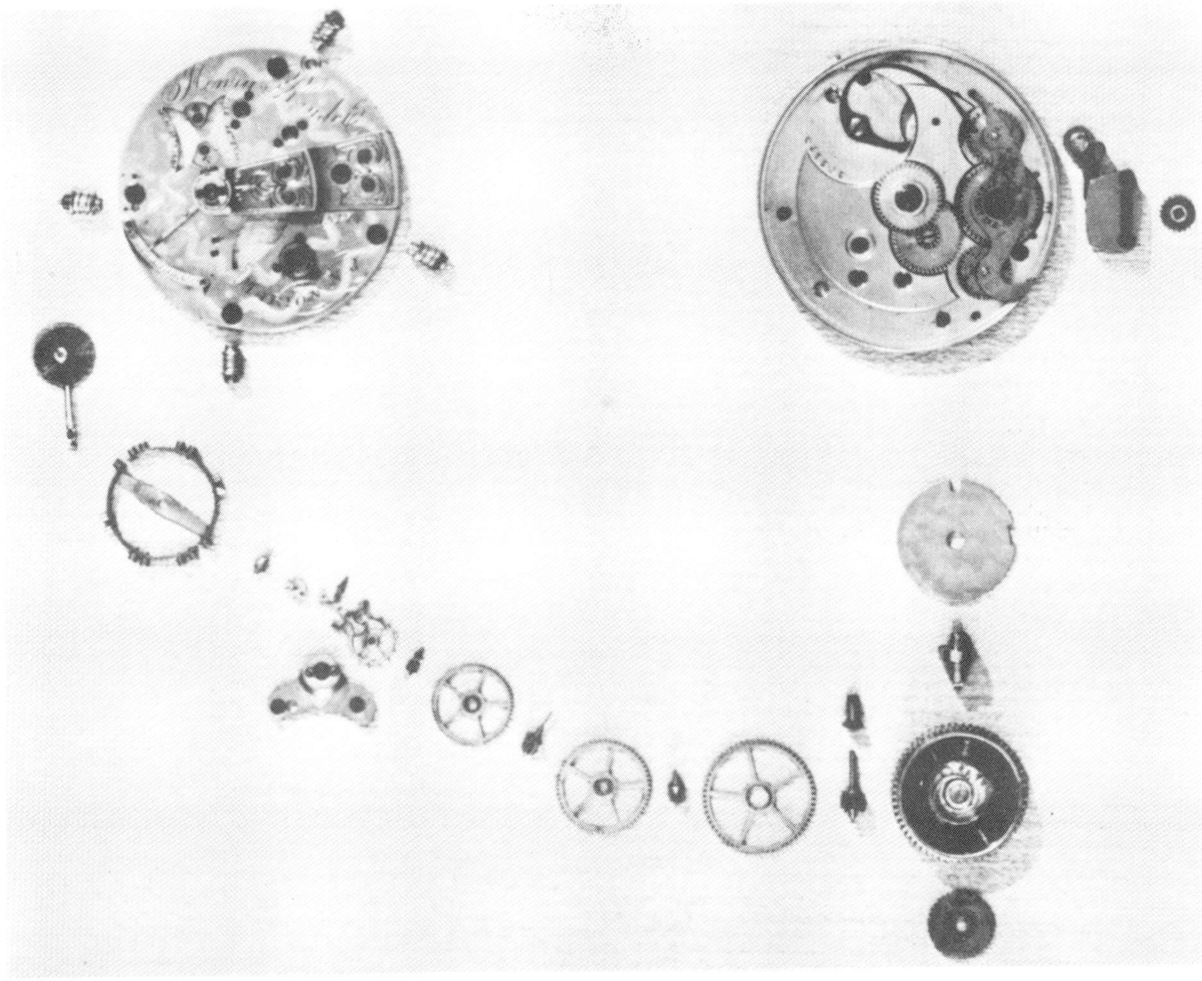

ABOVE. 165. Disassembled "Henry Randel" SN 205174, 18s, Model 54 (Marion Watch Co. dial). Note different damaskeening and engraving style from SN 205162. BELOW. 166. "Henry Randel" 18s, SN 52025, ¾ plate, second generation model, key wind and set, Model 55½. Circa late 1872.

	and cold. Sunk Seconds. Both Key wind and Stem wind movements.
No. 54½ (1871)	10 size and 16 size, Nickel, Quarter Plate and Bridge, same grade as No. 54 (1871). Patent Reversible Barrel and Patent Double Index Regulator. Both Key wind and Stem wind movements.
No. 55½ (1871)	14 size and 18 size, Nickel, Three-Quarter Plate, same grade as No. 54 (1871). Patent Reversible Barrel and Patent Double Index Regulator. Both Key wind and Stem wind movements.

Extras offered on the 1871 movements were:
If adjusted to positions $42.50
If enameled and frosted 12.50
If enameled and frosted on cock only 2.50
Patent Reversible Barrel added to
　full plate movement 3.00
Patent Double Index Regulator added to
　full plate movement 6.00
Patent Dust Band on full and ¾ plate ... GRATIS

"John W. Lewis"

This movement was originally produced prior to 1870. It was the lowest priced all nickel movement produced by the firm. It was probably named after one of the USWC major stock holders, but there is no documentation to support this deduction.

Factory descriptions follow:

No. 60 (1869)	18 size, Nickel, Full Plate Lever Movement, Straight Line Escapement, Hardened and Tempered Hair Springs, Exposed pallet jewels, Sprung over Expansion Balance, fifteen jewels. Sunk Seconds. Stem wind and set.
No. 61 (1869)	Same as No. 60 but Key wind and set.

Extras offered on the above two movements were:
  If adjusted to heat and cold $10.00
  If adjusted to position 50.00

No. 60 (1871)	18 size, Nickel, Full Plate Lever Movement, Straight Line Escapement, Hardened and Tempered Hair Springs, Exposed pallet jewels, Sprung over Expansion Balance, fifteen jewels. Sunk Seconds. Both Key wind and Stem wind movements.
No. 60½ (1871)	10 size and 16 size, Nickel, Quarter Plate and Bridge, Same grade as No. 60 (1871). Patent Reversible Barrel and Patent Double Index

168. "John W. Lewis," SN 50692, ¾ plate, 18s, 15j, second generation Model 61½. This is a very early example of the ¾ plate style, probably late 1872. Beautifully damaskeened and frosted. Patent double index regulator and patent applied for reversible barrel. Original USWC marked 18k hunting case.

167. "John W. Lewis," SN 21756, 18s, 15j, Model 60, first generation, frosted and damaskeened. This was lowest priced all nickel movement produced, yet the grade has many features of the higher priced models. This example is circa late 1869 or early 1870.

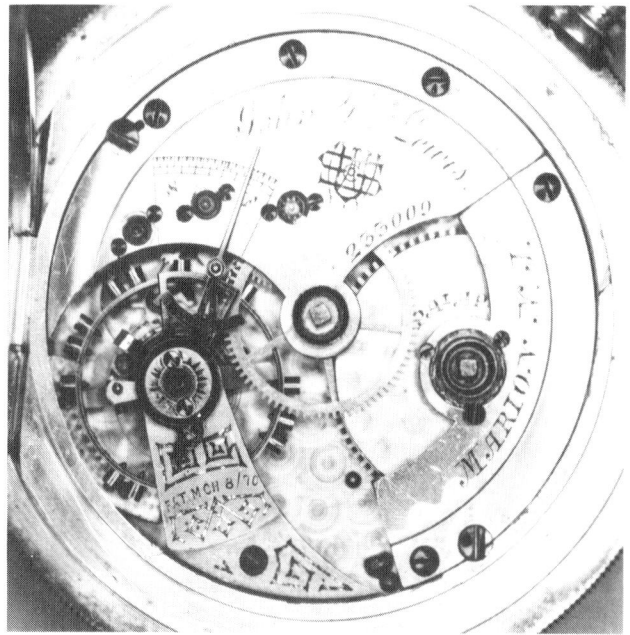

169. "John W. Lewis," SN 233009, ¼ plate and bridge, 16s, 15j, second generation Model 60½. Note patent applied for barrel, indicating this serial number reached before the December 30, 1873, patent granted. USWC logo on movement, frosted and damaskeened.

No. 61½ (1871) Regulator. Both Key wind and Stem wind movements.
14 size and 18 size, Nickel, Three-Quarter Plate, Same grade as No. 60 (1871). Patent Reversible Barrel and Patent Double Index Regulator. Both Key wind and Stem wind movements.

Extras offered on the 1871 movements:
If adjusted to heat and cold	$ 6.00
If adjusted to positions	30.00
If enameled and frosted	6.00
Patent Reversible Barrel added to full plate movement	3.00
Patent Double Index Regulator added to full plate movement	6.00
Patent Dust Band on full and three-quarter plate	GRATIS

"Asa Fuller"

This grade was originally produced prior to 1872. It was designed to compete with the less expensive watches being produced by USWC competitors. It offered a Nickel Barrel Bridge in the 18 size full plate movement, plus quarter plate and three-quarter plate movements.

Whether or not the USWC was the first American watch manufacturer to offer a combination nickel and gilt movement is open to some debate. It was an interesting variation, and some of these have even been reported with damaskeened nickel barrel bridges.

The "Asa Fuller" grade name resulted from an interesting combination of middle names from Frederick "Asa" Giles and George Channing "Fuller" Wright.

Factory descriptions follow:
No. 43½ (1871) 18 size, Nickel Barrel Bridge, Full Plate Lever Movement, Straight Line Escapement, Hardened and Tempered Hair Spring, Exposed pallet jewels, Sprung over Expansion Balance, eleven jewels. Both Key wind and Stem wind movements.
No. 44¼ (1871) 10 size and 16 size, Quarter Plate and Bridge, same grade as No. 43½ but fifteen jewels. Patent Reversible Barrel and Patent Double Index Regulator. Both Key wind and Stem wind movements.
No. 44½ (1871) 14 size and 18 size, Three-Quarter Plate, same grade as No. 43½. Patent Reversible Barrel and Patent Double Index Regulator. Both Key wind and Stem wind movements.

Extras offered on the above movements:
If with Sunk Seconds dial	$1.00
If adjusted to heat and cold	6.00
If Damaskeen finish	8.00
Patent Reversible Barrel added to No. 43½	3.00
Patent Double Index Regulator added to No. 43½	6.00
Patent Dust Band on full and three-quarter plate	GRATIS

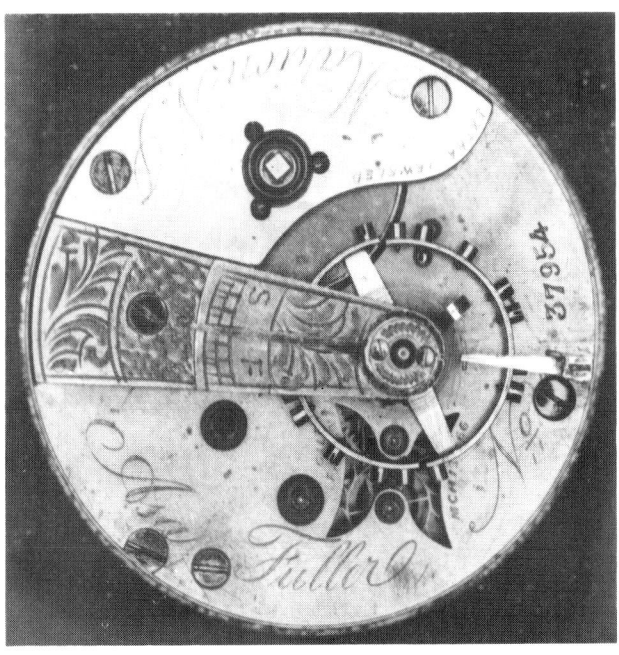

170. "Asa Fuller," SN 37954, 18s, 11j, gilt movement with nickel barrel bridge. Model 43½. Note first use of "extra jeweled" on barrel bridge. Some of this same model have been reported with a damaskeened barrel bridge (SN 34939 for example). Key wind and set.

171. "Asa Fuller," SN 71270, 18s, 11j, gilt, Model 44½ in ¾ plate movement style. USWC logo on movement. Interesting to note that this reversible barrel has the patent date of December 30, 1873. Because of this kind of variation, some USWC historians question the use of this barrel in dating the various products.

"R. F. Pratt"

This movement was also produced prior to 1870. It is similar to the "Chas. G. Knapp" grade and was produced in Ladies' size only. Both grades, Pratt and Knapp, were produced for USWC by the firm of A. Bourquin in Bienne, Switzerland. Evidence also exists that the firm of Augustin Perrenoud in Ponts, Switzerland, produced some of these for USWC.

The watch was apparently named after a relative of Edwin Rollo Pratt who was a long-time employee of both the Giles/Wales organization and the USWC. Another possibility is that R. F. Pratt is the Pratt from Read, Pratt & Co., stockholders of USWC, and comb manufacturers in Deep River, Connecticut.

Factory description follows:

No. 38 (1869)	10 size, Quarter Plate, Cock and Bridge, Lever Movement, four pair extra jewels, Straight Line Escapement, Exposed ruby pallets. Sunk Seconds. Key wind and set.

"Chas. G. Knapp"

This movement was produced prior to 1870 and was made in Ladies' size only.

Factory descriptions follow:

No. 39 (1869)	10 size, Quarter Plate, Cock and Bridge, Lever Movement, four pair extra jewels, Straight Line Escapement, Exposed ruby pallets. Sunk Seconds. Key wind and set.
No. 39½ (1869)	Same as No. 39, Damaskeen finish, extra $15.00

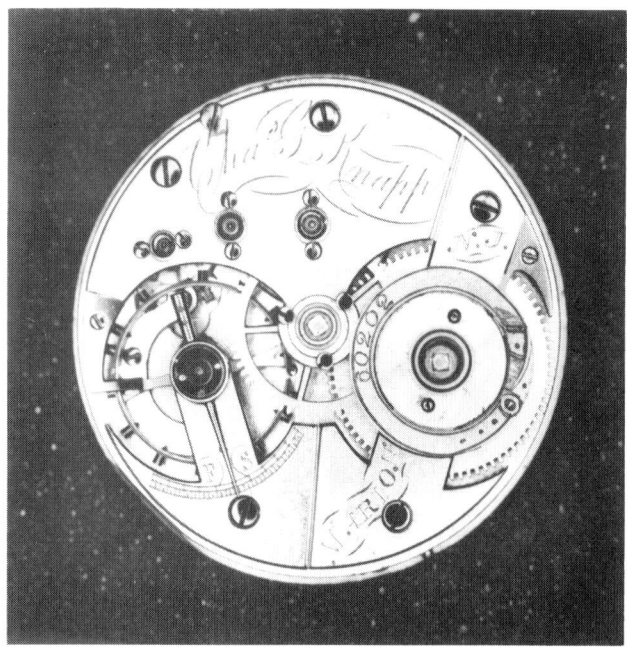

173. "Chas. G. Knapp," SN 60602, 10s, 15j, ¼ plate, cock and bridge also Swiss made. Most Knapps have screwed in settings while the Pratts jewels are friction set, but this is not always the case as some Pratts have survived with screwed in settings. Apparently, the USWC decided the Swiss connection would permit them to get a ladies' watch on the market sooner.

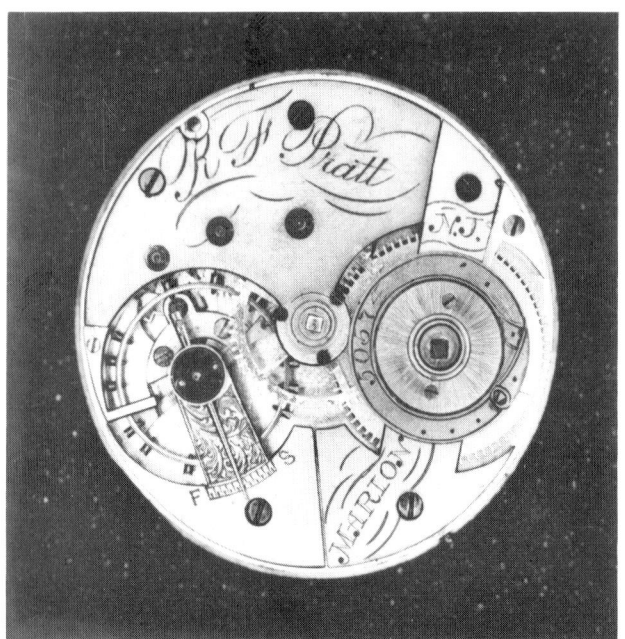

172. "R. F. Pratt," SN 50574, 10s, 15j, ¼ plate, cock and bridge designed by Fayette Stratton but made for USWC by the Swiss firms of A. Bourquin and A. Perrenoud. The Pratt and Knapp models were the first 10s ladies' watches marketed by USWC.

174. Dial from an unmarked movement that is the twin of "Chas. G. Knapp" model with screwed in settings and same design is marked "Augustin Perrenoud, Ponts." So, Perrenoud either made watches for USWC or purchased the same movements from Bourquin.

"Young America"

This movement was initially produced in 1872 and was sold as a Boy's Watch. It was made in 14 size only.

It was probably the USWC response to the Boy's Watch, "American Watch Co., Adams St.," a 14 size, ¾ plate introduced by the American Watch Company in 1870-71.

Factory descriptions follow:

No. 45⅓ (1872) 14 size, Three-Quarter Plate, Lever Movement, Straight Line Escapement, Hardened and Tempered Hair Springs, Exposed pallet jewels, Chronometer Balance, seven jewels, Patent Reversible Barrel. Both Key wind and Stem wind movements.

No. 45¾ (1872) Same as No. 45⅓, but with eleven jewels. Patent Reversible Barrel. Both Key wind and Stem wind movements.

Extra, if with Sunk Seconds dial $1.00

176. "J. W. Deacon," SN 96177, 18s, 11j, Model 45, key wind. Marion Watch Company dial. Note U-slot hair spring stud, typical of Deacon full plate models. Because of its relatively low cost, the models of this grade were a good seller for USWC. This watch in its original USWC marked silver hunting case.

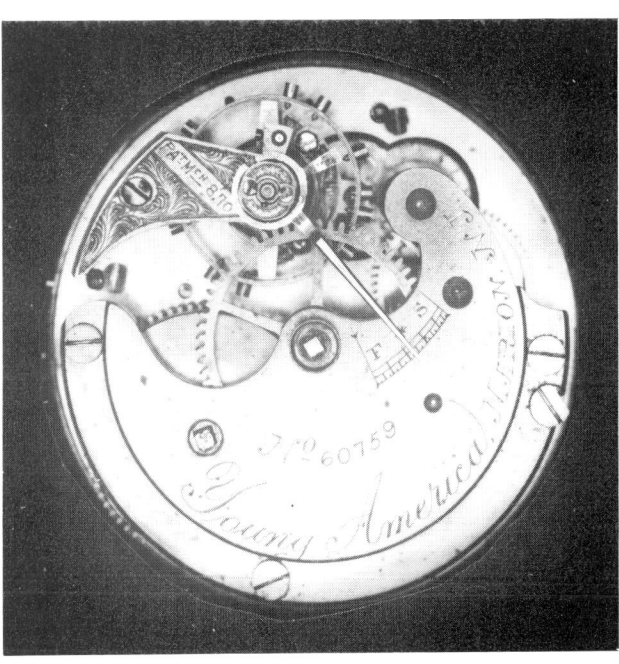

175. "Young America," SN 60759, 14s, 7j, ¾ plate, was produced as a boy's watch to compete with the "Adams Street" grade of Waltham. Model 45-1/3, key wind.

177. "J. W. Deacon," SN 62154, 14s, 11j, ¾ plate, Model 45¼, key wind. Standard regulators used on these grades, but the Elson patent date still showed on the balance cock. The patent reversible barrel was used on ¾ plate models.

"J. W. Deacon"

This movement was originally produced prior to 1872. The standard 18 size, full plate with butterfly opening is shown in Figure 35, the three-quarter plate movement in Figure 36.

Factory descriptions follow:

No. 45 (1871) 18 size, Frosted, Full Plate Lever Movement, Straight Line Escapement, Hardened and Tempered Hair Springs, Exposed pallet jewels, Sprung over Chronometer Balance,

No. 45⅛ (1871) eleven jewels. Both Key wind and Stem wind movements.

No. 45⅛ (1871) 10 size, Quarter Plate and Bridge, same grade as No. 45 but fifteen jewels. Patent Reversible Barrel and Patent Double Index Regulator. Both Key wind and Stem wind movements.

No. 45¼ (1871) 14 size and 18 size, Three-Quarter Plate, same grade as No. 45, eleven jewels. Patent Reversible Barrel. Both Key wind and Stem wind movements.

Extras offered on the above movements:
If with Sunk Seconds dial $1.00
Patent Reversible Barrel added to
 full plate movement 3.00
Patent Double Index Regulator added to
 full plate movement 6.00
Patent Dust Band on full and
 three-quarter plate .... GRATIS

"G. A. Read"

This movement was initially produced prior to 1872. It was designed to be the "value leader" in the USWC line — the watch meant to compete with the "good cheap American watch" which many competitors were now beginning to produce such as the popular "Wm. Ellery" model of the American Watch Company of Waltham, MA. The "G. A. Read" movement was made in 18 size only, in full and three-quarter plate. It has also been noted with "Marion Watch Co." dials.

It was named after G. A. Read, a stockholder in the

178. "G. A. Read," SN 158886, 18s, 7j, key wind, Model 46. Most popular watch of USWC, this key wind full plate model sold for $10.50. The Read grades were only available in 18 size, full and ¾ plate styles.

179. "G. A. Read," SN 79995, ¾ plate, 18s, 7j, Model 46½. This movement is gilt damaskeened. It has the Giles patent reversible barrel with December 30, 1873, date. Note no Elson patent date on balance cock. Marion Watch Company dial.

180. "G. A. Read," SN 282800, ¾ plate, 18s, 7j, Model 46½ without damaskeening. Note this late example has "patent applied for" on reversible barrel, with standard USWC dial. Some USWC historians believe this and the example above definitely indicate the problem of attempting USWC dating from the barrel patent date.

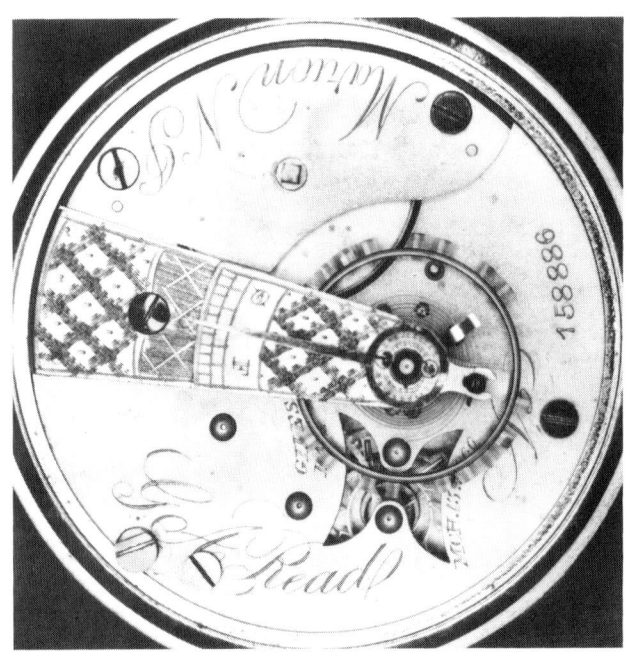

158

USWC, who was a comb manufacturer from Connecticut, of the firm Read, Pratt & Co.

G. A. Read factory descriptions follow:

No. 46 (1871)   18 size, Frosted, Full Plate Lever Movement, Straight Line Escapement, Hardened and Tempered Hair Springs, Exposed pallet jewels, Sprung over Chronometer Balance, seven jewels. Both Key wind and Stem wind movements.

No. 46½ (1871)   18 size, Three-Quarter Plate. Same grade as No. 46. Both Key wind and Stem wind movements.

Extras offered on the above movements:

If with Sunk Seconds dial $1.00
Patent Reversible Barrel added to No. 46 .... 3.00
Patent Double Index Regulator added to No. 46 6.00
Patent Dust Band on full and ¾ plate ... GRATIS

"North Star"

It is interesting that no factory descriptions, other than a material parts list, have turned up on the "North Star." Most USWC historians equate this grade to the "Home Watch Co." movement of American Watch Co. The "North Star" was probably the least expensive watch produced by USWC and the only one produced with an uncompensated balance wheel.

It has been noted both in full plate and three-quarter plate styles, in 14 and 18 sizes.

"I. H. Wright"

This grade was initially produced after 1872. This was the time of change — marked by the departure of William Wales and his replacement by George C. F. Wright as the Number Two man. Thus, it is logical to assume that the "I. H. Wright" movement was named after his older brother Henry Wright.

It has been recorded that I. H. Wright was "cashier and confidential agent" of the USWC.

Factory descriptions follow:

No. 33½ (1872)   18 size, Frosted, Full Plate Lever Movement, Straight Line Escapement, Hardened and Tempered Hair Springs, Exposed pallet jewels, eleven jewels, Sprung over Chronometer Balance, Sunk Seconds. Both Key wind and Stem wind movements.

No. 34½ (1872)   10 size and 16 size, Quarter Plate and Bridge, same grade as No. 33½. Patent Reversible Barrel and Patent Double Index Regulator. Both Key wind and Stem wind movements.

No. 34¾ (1872)   14 size and 18 size, Three-Quarter Plate, same grade as No. 33½. Patent Reversible Barrel and Patent Double Index Regulator. Both Key wind and Stem wind movements.

181 and 182. Dial and movement, "North Star" grade, SN 0204, 18s, 7j, stem wind but key set, gilt, open face with no seconds bit on the dial. A material parts list is the only documentation yet to turn up on this product. It seems likely that it was made by USWC in response to Waltham's "Home Watch Co., Boston" grade. Giles attempted to keep as much association as possible away from USWC with this inexpensive product; the Giles patent data is not even engraved around the butterfly cut out. Some USWC historians believe this product led to the adoption of the Empire City Watch Co. name as a corporation in 1876.

183. "I. H. Wright" grade, SN 8024, 18s, 11j, key wind and set, Model 33½. A lot of mystery surrounds Henry Wright, but he was loyal to USWC up to the very end.

Extras offered on the "I. H. Wright" movements:
If adjusted to heat and cold	$ 9.00
If adjusted to positions	42.50
If Damaskeen finish	12.50
Patent Reversible Barrel added to No. 33½	3.00
Patent Double Index Regulator added to No. 33½	6.00
Patent Dust Band on full and ¾ plate	GRATIS

"A. J. Wood"

This movement was produced initially in 1872 and was made in 10 size (Ladies') only. It was probably one of the last production models of the USWC and was produced in limited quantities.

The "A. J. Wood" was named in honor of stockholder Andrew J. Wood of Brick Church, New Jersey, a merchant in fats and oils.

Andrew Wood was another of those who did everything possible to support Frederick Giles throughout the tenure of USWC. Recall that in 1876 he was one of four individuals, along with Frederick, George Wright, and William Wyse, that incorporated the Empire City Watch Company.

Factory description follows:

No. 47 (1872) 10 size, Frosted, Quarter Plate and Bridge, Lever Movement, Straight Line Escapement, Hardened and Tempered Hair Springs, Exposed pallet jewels, Chronometer Balance, fifteen jewels, Patent Reversible Barrel. Sunk Seconds. Both Key wind and Stem wind movements.

184. Dial of "A. J. Wood," SN 263079, illustrates an interesting USWC monogram. Dial is single sunk with feet held in place by pins, no screws, 10s. This style monogram has also been noted on 14s dials.

185. Movement, "A. J. Wood," SN 263079, 10s, 15j, gilt, ¼ plate and bridge, Model 47. Key wind with standard regulator and reversible barrel marked patent applied for. This USWC-made 10s replaced the 10s Swiss-made Pratt and Knapp grades. A. J. Woods are relatively scarce, possibly because they came so late in the USWC life cycle.

Extras, Dials, Parts

A major "extra," common to all models, were the dials that one could buy to suit his own desires. All watches were provided with a plain seconds dial or a sunk seconds dial. In addition, special dials were also available at additional cost:

"Dials with Lord's Prayer on Seconds Bit .. $15.00
Dials, double sunk, White, each 6.50
  (same) Black, Green or Red, with
    White Figures, each 18.00
  (same) Black, Green or Red, with
    Gold Figures, each 21.00
Dials, single sunk, Black, Green or Red, with
    Gold Figures, each 13.50
  (same) Black, Green or Red, with
    White Figures, each 10.50
Dials with Old English Figures, extra 1.50
  (same) in Gold and Blue, extra 4.50
Dials, cap. letters in place of Figures, extra 1.00
Dials, Old English letters,
    in place of Figures, extra 2.50
Dials, with name in Old English,
    Gold and Blue, extra 2.50
Dials, Letters in place of Figures.
    Gold and Blue, extra 6.00
Dials with Monograms, Plain, extra....... 2.00
  (same) Gold and Blue, extra 4.00
Dials with Masonic: Square and Compass,
  Apron, Bible, Tiling, Carpet and Altar,
  or full Masonic Emblems, extra .. 2.00 - 30.00

Dials with any special design desired, for presentation or other purposes made to order, and charged for according to amount of work on them."

Data above is from a circa 1872 price list shown completely in Figure 193, pages 163-164.

186. 18s, single sunk dial. The standard dials used on most USWC products. Note thin Roman numerals.

187. 18s, double sunk dial. Used on higher grade USWC products. Note bold numerals and USWC printing. These double sunk dials are by far less common on USWC products than the typical single sunk variety. The technique for making them was even kept secret.

188. 18s, double sunk dial with one of several USWC monogram styles used by the company. These monogram dials were an extra cost option usually seen on the prestige USWC grades. Some USWC monograms are also found in the seconds bit in conjunction with another extra cost option above the center post.

ABOVE. 189. 18s, double sunk presentation dial. Everything about this dial is custom. Lord's Prayer hand painted in an area less than half of the seconds bit which requires magnification to read. Custom monogram with figures in Old English. Multi-colors, gold, red, green, and black. A very fine example of dial painting skill. BELOW. 190. 18s, double sunk combination Giles, Wales & Co. and USWC in a red banner. Outer chapter and seconds bit white, with a pastel blue in center bit. Another USWC show piece dial, only used on finer grades. Commonly called a USWC "banner" dial or "ribbon" dial. Different color center bits are seldom seen on USWC products.

ABOVE. 191. 18s, double sunk custom dial with unusual "peace dove" painted above center post and USWC monogram in the seconds bit. Green wreath with black dove and black monogram. This "Symbol of Peace" was especially popular in America in the years following the Civil War, but not seen often on any watch dials. BELOW. 192. 18s, double sunk "banner" dial with black printing in green banner in conjunction with a black USWC monogram in the seconds bit. Only a small quantity of any of these special dials by the USWC have survived the years. Those that do remain are a testimonial to the fine work of USWC dial painters, many of them women.

United States Watch Co.

GILES, WALES & CO.

Factory: MARION, New Jersey.

193. USWC Trade Price List of Material, late 1872 or early 1873. Prices are per dozen unless otherwise noted. Note column headings are initials for the various USWC grades. First column includes "G. A. Read," "J. W. Deacon," and "North Star."

PRICE LIST OF MATERIAL. Prices per Dozen.	G. A. R. J. W. D. & N. S.	A. F. & E. R.	G. C. F. S. M. W. CO. J. W. L. S. M. B. & W. A	Y. A.	F. A. & CO. A H. W. H. R.	U. S. W. CO.	Ladies and 16 Size. Fine Trains.
Arbor Cups, gilt	$1 50	$1 50	$1 50	$	$	$	$
Arbor Cup Steels	5 00	.....	.....	6 00	6 00	12 00	6 00
Barrel Arbors	7 50	7 50	9 00	9 00	9 00	18 00	9 00
Barrels, gilt	12 00	12 00	15 00	15 00	18 00	24 00	18 00
" Nickel	.....	.....	24 00	24 00	36 00	48 00	36 00
Barrels with Arbors fitted, gilt	21 00	21 00	25 50	25 50	27 00	56 00	27 00
" " nickel	.....	.....	36 00	36 00	45 00	60 00	45 00
Balances Chronometer	24 00	.....	.....	.....	.....	.....	.....
" expansion	.....	33 00	48 00	36 00	60 00	150 00	60 00
Balance Staffs	6 00	6 00	7 50	7 50	12 00	30 00	18 00
Cannon Pinions	6 00	6 00	7 50	7 50	9 00	24 00	12 00
do do Split for Stem Winders	9 00	9 00	12 00	12 00	15 00	30 00	18 00
Center Arbors, ¾ plate	4 50	.....	.....	5 00	.....	.....	9 00
Center Cup Steels, ¾ plate	4 00	.....	.....	6 00	.....	.....	12 00
do do ¾ plate for Stem Winders	6 00	.....	.....	9 00	.....	.....	16 50
Click Springs, long	1 25	1 25	1 50	1 50	1 75	4 50	3 00
Click and Spring Combined, new style, patent	2 50	3 00	4 50	3 00	4 50	12 00	6 00
Clicks	1 25	1 25	1 50	1 50	1 50	4 50	3 00
Collets	75	75	75	75	1 00	3 00	1 50
Dials, plain seconds	18 00	18 00	.....	.....	.....	.....	.....
" sunk, "	.....	21 00	24 00	24 00	30 00	48 00	36 00

Dials with Lords Prayer on Seconds Bit ....each.	15 00
Dials, double sunk, White .... "	6 50
" " Black, Green or Red, with White Figures .... "	18 00
Dials, double sunk, Black, Green or Red, with Gold Figures .... "	21 00
Dials, single sunk, Black, Green or Red, with Gold Figures .... "	13 50
Dials, single sunk, Black, Green or Red, with White Figures .... "	10 50
Dials with Old English Figures ....extra.	1 50
" " " in Gold and Blue .... "	4 50
Dials, cap. letters in place of Figures .... "	1 00
Dials, Old English letters, in place of Figures .... "	2 50
Dials with name in Old English, Gold and Blue .... "	2 50
Dials Letters in place of Figures, Gold and Blue .... "	6 00
Dials with Monograms, Plain .... "	2 00
" " " Gold and Blue .... "	4 00
Dials with Masonic: Square and Compass, Apron, Bible, Tiling, Carpet and Alter, or full Masonic Emblems ....extra.	$2 00 to 30 00

Dials with any special design desired, for presentation or other purposes made to order, and charged for according to amount of work on them.

163

	G.A.R. J.W.D. & N.S.	A.P. & E.R.	G.C. F.S. M.W.CO. S.M.B. J.W.L. & W.A.	Y.A.	F.A. & CO. A.H.W. H.R.	U.S.W. CO.	Ladies and 16 Size Fine Trains
Ratchets	1 25	1 25	1 50	1 50	2 00	4 50	3 00
Regulators	4 50	5 00	6 00	6 00	9 00	18 00	9 00
" Patent							60 00
Rollers	2 00	2 25	3 00	3 00	4 50	9 00	4 50
Rollers and Jewel Pin	3 25	3 50	4 50	4 50	6 00	12 00	6 00
Screws long Case	25	25	38	38	38	1 50	38
" Banking	25	25	25	25	25	38	
" Pillar, Potance							
" 3d Bridge, Index	25						
" Hair Spring Stud, short Case	$2 00 per gross, assorted. do do gilded $12 00 per gross.						
" Plate Jewel							
" Round head Jewel							
" Expansion Balance	50	50	75	75		1 50	75
Stop Works, per doz. pieces			3 00		3 00	6 00	4 50
Wheels, gilt center	3 00	3 00	4 00	4 00	4 50	6 00	4 50
" 3ls and 4ths	3 00	4 00	4 00	4 50	4 50	6 00	4 50
" 'scapes	4 50	4 50	6 00	4 50	12 00	18 00	12 00
" hour	3 75	3 75	4 50	4 50	4 50	6 00	4 50
" minute	3 75	3 75	4 50	4 50	4 50	6 00	4 50
Wheels and Pinions, center	12 00	12 00	15 00	15 00	18 00	30 00	30 00
" 3ls and 4ths	9 00	9 00	12 00	12 00	13 50	27 00	21 00
" 'scapes	10 50	10 50	13 50	13 50	15 00	30 00	21 00
Winding Arbor, (Stem Winder.)	50	50	50	50	50	1 50	75
" Wheel contrite "	4 50	4 50	6 00	6 00	6 00	15 00	7 50
" intermediate (Stem Winder.)	1 25	1 25	1 50	1 50	2 00	4 50	3 00
Winding Wheel, Bevel, (Stem Winder.)	4 50	4 50	6 00	6 00	6 00	15 00	7 50
Winding Wheel, setter, (Stem Winder.)	1 25	1 25	1 50	1 50	2 00	4 50	3 00
Winding Yoke, (Stem Winder.)	12 00	12 00	15 00	12 00	18 00	48 00	18 00

	G.A.R. J.W.D. & N.S.	A.F. & E.R.	G.C. F.S. M.W.CO. S.M.B. J.W.L. & W.A.	Y.A.	F.A. & CO. A.H.W. H.R.	U.S.W. CO.	Ladies and 16 Size Fine Trains
Forks	3 00	4 50	6 00	6 00	9 00	24 00	15 00
Hair Spring Studs	3 00	5 00	6 00	6 00	9 00	24 00	15 00
" ⅜ plate	1 50						
Hair Springs, tempered	2 50	2 50	2 50	2 50	4 50	24 00	3 00
" Breguet							27 00
Hand, steel	1 50	1 50	2 00	2 25	4 50	27 00	6 00
" fleur de lis			2 25	3 50	1 75	12 00	
" seconds	25	50	1 75	1 75	1 75	4 50	3 00
Indexes, Silver			3 00		3 00		
" Gold						12 00	12 00
Jewels, chrysolite, opened, viz.: 3d, 4th and 'scape, upper	4 00	4 50	6 00	4 00	9 00		24 00
" lower	4 50	3 00	4 00	4 00	6 00		
Cock and foot holes, turned down	4 50	4 50	5 25	5 25	5 25		
Jewels, End Stones, in Brass, turned down	2 75	2 75	2 75	2 75	2 75		
Jewels, Aqua Marine, opened, viz.: 3d, 4th and 'scape, upper	3 00	3 00					
" lower	2 50	2 50					
Cap Jewels in Brass	2 75	2 75	2 75	2 75	2 75		6 00
Ruby, 3d, 4th and 'scape upper						12 00	6 00
" lower					4 50	9 00	4 50
" End Stones, Balance					4 50	7 50	4 50
Roller Pins	1 00	1 00	1 00	1 00	1 50	3 00	1 50
Main Springs	2 25	2 25	3 75	3 75	3 75	3 75	3 75
Pallet Arbors	4 00	4 00	4 50	5 00	6 00		9 00
" Conical Pivots					12 00	30 00	18 00
Pallets	9 00	10 50	12 00	12 00	24 00	60 00	36 00
Pinions, Center	7 00	7 00	7 50	9 00	12 00	21 00	
" Hollow ¾ plate	12 00			15 00		36 00	24 00
" 3d, 4th and 'scape	5 00	5 00	6 00	7 50	10 50	21 00	15 00

SALESROOMS:

GILES, WALES & CO., 13 Maiden Lane, New York.
GILES, BRO. & CO., 384 Wabash Avenue Chicago, Ill.

193. *USWC Trade Price List of Material (continued). Note that all train wheels shown as gilt. This plus the USWC surviving movement price schedules confirm that no grades were completed with a gold train.*

Chapter 20

Honors and Testimonials

The USWC was an active participant at various Fairs and Exhibitions, entering their watches into competition against all comers. Many honors, awards and testimonials have been recorded.

First Premium (first prize) Awards for perfection over all competitors were awarded at the following Fairs and Exhibitions:

1869 — Fair of Cincinnati Industrial Exposition, Ohio
1870 — American Institute, New York City
1870 — Ohio Mechanics Institute, Cincinnati, Ohio
1870 — Louisiana State Fair, New Orleans
1871 — Texas State Fair, Houston
1871 — International Industrial Exposition of the Mechanics Institute, Buffalo
1872 — New Jersey State Fair
1872 — Iowa State Fair
1873 — American Institute, New York City

The American Horological Journal for November 1870 reported the following from the Twenty-third Exhibition of the American Institute of New York:

"WATCHES — The United States Watch Co., Marion, N. J. — Giles, Wales & Co., 13 Maiden Lane, New York, General Agents — display a fine variety of ladies' and gentlemen's watches, with full and three-quarter plate movements, exposed pallets, straight-line escapements, and stem winders — the Company claiming strength and simplicity as the peculiar feature of the winding mechanism. They exhibit over fifty different styles, many of them in solid nickel, and beautifully damaskeened. Their style of casing is especially worthy of notice, both in design and finish, some of them being very elaborately enameled, and especially adapted for presentation purposes. Among the ladies' watches was one set with diamonds, valued at $1,800, and one valued at $700. They also exhibit a number of dials with masonic and other emblems, which were very beautifully executed."

RIGHT. 194 and 195. Obverse and reverse views of the actual silver medal awarded to USWC for first premium award at the New York American Institute Fair in 1873. Actual dimensions are two inches diameter and 1/8 inch thickness. This award was received ". . . for the best watches, watch movements and materials." Surviving data indicates that USWC received first premium awards at nine exhibitions during the five-year period from 1869 through 1873. Surviving watches indicate plans had been made to participate in the 1876 Centennial Exhibition held in Philadelphia.

It was at this 1870 Exhibition that the USWC was awarded "First Premium." The medal is shown at the top of the advertisement in Figure 196.

Table V
Some Original Owners of USWC Watches
1867-1872

Ser. No.	Grade	Original Owner
1006	Frederic Atherton & Co.	Wm. Mitchell, Conductor, P. & K. R.R.
1037	Frederic Atherton & Co. June 1867	Henry Smith, Treasurer, Panama R.R. Co., 88 Wall St., N.Y.
1064	Frederic Atherton & Co. September 1868	William Derby, of Derby, Snow & Prentiss, Jersey City, N.J.
1081	Frederic Atherton & Co. July 1868	John D. Egbert, 5 College Place, Room 8, New York
1089	Frederic Atherton & Co. December 1868	L. E. Chittenden, Late Treasurer, U. S. Treasury, N.Y.
1105	Frederic Atherton & Co. January 1869	A. H. King, Elastic Cone Spring Co., 7 Park Place, New York
1117	Frederic Atherton & Co. July 1868	B. F. Phelps, Conductor, N. J. Central R.R.
1124	Frederic Atherton & Co. June 1869	A. L. Dennis, President, N. J. R.R. & T. Co.
1125	Frederic Atherton & Co. March 1869	H. Lassing, Knickerbocker Ins. Co., 161 Broadway, New York
1143	Frederic Atherton & Co.	James B. Ryer, of Kelty & Co., 447 Broadway, New York
1154	Frederic Atherton & Co.	Horace Hatch, M.D., 25 W. 38th St., New York
1176	Frederic Atherton & Co. February 1869	Henry DeLancey, Engineer, Phila. & Erie R.R.
1244	Frederic Atherton & Co. November 1869	E. C. Keys, Pittsburgh, Pa.
1251	Frederic Atherton & Co. August 1869	F. A. Haskell, Conductor, Hudson River R.R.
1259	Frederic Atherton & Co. June 1869	E. Rice, of Whitney & Rice, 179 Broadway, N.Y.
1320	Frederic Atherton & Co.	W. H. Hawkins, Chicago, Burlington & Quincy R.R.
1650	Frederic Atherton & Co.	FIRST PREMIUM AWARD — 1870 Cincinnati Industrial Exposition
1658	Frederic Atherton & Co. October 1869	John Lindstroom, 344 Atlantic St., Brooklyn, N.Y.
1706	Frederic Atherton & Co. October 1869	Jno. W. Smith, Amsterdam Ins. Co., Dubuque, Iowa
1788	Frederic Atherton & Co.	Henry Morford, Equitable Insurance Co., 120 Broadway, New York
1835	Frederic Atherton & Co.	S. M. Moore, of S. M. Moore & Co., Chicago, Ill.
1894	Frederic Atherton & Co.	H. Cottrell, 128 Front St., N.Y.
2100	Fayette Stratton January 1868	Walter H. Kirkpatrick, Conductor, Penn. C. R.R.
2183	Fayette Stratton	Wm. Dunne, Baggage Express, Utica, N.Y.
2226	Fayette Stratton	Oscar M. Sanford, Utica, N.Y.
2260	Fayette Stratton	A. M. Osgood, Ilion, N.Y.
2291	Fayette Stratton November 1868	E. O. Whipple, Conductor, U. & B. R.R.
2617	Fayette Stratton May 1869	I. Vrooman, Engineer, N. Y. C. & H. R.R
2656	Fayette Stratton	Jacob Weart, Collector of Internal Revenue, Jersey City, N.J.
2755	Fayette Stratton	Jas. B. Weaver, 111 & 115 William St., New York
2798	Fayette Stratton	John M. Woolhause, Conductor, C. & N. W. Railway, Chicago
4026	Edwin Rollo November 1869	Joshua I. Bragg, Conductor, N. J. R.R.
4130	Edwin Rollo	Thos. E. Miner, Pier No. 5, Elizabethport, N.J.
4284	Edwin Rollo	David H. Peck, Ferry Master, Central R.R. of N.J.
10548	Frederic Atherton & Co. September 1868	Z. C. Priest, Asst. Supt., N. Y. C. & H. R.R.
12003	United States Watch Co.	W. S. Dunn, with H. B. Claflin & Co., N.Y.

12006	United States Watch Co. February 1869	Judge Chas. H. Voorhis, Hackensack, N.J.
12012	United States Watch Co. August 1869	Geo. Lovis, Passenger Agent, Toledo, Wabash & Western Railway
?	"Gold, stemwinder" (?) October 1, 1870	President Ulysses S. Grant
20019	Wm. Alexander	Geo. W. McDonald, Supt., Brooklyn Water Works
20293	Wm. Alexander August 1872	James E. McLean, Chicago, Ill.
21039	S. M. Beard December 1871	S. M. Beard, of Beard & Cummings, 128 Front St., New York
21767	Fayette Stratton	Sam'l. Merrill, Governor of Iowa (see Fig. 46)
22105	S. M. Beard	S. F. Wilson, D.D.
24008	United States Watch Co.	Chas. H. Wolf, Cincinnati, Ohio

OPPOSITE PAGE AND ABOVE. Table V. Compiled from many USWC advertisements with testimonials, this table shows original owners, serial number, grade and date of purchase (if known).

RIGHT. 196. "Harper's Weekly" ad, April 13, 1872, showing first premium awards presented to the USWC at the New York City and Cincinnati Expositions in 1870. Note the typical inclusion of a few testimonials at the bottom of ad.

LEFT. 197. Dial view "Pennsylvania Rail Road," 18s, SN 21526. Waltham and Elgin had already sold a number of their watches to the Pennsylvania R.R. when the USWC approached them with a novel idea . . . buy USWC products and get the PRR logo on the dial with the manufacturer's logo in the seconds bit. Both competitive products only had the plain rail road name on the dial. But the proposition didn't end there. . . .

RIGHT. 198. Movement view, "Pennsylvania Rail Road," SN 21526. In addition to the fancy dial, the USWC would engrave Pennsylvania R.R. on the movement with the manufacturer's name below on the barrel plate. But there was still more . . . buy USWC and get a quality frosted nickel, 15j movement, contrasted to the Elgin and Waltham gilt movements. So, Frederick Giles established three interesting firsts: (1) first RR logo on a dial, (2) first RR marked movement, and (3) first nickel movement sold to a RR. This watch is believed to be late 1871 or early 1872.

Chapter 21
Special Order Watch Production

A small but important portion of any watch company's business is that referred to as "special orders." In recent years, these have been called "private label" or "jeweler's contract" watches. The USWC was one of the first to actively pursue this type of business. In addition to jewelry stores, the USWC sold their products to at least four railroads and five different newspapers. Unless some special options were desired, these names were engraved on the movements with location at no additional charge. And, in some instances, the name, or logo of the purchaser, was also painted on the dials.

Orders for such work varied in amount, from one to several dozen. The single order was likely for a retirement ceremony, sales contest winner, etc., with the larger orders limited only by the retail volume of the jeweler's business.

Movements for these "special order" watches were taken from stock or a current production run and just about all types and sizes of movements in the USWC inventory were utilized.

A listing of some of these jeweler's contract watches is shown in Table VI.

199. "Mason Ware & Co., San Francisco, Cal.," SN 15270, 18s, 15j. One of the better grades used by jewelers, this example is also one of the earliest made up on special order. Circa 1869. Three examples of this jeweler's products have turned up to date.

200 and 201. Two examples of 18s, 15j, ¾ plate special orders. "John S. Robson, Brooklyn, NY," SN 51022, is nickel damaskeened, while "H. G. Cook, Lanesboro, Minn.," SN 71203, is gilt. Both have USWC logos engraved on the movement.

Table VI
Some Jeweler's Contract USWC Products

Ser. No.	Jeweler & Location	Details
14923	Mason, Ware & Co., San Francisco	18s, KW, Giles 1866 patent
15267	Mason, Ware & Co., San Francisco	18s, KW, Giles 1866 patent
15270	Mason, Ware & Co., San Francisco	18s, KW, Giles 1866 patent
17595	H. H. Reed, Brooklyn, NY	18s, KW, Giles 1866 patent
19106	D. Dunseath, Pittsburgh, PA	18s, KW, Giles 1866 patent
20402 & 20405	D. Dunseath, Sr., Pittsburgh, PA	18s, 19J, Giles 1866 patent
29009	Wm. Edwards, New York	10s, 15J, KW
30340	A. Frankfield & Co., NY	
36303	Ben Franklin, Utica, NY	18s, 15J, full plate, gilt, probable Fuller movement
37477	McGonigle, S. Amboy, NY	18s, KW, Giles 1866 patent
40935	Daniel Smith	18s, 11J, full plate, gilt
51022	John S. Robson, Brooklyn, NY	18s, 15J, three-quarter plate, nickel
55302	Geo. E. O. Chace, New York	16s, 15J, quarter plate & bridge, nickel
60120	Wm. Edwards, New York	10s, 15J, KW
60434	Benjamin & Ford, New Haven	10s, 15J, KW
62053	S. Conradi, Houston	14s, KW
69106	G. D. Parks, Yorkville, NY	14s, 11J, three-quarter plate
69168	Wilson & McGraw, Cincinnati, Ohio	14s
69189	Wirth Bros., Brooklyn, NY	14s, 11J, KW, three-quarter plate
71203	H. G. Cook, Lanesboro, Minn.	18s, 15J, three-quarter plate
74037	S. J. Delan, New York	18s, 15J, KW
110092	D. B. Judd, Chester, Mass.	18s, Giles 1866 patent
112714	A. Frankenfeld & Co.	18s, 15J, KW
112839	M. C. Haight, Geneva, NY	
140739	A. Frenkfield & Co., New York	18s, KW, 15J
140961	D. B. Judd, Chester, Mass.	18s, KW, Giles 1866 patent
142723	D. B. Judd, Chester, Mass.	
142901	H. H. Reed, Brooklyn, NY	18s, KW, Giles 1866 patent
155772	John McPherson, Vassar, Michigan	18s, KW
157411	T. E. Parker, Lydonville (sic), Vt.	
241012	Fisher & Thatcher, L.A., Calif.	10s, 15J, quarter plate & bridge, gilt
263080	Oliver Arnzen, Brooklyn, E.D.	10s, 15J, quarter plate & bridge, gilt
282882	D. R. Brown, Stroudsburg, Pa.	18s, KW, three-quarter plate
283383	A. M. Murphey, Tyler, Texas	18s, KW, three-quarter plate

202. "Benjamin & Ford, New Haven, Conn.," SN 60434, Swiss made for USWC, 10s, 15j, ¼ plate, cock and bridge movement style, key wind and set.

203. "Oliver Arnzen, Brooklyn, E.D. (Eastern District)," SN 263080, USWC made, 10s, 15j, ¼ plate and bridge movement style, key wind and set.

204. "D. R. Brown, Stroudsburg, PA.," 18s, ¾ plate, 7j, gilt, SN 282882. Relatively late USWC jeweler's contract, key wind and set, standard regulator. December 30, 1873, patent date on reversible barrel.

Other "special order" watches were recorded and are listed below under appropriate headings.

NEWSPAPERS
    Rural New Yorker — 18s, 15J, full plate, KW
    St. Paul Press — 18s, 15J, full plate, stemwind
    American Working People — 18s, 15J, full plate, KW
    Cincinnati Gazette — 18s, 11J, full plate, KW
    Cincinnati Enquirer — 18s, 11J, full plate, KW

RAILROADS
    Pennsylvania R.R. — 18s, 15J, full plate, KW
    Union Pacific R.R. — 18s, 15J, full plate, KW
    B.C.&M. R.R. — (Boston Concord & Montreal), 18s, full plate, KW
    N.J. R.R. & T. Co. — (New Jersey), 18s, full plate, KW

OTHER
    George C.F. Wright — 16s, 19J, ¼ plate & bridge, flat hairspring, Elson double index regulator

The USWC seems to be one of very few, or the only watch company to go after "special order" business from the newspaper trade.

An unexplained alphabet, or letter code, has been noted on some special order newspaper movements and on a few jeweler's contract watches. Capital letters from A through H either engraved above or below the newspaper or jeweler's name have been recorded to date. One speculation is that the letters were used on certain, small order items to designate the number in the lot . . . A for 1 in the lot, B for 2 in the lot, etc. Numbers could not be used because of the possible confusion with the serial number.

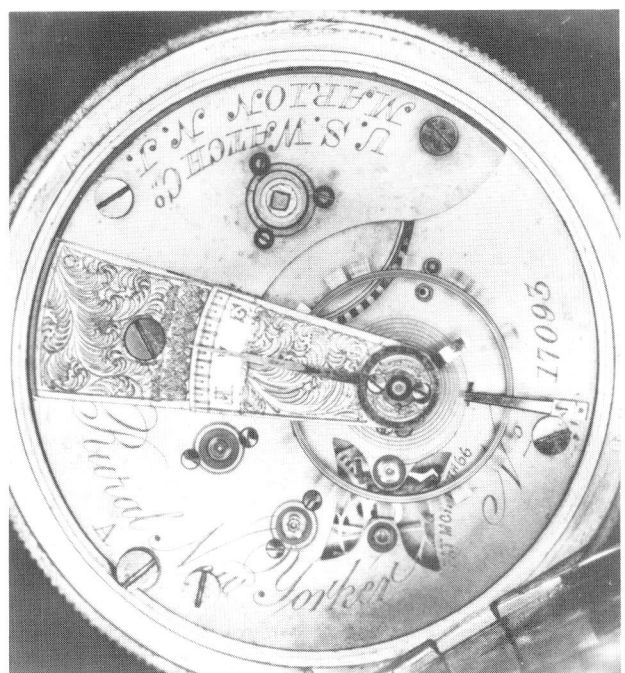

205. "Rural New Yorker," SN 17093, 18s, 15j, with letter A under Rural. This newspaper was a big user of USWC watches contrasted to some of the other newspapers. The USWC also reciprocated by using the Rural New Yorker for many of their advertisements, and this fact could explain the reason more of these have turned up than any of the others.

206. "St Paul (Minn) Press," SN 110459, 18s, 15j, with letter A above Paul. This paper is still in business and could only speculate that this may have been a retirement presentation, or possibly the prize in a contest. Only one example has turned up.

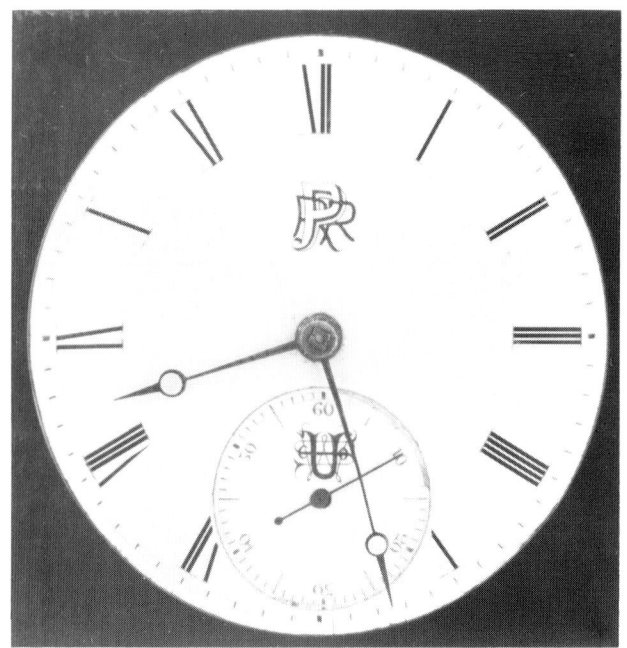 

207 and 208. Dial and movement views, "Pennsylvania Rail Road," SN 21518, 18s, 15j, key wind and set. Very close to SN 21526 illustrated in Figures 197 and 198 and in the same lot. Note the differences . . . USWC logos in seconds bit are very different, and the engraving style varies on the two balance cocks, yet the PRR logo is the same.

209. "Union Pacific," SN 142375, 18s, 15j. Another USWC special order railroad watch. The few surviving examples are all gilt and apparently the only nickel movements were those made for PRR. There were at least two runs produced of the Union Pacific based on surviving examples. There was only one run for PRR, BC & M RR, and the New Jersey RR & T based on examples turned up to date.

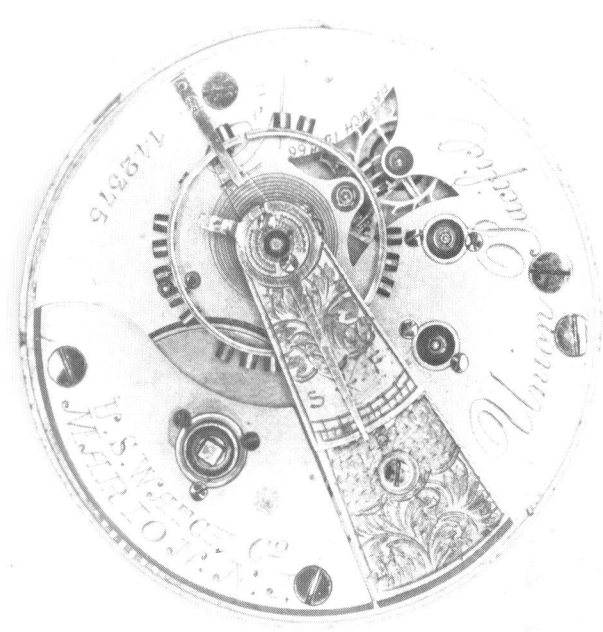

Many watch companies catered to the railroad trade including USWC. . . . if the watch could take the environment of a locomotive, surely it was good for everyday use! Consequently, many USWC testimonials came from railroad personnel.

Frederick's first sale of some 50 watches (based on surviving examples) to the Pennsylvania R.R., led to subsequent sales to the Union Pacific R.R., the Boston, Concord & Montreal R.R., and the New Jersey R.R. and T. Not many examples of these seem to have survived, indicating that probably relatively small quantities were ever sold. And, none of the other railroad examples have been noted with logos on the dial like the PRR, although it would seem there may be others.

At some time during the 1870's it is thought that the railroads abandoned purchase of company watches and required railroad personnel to provide their own.

The "George C. F. Wright, New York" watch was illustrated in Figure 111. It is 16 size, ¼ plate and bridge style with 19 jewels.

This watch, with the name of an original partner of Giles, Wales and Co. and of the USWC, may have been a special presentation piece or his personal watch. Secondly, it could have been made by Wright, after the demise of the USWC when he was in business as a watchmaker/jeweler. He was in a position where he could acquire movements, parts, etc., at the time of final bankruptcy, and then market these watches under his own name. This latter possibility seems most likely since the location on the movement is shown as "New York," where George set up his business on Maiden Lane after leaving USWC.

Chapter 22

The Centennial Commemoratives

The subject of USWC Centennial Commemoratives was initially reported and discussed by Eugene T. Fuller in an exploratory article published in the October 1977 issue of the NAWCC BULLETIN, entitled: "U. S. Watch Company Centennial Commemoratives."

Since the publication of the Fuller article, new and probably significant information, plus three additional Commemorative watches have been reported.

It now appears likely that these Centennial Commemorative watches were products of the USWC and made specifically for exhibition or for sale at the Philadelphia Centennial Exhibition of 1876.

First, a short re-cap of the three watches reported in the Fuller article, plus the recently reported other three examples.

"Independence" — The "Independence" model, probably the most elaborate of the six, is shown in Figure 210. It is the standard 18 size, full plate model, with the butterfly opening, except in this case, the cut out is a double butterfly with rounded wing tips which exposes more of the train wheels. Actually, a skeletonized top plate and balance cock with an engraved American shield. The movement is nickel with beautiful damaskeening and jeweling with gold settings.

210. "Independence America," SN 5291, 18s, 15j, richly damaskeened and frosted, skeletonized top plate in double butterfly style (see Figure 60), skeletonized balance cock with American shield engraving. Entire engraving in Old English style surrounded by fancy scrolling, similar to that used on "United States" grades.

The overall quality, engraving and finish of this watch equates to the "United States Watch Co." grade but is only 15 jewels. The serial number on the top plate, 5291, has no relationship to the serial number on the pillar plate, 21004 (from an S. M. Beard run). This has all indications of being a specially designed movement from existing material stock for display or exhibit at the Centennial.

The "George Washington Exhibition Watch" is shown in Figure 211. This movement is similar to the "Independence" model except that it has the standard butterfly cut out with pointed wing tips. The movement is nickel with beautiful damaskeening and has the extra jeweling which is typical of the USWC's better quality products.

The "Philadelphia Exhibition Watch" is shown in Figure 212. It too has a nickel movement with beautiful damaskeening and the standard butterfly cut out. One major difference is the placement of the Slow/Fast regulator index. It appears on the balance bridge rather than on the top plate as shown in the two previous movements. Usually, a movement with the Slow/Fast indicator on the balance bridge also has the serial number engraved on the top plate at the left, preceding "Marion, N. J." on the standard models. As can be seen in the photograph, this is not the case. The serial number is below the balance bridge, as in the previous two examples. This arrangement is similar to the second generation of the "George Channing" model, produced in 1872 or later.

These first three examples in nickel would appear to have been produced for display or exhibition purposes. The next three in gilt would have been more likely to be produced for sale in conjunction with the Centennial. Further examples to turn up in the future similar to our last three gilt examples would tend to confirm this deduction.

The fourth Centennial Commemorative watch, having turned up since the Fuller article, is the "Centennial" shown in Figures 213 and 214. This also is a standard 18 size, full plate with butterfly opening. The movement is gilt, frosted, with a nickel balance bridge. The Slow/Fast indicator is on the balance bridge and the serial number is placed on the left side. Also inscribed on the back plate are the words "Extra Jeweled." This particular movement appears to be identical to the regularly produced USWC "Asa Fuller" grade, an example of which is illustrated in Figure 170.

The fifth watch that has turned up since the Fuller article is shown in Figure 216. It is 14 size, gilt finish, and the first to appear in three-quarter plate. Like our fourth example, the movement is also engraved "Centennial, Philadelphia." But unlike the fourth example, the dial is marked "Centennial." It is in the original silver hunting key-wind case marked "Centennial" inside the front and back covers. The regulator is standard without the Elson double index, and the movement apparently is a basic three-quarter plate "J. W. Deacon" model.

The sixth example reported is the first duplicate to

211. "George Washington Exhibition Watch, Philadelphia, 1876," SN 22432, 18s, 15j, beautifully damaskeened and frosted. Excellent condition gold hunting case, showing little indication of any use. Serial numbers match, making this a probable "Wm. Alexander" movement. Dial signed "Philadelphia."

212. "The Philadelphia Exhibition Watch, 1876," SN 160974, 18s, 15j, nicely damaskeened and frosted. Serial numbers match, making this a probable second generation "George Channing" movement. The engraving designation on this watch and the previous "George Washington" leave little doubt as to their purpose.

ABOVE. 213 and 214. Dial and movement views, "Centennial, Philadelphia," SN 175533. This is the only example to turn up with "Exhibition" label on dial, even though the movement is engraved with the same wording as the 14s examples. This watch is 18s, 11j, gilt with the nickel barrel bridge and is easily associated with the "Asa Fuller" grade. Key wind and set. BELOW. 215 and 216. "Centennial Philadelphia," SN 1004 and 1005, both 14s, gilt, 11j, ¾ plate style, probably from a run of "J. W. Deacons." Serial numbers assigned out of sequence on these two examples, but all numbers throughout movement match the top plate number. Key wind and set. These two 14s gilt and the 18s gilt/nickel with same movement engraving apparently were meant to be commercial, or sales items, for the Centennial. The two 14s both are in original silver cases marked "Centennial."

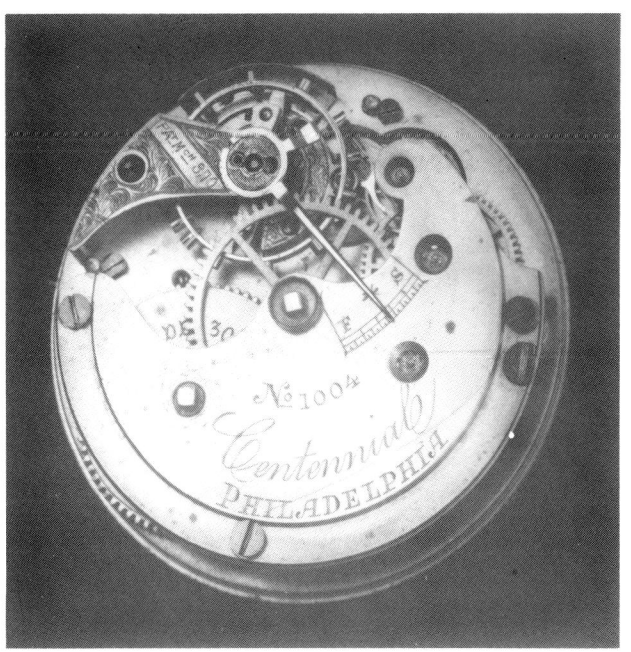

175

Table VII

USWC 1876 Centennial Product Contrast

Model Name:	Centennial Philadelphia	Independence America	George Washington Exhibition Watch Philadelphia 1876	The Philadelphia Exhibition Watch 1876	Centennial Philadelphia
Figure:	215 and 216	210	211	212	214
Serial Number:	1004 and 1005	5291	22432	160974	175533
Placement:	Center	Below Balance Bridge	Below Balance Bridge	Below Balance Bridge	Left Side
Slow/Fast Indicator:	On Top Plate	On Top Plate at Left	On Top Plate at Left	On Balance Bridge	On Balance Bridge
Other:	Standard	Skeletonized	Standard	Standard	Standard "Extra Jeweled"
Movement:	Gilt	Nickel	Nickel	Nickel	Gilt
Finish:	Standard	Damaskeen and Frosted	Damaskeen and Frosted	Damaskeen and Frosted	Frosted with Nickel Barrel Bridge
Jewels:	11j	15j	15j	15j	11j
Name on Dial:	Centennial	United States Watch Co.	Philadelphia	United States Watch Co.	Exhibition
Movement Style:	Probable "J. W. Deacon" movements	Special Design "S. M. Beard" movement	Probable "Wm. Alexander" movement	Probable "George Channing" movement	Probable "Asa Fuller" movement
Size:	14 size	18 size	18 size	18 size	18 size
Type:	Three-Quarter Plate	Full Plate	Full Plate	Full Plate	Full Plate
Winding:	Key Wind	Stem and Key Wind	Key Wind	Key Wind	Key Wind

217 and 218. Dial and front cover case, "Centennial, Philadelphia," SN 1004, 14s, ¾ plate. Dial and case are the same on SN 1005, with the dial marked "Centennial" and both front and rear covers of a coin silver, engine turned hunting case that are also marked "Centennial." Close inspection of these two examples in gilt with matching dial and cases certainly points to a product that was made to be commercial. It is very interesting that the only two examples of this 14s known have adjacent serial numbers.

any of the other five — it is the same as the fifth example with consecutive serial numbers, 1004 and 1005. These two examples in 14 size are both in original coin silver cases marked "Centennial" inside front and back covers.

It would appear that these Centennial Commemorative watches are representative of various phases of USWC quality and workmanship. Fortunately, the six that have been recorded to date, quite easily align themselves within the normal development of the USWC. This is best shown in Table VII, "USWC 1876 Centennial Product Contrast."

One possible explanation for at least some of these examples may be linked to a Hugh Mulligan in Philadelphia. In 1875, this gentleman registered trademarks for (1) "General George Washington Exhibition Watch, Philadelphia, 1876" — No. 2535, May 18 and (2) "General Andrew Jackson Exhibition Watch, Philadelphia, 1876" — No. 2820, August 10. Mulligan may have "special ordered" a small quantity, or trial run, of Centennial products from the USWC sometime in 1874 or 1875 to establish trade-mark *usage*.

An interesting coincidence surrounds two of these watches. The "George Washington Exhibition Watch" was acquired in Montague some years ago and the "Independence America" was obtained some 40 miles away in the Berkshires area. According to the late Christabel G. Burns (Frederick Giles' granddaughter), Frederick probably brought a few watches and USWC award medals with him as keepsakes when he returned to Montague in 1878. The "Philadelphia Exhibition Watch" is part of the Vogel collection owned by the American Clock & Watch Museum and its origin cannot be traced.

In summary, six Centennial oriented products from the USWC have now been reported. All six have different, yet very similar characteristics and engraving styles. All six watches were probably produced prior to the demise of the USWC in anticipation of a display and/or sale at the 1876 Centennial Exhibition in Philadelphia. Proof is still lacking, but deduction at this time would indicate the three nickel examples were produced for display or competition, while the three gilt examples were intended to be commercial, or sales items. We do know for a fact that at earlier fairs and displays USWC exhibited "over 50 watches in both nickel and gilt finishes." As discussed in the Fuller article, considerable advance planning was obviously necessary for the Centennial with their policy for *free* exhibit space announced early in 1873.

The USWC was an active participant on the exhibition circuit and Frederick Giles does not seem to have been the kind of person that would have passed up a major exhibit like the Centennial which might have given new life to his faltering company. When the 1876 Centennial arrived, the USWC was going through their final financial trauma, and reorganization into the Empire City Watch Co. was probably more pressing than completing the Centennial exhibit. The great Centennial spawned a variety of commemorative products, but very few from American watch companies. In fact, Giles and the USWC *may have the distinction of being the first American watch company to produce a commemorative watch for a major exhibition.*

PRICE LIST OF MOVEMENTS.

"EMPIRE CITY WATCH CO."

		Key Winders	Pendant Winders
No. 1	Quarter Plate and Bridge.	NICKEL, 10 (Ladies) and 18 size, Damaskeen; and Frosted, ENAMELED, Gold; Index Plate, Gold Jewel Settings throughout. Best quality, Lever Movement, *St. Line* Escapement, *Exposed ruby pallets*, 19 fine ruby jewels, 3 pairs *conical* pivots, *cap jeweled*, with fine rubies, Breguet hair spring, hardened and tempered, Compensation Balance, accurately adjusted to heat and cold, Isochronism and position. *Patent Reversible Barrel* and *Patent Double Index Regulator*. Named "Empire City Watch Co." New York.	$450 00	$450 00
No. 2	Three-Quarter Plate.	NICKEL, 14 and 18 size, same quality as above, *Patent Reversible Barrel* and *Patent Double Index Regulator*. Named "Empire City Watch Co." New York.	400 00	430 00
No. 3	Full Plate.	NICKEL, 18 size. Same quality as above.	370 00	400 00

"W. S. WYSE."

No. 4	Full Plate.	NICKEL, 18 size, Lever Movement, *St. Line* Escapement, *Exposed pallet jewels*, 6 pair extra jewels, and 2 pair extra *conical* pivots, *cap jeweled* in *gold* settings, Sprung over Expansion Balance, adjusted to heat and cold, Sunk Seconds. *Patent Double Index Regulator* and *Reversible Barrel*. Named "W. S. Wyse," New York.	275 00	300 00
No. 5	Quarter Plate and Bridge.	NICKEL, 10 (Ladies) and 16 size. Same quality as 4. *Patent Reversible Barrel* and *Patent Double Index Regulator*.	310 00	335 00
No. 6	Three-Quarter Plate.	NICKEL, 14 and 18 size. Same grade as 4. *Patent Reversible Barrel* and *Patent Double Index Regulator*.	285 00	310 00
		Extra on 4, 5, or 6, if adjusted to positions.	50 00	50 00

"L. W. FROST."

No. 7	Full Plate.	NICKEL, 18 size, Lever Movement, *St. Line* Escapement, *Exposed pallet jewels*, 4 pair extra jewels, Sprung over Expansion Balance, adjusted to heat and cold. Sunk Seconds. *Patent Reversible Barrel* and *Patent Double Index Regulator*. Named "L. W. Frost," New York.	175 00	200 00
No. 8	Quarter Plate and Bridge.	NICKEL, 10 (Ladies) and 16 size, same grade as 7. *Patent Reversible Barrel* and *Patent Double Index Regulator*.	200 00	225 00
No. 9	Three-Quarter Plate.	NICKEL, 14 and 18 size, same grade as 7. *Patent Reversible Barrel* and *Patent Double Index Regulator*.	185 00	210 00
		Extra on 7, 8, or 9, if adjusted to positions.	50 00	50 00

"CYRUS H. LOUTREL."

No. 10	Full Plate.	NICKEL, 18 size, Lever Movement, *St. Line* Escapement, *Exposed pallet jewels*, 4 pair extra jewels, Sprung over Expansion Balance, adjusted to heat and cold. Sunk Seconds. *Patent Reversible Barrel* and *Patent Double Index Regulator*. Named "Cyrus H. Loutrel," New York.	110 00	125 00
No. 11	Quarter Plate and Bridge.	NICKEL, 10 (Ladies) and 16 size. Same grade as 10. *Patent Reversible Barrel* and *Patent Double Index Regulator*.	150 00	165 00
No. 12	Three-Quarter Plate.	NICKEL, 14 and 18 size. Same grade as 10. *Patent Reversible Barrel* and *Patent Double Index Regulator*.	125 00	140 00
		Extra on 10, 11, or 12, if adjusted to positions.	50 00	50 00

"J. L. OGDEN."

No. 13	Full Plate.	NICKEL, 18 size, Lever Movement, *St. Line* Escapement, *Exposed pallet jewels*, 4 pair extra jewels, Sprung over Expansion Balance, adjusted to heat and cold. *Patent Reversible Barrel* and *Patent Double Index Regulator*. Sunk Seconds. Named "J. L. Ogden," New York.	70 00	85 00
No. 14	Quarter Plate and Bridge.	NICKEL, 10 (Ladies) and 16 size. Same grade as 13. *Patent Reversible Barrel* and *Patent Double Index Regulator*.	85 00	100 00
No. 15	Three-Quarter Plate.	NICKEL, 14 and 18 size. Same grade as 13. *Patent Reversible Barrel* and *Patent Double Index Regulator*.	75 00	90 00
		Extra on 13, 14, or 15, if adjusted to positions.	42 00	42 00

"E. F. C. YOUNG."

No. 16	Full Plate.	NICKEL, 18 size, Lever Movement, *St. Line* Escapement, Sprung over Expansion Balance, 15 jewels. Sunk Seconds. *Patent Reversible Barrel* and *Patent Double Index Regulator*. Named "E. F. C. Young," New York.	35 00	50 00
No. 17	Quarter Plate and Bridge.	NICKEL, 10 (Ladies) and 16 size. Same grade as 16. *Patent Reversible Barrel* and *Patent Double Index Regulator*.	50 00	65 00
No. 18	Three-Quarter Plate.	NICKEL, 14 and 18 size. Same grade as 16. *Patent Reversible Barrel* and *Patent Double Index Regulator*.	42 50	57 50
		Extra on 16, 17, or 18, if adjusted to heat and cold	6 00	6 00
		Extra on 16, 17, or 18, if adjusted to positions.	30 00	30 00

"D. C. WILCOX."

No. 19	Full Plate.	NICKEL Top Plate, Barrel Bridge and Cock, Lever Movement, 18 size, *St. Line* Escapement, 4 pair extra jewels, Sprung over Expansion Balance. *Patent Reversible Barrel* and *Patent Double Index Regulator*. Sunk Seconds. Named "D. C. Wilcox," New York.	30 00	45 00
No. 20	Quarter Plate and Bridge.	NICKEL, 10 (Ladies) and 16 size. Same grade as 19. *Patent Reversible Barrel* and *Patent Double Index Regulator*.	45 00	60 00
No. 21	Three-Quarter Plate.	NICKEL, 14 and 18 size. Same grade as 19. *Patent Reversible Barrel* and *Patent Double Index Regulator*.	37 50	52 50
		Extra on 19, 20, or 21, if adjusted to heat and cold	6 00	6 00

"HENRY HARPER."

No. 22	Full Plate.	NICKEL COCK, Lever Movement, Frosted, 18 size, *St. Line* Escapement, Sprung over Expansion Balance, 15 jewels. *Patent Reversible Barrel* and *Patent Double Index Regulator*. Sunk Seconds. Named "Henry Harper," New York.	26 50	41 50
No. 23	Quarter Plate and Bridge.	NICKEL, 10 (Ladies) and 16 size. Same grade as 22. Sunk Seconds. *Patent Reversible Barrel* and *Patent Double Index Regulator*.	38 50	53 50
No. 24	Three-Quarter Plate.	NICKEL, 14 and 18 size. Same grade as 22. Sunk Seconds. *Patent Reversible Barrel* and *Patent Double Index Regulator*.	31 50	46 50
		Extra on 22, 23, or 24, if adjusted to heat and cold.	6 00	6 00

"JESSE A. DODD."

No. 25	Full Plate.	NICKEL, Lever Movement, 18 size, *St. Line* Escapement, Sprung over Expansion Balance, 11 jewels. *Patent Reversible Barrel* and *Patent Double Index Regulator*. Named "Jesse A. Dodd," New York.	21 50	29 50
No. 26	Quarter Plate and Bridge.	NICKEL, 10 (Ladies) and 16 size. Same grade as 25 but 15 jewels. *Patent Reversible Barrel* and *Patent Double Index Regulator*.	33 50	46 50
No. 27	Three-Quarter Plate.	NICKEL, 14 and 18 size. Same grade as 25. *Patent Reversible Barrel* and *Patent Double Index Regulator*.	27 50	39 50
		Extra on either of the above, if adjusted to heat and cold.	6 00	6 00

"E. C. HINE."

No. 28	Full Plate.	NICKEL, Lever Movement, Frosted, 18 size, *St. Line* Escapement, Sprung over Chronometer Balance, 11 jewels. *Patent Reversible Barrel*. Named "E. C. Hine," New York.	15 50	23 50
No. 29	Quarter Plate and Bridge.	NICKEL, 10 (Ladies) and 16 size. Same grade as 28 but 15 jewels. *Patent Reversible Barrel*.	26 50	37 50
No. 30	Three-Quarter Plate.	NICKEL, 14 and 18 size. Same grade as 28, 11 jewels. *Patent Reversible Barrel*.	21 50	29 50
		Extra on either of above if with Sunk Second Dial	1 00	1 00

"NEW YORK BELLE."

| No. 31 | Quarter Plate and Bridge. | NICKEL, 10 (Ladies) size, Lever Movement, Frosted, *St. Line* Escapement, 7 jewels. *Patent Reversible Barrel*. Named "New York Belle," New York. | 19 00 | 30 00 |

"THE CHAMPION."

No. 32	Full Plate.	NICKEL, Lever Movement, Frosted, 18 size, *St. Line* Escapement, Sprung over Plain Balance, 7 jewels. Named "The Champion," New York.	9 00	15 00
No. 33		The same, with Chronometer Balance.	10 50	16 50
No. 34	Three-Quarter Plate.	18 size. Same grade as 32	11 75	17 75
No. 35		The same, with Chronometer Balance	13 25	19 25
		Extra on either of the above, if with Sunk Second Dial	1 00	1 00

"BLACK DIAMOND."
(BOY'S WATCH.)

No. 36	Three-Quarter Plate.	NICKEL, 14 size, Lever Movement, *St. Line* Escapement, *Exposed pallet jewels*, Chronometer Balance, 7 jewels. *Patent Reversible Barrel*. Named "Black Diamond," New York.	11 75	19 75
No. 37	Three-Quarter Plate.	Same as 34, but with 11 jewels. *Patent Reversible Barrel*.	13 25	21 25
		Extra on either of above, if with Sunk Second Dial	1 00	1 00

Plain Dials, Plain Seconds.. $2 00
Plain Dial, Sunk Seconds.. 6 00
Double Sunk Dials... 6 30
Masonic and Presentation Dials of any design, to order......... $3 00 to 50 00

All the above Movements have the *Hardened and Tempered Hair Springs*. All except 32 and 33, have **PATENT REVERSIBLE BARREL**, unequaled for simplicity and perfection, it being a preventive damage to train in case of breaking of main spring. All Full Plates have **PATENT DUST BANDS**; all up to and including 27 have **PATENT DOUBLE INDEX REGULATOR**, without extra charge, the same being included in the list price of movements.

All except the 32, 33, 34 and 35 are warranted by special certificate.

Salesroom, No. 13 Maiden Lane.

Chapter 23
Related Watch Companies

Empire City Watch Company

Incorporated on January 5, 1876, this reorganization was the last attempt to salvage the USWC. Unfortunately nothing, certainly not another change of name, could alter the fact that Frederick A. Giles and his associates had run out of time and money.

The "North Star" grade with Empire City dial was a product of the earlier USWC group made for the "low end" market. The fact that it is not included in the Empire City price list gives credence to the theory that this earlier product was the original source for the company name adopted in 1876.

A listing of Empire City grades with their earlier equivalent USWC grades is shown in the table below:

Table VIII
Empire City and Equivalent USWC Grades

Empire City Watch Company	United States Watch Company
W. S. Wyse	A. H. Wallis
I. W. Frost	Henry Randel
Cyrus H. Loutrel	Wm. Alexander
J. L. Ogden	S. M. Beard
E. F. C. Young	John W. Lewis
D. C. Wilcox	George Channing
Henry Harper	Asa Fuller
Jesse A. Dodd	Edwin Rollo
E. C. Hine	J. W. Deacon
New York Belle	A. J. Wood
The Champion	G. A. Read
Black Diamond	Young America

The evidence suggests that naming of Empire City grades was done in terms of assistance expected rather than received. Only one was named after a stockholder, William S. Wyse, an advertising man. Cyrus H. Loutrel, a wholesale stationer, and Jesse A. Dodd, a diamond merchant, were important creditors. Henry Harper, publisher, had been a stockholder of USWC. The name Edward F. C. Young perhaps suggests the degree ethics had declined in Marion; he was cashier of the First National Bank of Jersey City. This institution had the primary voice in determining the negotiability of Giles and Wright's paper.

OPPOSITE PAGE. 219. "Empire City Watch Co." price list, circa 1876. Note two very interesting points: (1) prices for Empire City products were higher than their equivalent USWC grades, and (2) the Empire City model number system is new and very simple when contrasted to the earlier USWC system.

220 and 221. Dial and movement views, "E. F. C. Young," 18s, 15j, Model 16. Patent reversible barrel and double index regulator. Stem wind, lever set. Checker board damaskeening. Contrast this dial to "North Star" dial, Figure 181.

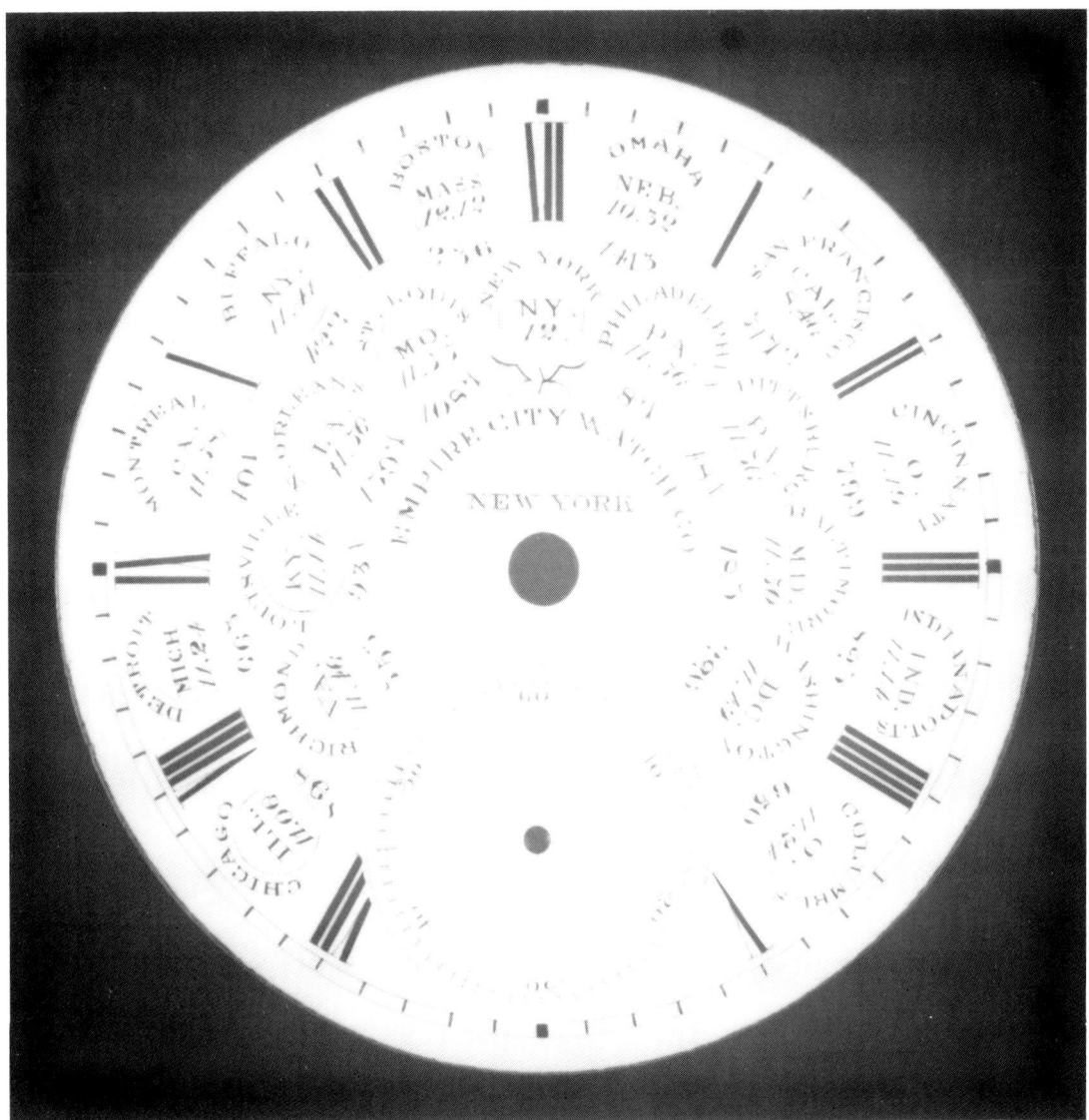

222. 18s, "Empire Combination Timer" dial. Possibly one of Frederick's final attempts to save his company, this interesting concept showed sun time in 18 North American cities when mean noon in New York City plus distances to each city from New York. It was the last fine work of the USWC dial department. Initials of the dial painter, "AM," are on the back.

Empire City products were produced in full plate, three-quarter, and quarter plate styles; the butterfly opening was *eliminated* on full plate models.

The "Empire Combination Timer" is an unlisted Empire City model and may represent one of Frederick Giles' final ideas to salvage his faltering company. Using a conventional 18 size, full plate or three-quarter plate movement, the unconventional dial showed "sun" time in 18 cities plus their distances from New York. Even if the idea had worked, the adoption of "standard" time in 1883 would have eliminated the need for this otherwise unique dial concept.

The "Empire Combination Timer" dial, circa 1876 or 1877, was an interesting revival of the unique and very finely painted dials characteristic of USWC in their earlier years. Only three of these particular Empire City models have turned up to date and all show New York City as the "keystone" with time at 12:00. All three are 18 size and the movements are marked "Empire Combination Timer." (See Figures 222-224 and 247.) These watches apparently were primarily designed to appeal to New York City "traveling men" or railroad personnel who would be especially interested in the times and distances to the various cities shown on the dial. Because only three have surfaced, their market appeal must have been very limited and they probably rank among the scarcest models produced by the USWC. The "Empire Combination Timer" did not save the struggling company, even though the complexity of sun times shown on the dial may have possibly "hastened the adoption" of standard time.

ABOVE. 223 and 224. "Empire Combination Timer," 18s, full plate (SN 11532) and ¾ plate (SN 51026). SN 11532 is gilt, 11j, patent reversible barrel and standard regulator. SN 51026 is nickel, 15j, patent reversible barrel and patent double index regulator. Note wide spread in serial numbers, even though both products are circa 1876 or 1877. Dial from SN 11532 is illustrated in Figure 222. Both dials show exactly the same data in almost identical format. Movements are both key wind and key set.   BELOW. 225 and 226. "D. C. Wilcox," 18s, full plate, SN 76052 and "Jesse A. Dodd," 18s, ¾ plate, SN 74009. The Wilcox full plate is nickel damaskeened, stem wind, 15j, Model 19 on price list. The Dodd ¾ plate is gilt, stem wind, 11j, Model 27. Note the Wilcox is missing its regulator; the Dodd is the patent double index regulator. Both of these movements have the Giles patent reversible barrel. Typical Empire City products.

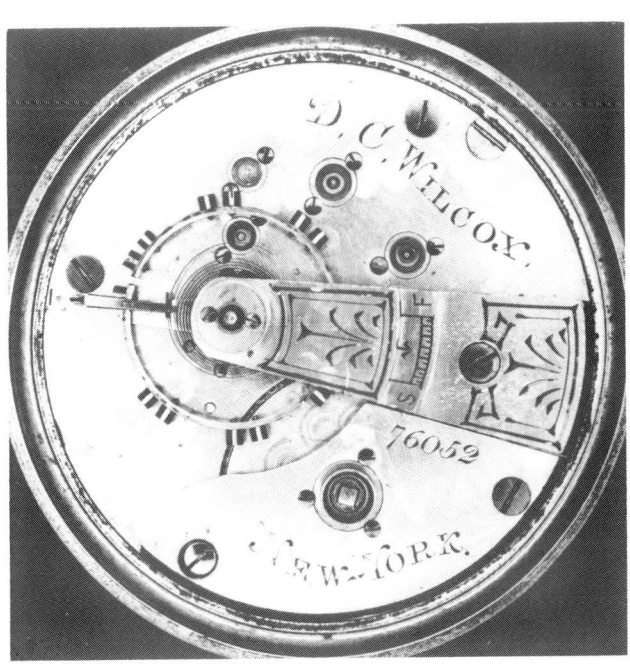

ABOVE. 227 and 228. Dial and movement views, "Royal Gold American Watch," 18s, typical single sunk dial, SN 16052, 11j, gilt, stem wind and lever set. BELOW. 229 and 230. Under dial close-up views, SN 16052, with detail of late setting mechanism. Note evidence of rework on the yoke. Downward pressure on lever at 4 o'clock position disengages winding and engages setting mechanism.

Royal Gold American Watch

Not a company, not a model, the only word that really applies is group. These watches were the result of the previously described "infamous" business arrangement between E. H. Elias and Frederick Giles. They are inscribed "Royal Gold American Watch, New York."

Both the full plate and three-quarter plate movements, inscribed as above, are standard USWC movements. The additional marking on the back plate, "Extra jeweled," is done in the manner of the "Asa Fuller" model, and are, of course, surplus movements sold under the Royal Gold name. Both full and three-quarter plate movements are shown in Figures 227 through 232.

The basic idea of Royal Gold American watches was to provide the cash flow necessary to support production and promotion of Empire City Watch Company watches. Elias et al certainly did not require quality, so the USWC "string saving stock" of reject and imperfect movements provided the perfect pool for rework and addition of Royal Gold American dials and engraving. At the same time Empire City watches were primarily drawn from USWC stock of unsold, or partially complete, good movements then finished out with Empire City Watch Company dials and engraving. It can be seen that this final production phase of USWC was largely a rework of existing stock. Based on surviving examples noted to date, there were relatively more

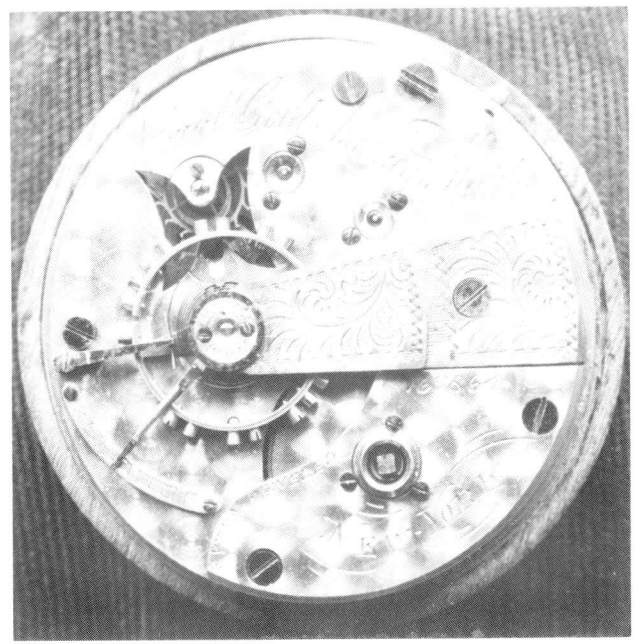

231 and 232. "Royal Gold American Watch," 18s, full plate, SN 169864, in unusual gilt damaskeened finish and 14s, ¾ plate, SN 260187, in typical gilt finish. The wide serial number spread in these previous three examples prove random selection from surplus stock.

233 and 234. Howard brothers products, "Independent," 18s, SN 191053, nickel damaskeened, unusual straight line pattern across plates and balance cock with distorted butterfly cut out; "Little Jewel," 10s, SN 195249, ¼ plate and bridge, nicely gilt damaskeened.

Royal Gold watches completed than Empire City models. Neither were produced in large quantities, pointing to a short duration for this phase of USWC operations. The scattered serial number sequence of these watches makes precise production estimates extremely difficult.

Howard Brothers

The Howard Brothers of Fredonia, New York, were buyers of USWC movements. They were established businessmen in the patent medicine field, selling their wares via mail order. In the early 1870s, they began selling watches, also via mail order. They originally purchased Swiss movements, upon which they engraved their own names, and sold them as if they were the manufacturers. The venture proved a great success and they switched from the imported Swiss movements to American-made movements. The USWC, among others (Hampden, Illinois, Cornell), supposedly sold hundreds of movements to the Howards, which again, they cased and sold under their own name. These USWC-made

235. Howard Brothers "Independent," SN 191099, 18s, very close serial number to previous 18s example with the same distorted butterfly cut out, but with different damaskeening wave pattern running across entire plates and balance cock. Friction set jewels. These products were nicely finished, yet relatively inexpensive.

movements usually carried the name of "E. D. Howard" or "Independent Watch Co."

Upon closing of the factory, the Howards purchased large quantities of unfinished movements, materials, and some machinery. Most of these were 18 size full plate with butterfly opening. In some instances, this opening was then cut out in different forms to eliminate the butterfly "trade-mark." The Howards were unable to sell these hybrid watches in large quantities and were forced to reorganize as the Fredonia Watch Company. The new Fredonia watch was a cross between the best of the Newark-Cornell-California watch and the best of the Marion full plate products. Although the design was a bit dated, it was in many respects a superior and sturdy beast, but alas, it did not sell well either. (More data on the Howard ventures is found in the Muir article, "Peoria and Non-Magnetic," NAWCC BULLETIN #178, October, 1975, pp. 490-491.)

Chapter 24

Production Estimates

An effort was made to reconstruct the production schedule of the USWC. There appeared to be a reasonable chance of at least a framework for such a schedule with the information available at this time.

Several "hard" inputs were considered and evaluated, the most important one being movement serial numbers, which are the primary basis for statistically reliable production estimates.

Two other inputs were considered in conjunction with the serial numbers, those being known patent dates of inventions and known release dates of watch models to the general public. Two "soft" inputs were also considered; the information that has been passed on to us during the past hundred years, plus some good, cold, "guesstimates," if there is such a thing. Thus, it was felt that these five inputs, when played one against the other, could result in a "first draft" production schedule of the USWC.

Movement serial numbers are the best source of data, and many times, the only source of information needed if available in quantity. However, in the case at hand, approximately 700 movements have been recorded out of a possible total estimated production run of some 50,000 for the period 1867-1874, a relatively small sized sample of less than 2%. Suffice to say that sufficient numerical sequences are noted in various blocks that strongly suggest several different production runs on the more popular models.

In arriving at preliminary production estimates, it should be noted that quantities are likely to be on the *high* side for the more expensive, nickel grades, and *low* for the cheaper gilt grades. This is the result of a somewhat biased random sample because of (1) the fact that over the years, higher quality movements tend to survive longer than the lower quality items, and (2) higher quality items tend to be reported more readily than lower quality items, i.e. one tends to be prouder of a "United States" or "A. H. Wallis" than of a "Read" or "Rollo."

In the reporting of serial numbers three *errors* are worthy of note. The first might be an incomplete description and the model recorded as a "United States" model when it is actually some other model. The second error concerns movement style/design. . . . USWC quarter plate and bridge are sometimes called three quarter or split three quarter plate by those not familiar with the differences. Third, Frederick Giles labeled his ¼ plate and bridge as 10 size and 16 size. The labeled 16 size actually measures a good 17 size and so sometimes are called 18 size. Wherever possible, these errors have been eliminated from the survey.

Based on the numbers available at this time, it appears that as a particular model or movement size was about to be put into production, a block of numbers was assigned to that product. This is especially true of the earlier production runs. Later evidence shows that numbers within a block were not used consecutively for a given model, but that several different named models could share the same block of numbers. Additionally, serial numbers were allocated out of sequence. That is, some of the later (1873-1876) production runs were allocated low serial numbers and thus, there is no production/serial number relationship.

Only two known release dates of watches for retail sale are available. The "Atherton" model was released in June 1867 and the "United States" model in February 1869. These dates, in conjunction with known serial numbers, aided in the placement of other watch models, both between and after the known release dates.

Patent dates were also useful. Since it was then required that these dates appear on the patented device, generally they provide some notion as to the earliest reasonable date of production. Unfortunately, this information must be used with some care. One must remember that parts — even plates and bridges — have been, and will continue to be, switched from watch to watch. The only real test is a complete check of the individual parts; even then, part numbers can be altered. In this respect, the Marion part numbering stamps have been to some extent identified and could be in time an important aspect in cataloging the company's watches. The USWC is known to have used at least two sizes of numbering punches. Roughly speaking the earlier was a large sized set which wore out and was replaced by a small sized set. The large numbers are particularly useful since some numbers lost segments, but were apparently not replaced until the entire set became useless.

The Giles patent reversible barrel presents a unique challenge. Generally speaking Marion-built watches fall into three groups: (1) pre-reversible barrel, (2) reversible barrel marked "patent applied for," and (3) reversible barrel marked "Pat. Dec. 30, 1873." The patent applied for version was introduced late in 1872, but how long this imprint appeared on barrels after introduction of the patent Dec. 30, 1873, version is subject to question. Barrel switching plus the large stock of USWC parts maintained for many years by supply houses after the factory's demise, complicate matters further. Even so, a large number of well-documented examples does help to reduce these questions to manageable proportions.

Marion-built watches with a patent date of March 8, 1870, on the balance cock refer to the Julius Elson

185

double index regulator introduced in 1872. Stem-winding yokes are normally marked with both the Giles August 15, 1865, patent and the December 22, 1868, improvement dates; very early yokes will have only the former mark.

An in-depth study of some 700 serial numbers suggests an allocation system of serial numbers to watch grades, at least to the extent that blocks of numbers were initially assigned to specific models. However, available evidence further suggests that, in most cases, a particular model did not use all of the numbers in the pre-assigned block, and at a later date, other models, special order watches, and those of Empire City and Royal Gold among others, were interspersed in these blocks in an apparent haphazard sequence. This lack of serial number sequence confirms that these later products were largely taken from existing USWC stock and completed with Empire City and Royal Gold engraving and dials.

A special effort was made in this preliminary serial number allocation to identify 19-jewel and 15-jewel versions of the "United States" grades. These same type of jewel count breakdowns are needed for the "Frederic Atherton" and "A. H. Wallis" grades as well as other models where the jewel count was lowered over their USWC lifetime. Future work is also needed to more accurately define other feature breakdowns for the various USWC models. Enough data was gathered in the present survey to develop the following table with generalizations concerning USWC movements:

Table IX
Relative USWC Feature Scarcity

Feature	Common	Scarcer	Scarce
Size	18s	10s-14s	16s
Style	Full Plate	¾ Plate	¼ Plate
Jewels	Less than 15J	15J-17J	19J
Settings	Friction-Brass	Nickel	Gold
Finish	Gilt	Gilt & Nickel	Nickel
Damaskeening	Nickel	Frosted	Gilt
Winding	Key	Stem, '68 Imp.	Stem, '65 Patent
Setting	Key	Lever	Button
Hairspring	Flat	—	Breguet
Steel Work	Plain	Blued	Gilded
Engraving	Plain	—	Enameled
Patents	Giles Butterfly	Giles Barrel	Elson Regulator
Dial	Single Sunk	Double Sunk	Special Design

Listed on the following pages are two numerical break-downs, based on movement serial numbers, showing estimated production quantities for all USWC models. The first (Table X) shows a summary of model names, lowest and highest serial numbers recorded and estimated production *ranked* by relative scarcity. The second (Table XI) amplifies the information and provides more detail.

Needless to say, any or all of this information is subject to further study and evaluation as additional information becomes available. At best, it is offered as a preliminary guide in an area where little or no information has heretofore been published.

Further research and evaluation by those who follow will surely expand these allocations and add further insight to the master plan as devised by the USWC.

Significant gaps will be noted between various blocks in the following tables. This is due to little or no information, mostly the latter; thus, the possibility exists that a significant number of serial numbers were never allocated for use and that the total production of the USWC falls far short of the oft mentioned figure of 300,000. Assuming full production within the serial number blocks indicated by the survey, then total production can be estimated at 56,524.

236. "Asa Fuller" SN 268029, a 10s Model 44¼ in ¼ plate and bridge movement style, key wind and set. More confusion has surrounded this movement style than any other. All USWC grades in this ¼ plate and bridge style are relatively scarce because of many production delays in getting them on the market before hard times hit with the Panic of 1873. Furthermore, while this 10s example is true in size, the 16s actually measures a good 17s. Finally, many people incorrectly call this movement style a ¾ or split ¾. On this particular example, note "patent applied for" designation on the barrel. This suggests that this late serial number was finished prior to the December 30, 1873, patent date. And, this is supported by the relatively nice finish and extra floral trim engraving on the plates plus a USWC logo. The ¾ plate "Asa Fuller" SN 71270 shown in Figure 171 has the December 30, 1873, patent date on barrel illustrating the problem with establishing an exact serial number relationship to a chronological dating.

In general, the greater the sample percentage (number surveyed) within each stratum (26 model/grade groups), the more reliable the total production estimate is for that particular group. Overall, the combined strata provide a relatively small sample of less than 2% (680 out of 56,524) for the grand total production estimate.

Furthermore, the grand total production estimate must be considered as a maximum estimate because the projection is based on full production within serial number blocks established by the sample. While this is highly unlikely, an offsetting factor (correcting error) is the discovery of new serial numbers in the future which could establish new serial number blocks.

It is significant to note that estimates for the Centennial Commemoratives, Empire City, and Royal Gold are extremely difficult because of the random, scattered sequence of their serial numbers. At best, these can be considered a judgement call.

An effort was made in this preliminary study to break down the "United States Watch Co." grades into their 19- and 15-jewel variations. With over 50 different variations of size, winding, and jewel types of this prestige grade possibly available, the surviving sample definitely indicates that many of these were never made. And, some USWC historians believe that when the 15-jewel models were introduced early in 1873, the 19-jewel models were no longer available or in sufficient demand. This is especially true of the 10 size and 16 size ¼ plate movement style, because of the delays in getting them to the market prior to the late fall, 1873 Panic which "crippled" the USWC.

Table X
Model/Grade Groups Ranked by Relative Production Scarcity

Rank	Model/Grade Groups	Number Surveyed	Lowest and Highest Serial Numbers	Estimated Total Production
1	Centennial Commemoratives	6	1004-175533	100
2	A. J. Wood	3	241012-263079	150
3	United States Watch Co. (19J)	15	12003-206330	200
4	Railroad Special Order	10	21515-142375	200
5	Newspaper Special Order	11	13291-158840	200
6	Young America	3	60717-60872	200
7	United States Watch Co. (15J)	11	19923-235003	275
8	R. F. Pratt	7	50100-50574	325
9	I. H. Wright	10	7733-268070	400
10	Henry Randel	13	9862-225039	600
11	Marion Watch Co.	9	7013-170019	600
12	Chas. G. Knapp	15	50420-60574	625
13	Royal Gold American	14	0924-260187	700
14	S. M. Beard	16	21023-22261	750
15	North Star/Empire City Watch	10	0204-263374	794
16	Wm. Alexander	15	20019-248704	1,040
17	A. H. Wallis	22	19034-255201	1,410
18	Fayette Stratton	39	2057-162275	1,500
19	John W. Lewis	16	23321-200170	1,550
20	Frederic Atherton & Co.	56	1006-18420	1,700
21	Others	70	—	4,000
22	George Channing	39	3026-286088	4,400
23	J. W. Deacon	36	45056-284037	6,575
24	Asa Fuller	53	23898-80466	8,050
25	Edwin Rollo	88	4005-286366	8,575
26	G. A. Read	93	10256-282800	11,755
	Totals	680	0204-286366	56,524

Note: It is important to review the qualifications for these estimates on pages 185, 186, and the top of this page. Additional data on each of the 26 model/grade groups are in Table XI, pages 188-192.

Table XI
Model/Grade Groups with Serial Number Allocations

Model/Grade Group	Total Number Surveyed	Total Estimated Production	Estimated Allocation	Estimated Production	Number Surveyed	Notes
Frederic Atherton & Co.	56	1,700	1001-2000	1000	50	18s, 17J-21J, both KW and SW models, "Giles Patent, 13 March 1866"
			9101-9300	200	2	18s, 17J-19J, KW
			10501-10600	100	2	18s, SW, Giles 1866 patent
			18051-18450	400	2	18s, 19J, KW, Giles 1866 patent
Fayette Stratton	39	1,500	2001-3000	1000	32	18s, 15J-17J, both KW and SW models, Giles 1866 patent
			10401-10450	50	1	18s, 17J, Giles 1866 patent (with butterfly)
			10651-11000	350	3	18s, 11J-19J, Giles 1866 patent
			21751-21800	50	1	
			162251-162300	50	2	Giles 1866 Patent
George Channing	39	4,400	3001-4000	1000	10	18s, 15J, KW, Giles 1866 patent
			14601-16700	2100	13	18s, 15J, KW, Giles 1866 patent
			160251-161250	1000	8	18s, 15J, KW, Giles 1866 patent, nickel movement
			161401-161500	100	2	18s, 15J, Giles 1866 patent, nickel movement
			161551-161600	50	1	
			161751-161800	50	1	
			286001-286100	100	4	18s, 15J-17J, KW, ¾ plate, nickel movement
Edwin Rollo	88	8,575	4001-6900	2900	33	18s, 11J-15J, KW, Giles 1866 patent
			11051-11100	50	1	18s, Giles 1866 patent
			30001-32400	2400	14	18s, 15J, KW, Giles 1866 patent
			50451-50475	25	1	18s, Elson Regulator
			65001-65300	300	2	
			110051-110150	100	2	18s, 15J
			110551-110600	50	1	18s, 15J, KW, Giles 1866 patent
			110801-110850	50	2	18s, 15J, KW, Giles 1866 patent
			140101-140850	750	10	18s, 15J, KW, Giles 1866 patent
			141701-143000	1300	10	18s, 11J-15J, KW, Giles 1866 patent
			147251-147300	50	1	18s, 15J, Giles 1866 patent
			228001-228050	50	3	16s, 15J, KW, ¼ plate and bridge
			238201-238850	150	2	14s, 15J
			270001-270050	50	1	10s, 15J, KW
			285101-285400	300	4	18s, 15J, KW, ¾ plate
			286351-286400	50	1	18s, 15J, ¾ plate
Marion Watch Co.	9	600	7001-7050	50	1	18s, Giles 1866 patent
			7301-7650	350	2	18s, 11J-15J, KW
			8401-8450	50	3	18s, 15J
			51601-51650	50	1	

Maker	Total	Serial Numbers	Qty	Count	Description
United States Watch Co. (19J)	15	77001-77050	50	1	18s, 15J, ¾ plate
		170001-170050	50	1	18s, 19J, KW, Giles 1866 patent
	200	12001-12025	25	4	18s, 19J, gilt, full plate with butterfly
		24001-24100	100	7	18s, 19J, nickel, full plate with butterfly
		25501-25525	25	1	18s, 19J, full plate with butterfly, nickel, KW/set only
		54401-54425	25	2	18s, 19J, nickel frosted, ¾ plate with Elson regulator — some are 15J
		206326-206350	25	1	18s, 19J, nickel, full plate with butterfly — some are 15J
United States Watch Co. (15J)	11	19901-19925	25	1	18s, 15J, nickel frosted, full plate with butterfly
		50051-50075	25	1	18s, 15J, nickel, ¾ plate with Elson regulator
		53526-53550	25	1	18s, 15J, nickel, ¾ plate with Elson regulator
	275	76501-76550	50	2	18s, 15J, gilt ¾ plate with Elson regulator
		202601-202625	25	1	18s, 15J, full plate, no butterfly, nickel
		206101-206125	25	1	18s, 15J, full plate
		206201-206250	50	2	18s, 15J, nickel frosted, full plate
		225301-225325	25	1	18s, 15J, nickel, ¾ plate with Elson regulator
		235001-235025	25	1	16s, 15J, ¼ plate and bridge, nickel
A. H. Wallis	22	19001-19800	800	13	18s, 15J, 17J-19J, nickel movement, Giles 1866 patent
		20541-20600	60	2	18s, 15J, nickel movement, KW/SW
	1,410	24951-25000	50	1	18s, 17J, nickel movement
		25051-25100	50	1	18s, 17J, nickel movement
		203001-203300	300	2	18s, 15J, nickel movement
		205201-205250	50	1	18s, nickel movement
		255110-255150	50	1	
		255201-255250	50	1	18s, 15J
Wm. Alexander	15	20001-20300	300	8	18s, 15J, Giles 1866 patent
		21661-21700	40	2	18s, 15J
	1,040	22151-22200	50	1	18s, 15J, nickel movement, Giles 1866 patent
		22651-23200	550	2	18s, 15J, nickel movement
		200001-200050	50	1	18s, nickel movement
		248701-248750	50	1	10s, 15J, KW, nickel movement
S. M. Beard	16	21001-21650	650	13	18s, 15J, both KW and SW models, nickel movement, Giles 1866 patent
	750	22101-22150	50	2	18s, 15J, Giles 1866 patent
		22251-22300	50	1	18s, 15J, nickel movement, KW, Giles 1866 patent
Henry Randel	13	9851-9900	50	1	
	600	20501-20540	40	1	18s, 15J, KW, nickel movement, Giles 1866 patent

Table XI (Continued)

Model/Grade Group	Total Number Surveyed	Total Estimated Production	Estimated Allocation	Estimated Production	Number Surveyed	Notes
Henry Randel (continued)			21651-21660	10	1	18s, 15J
			22551-22650	100	2	18s, 15J, nickel movement
			24501-24700	200	2	18s, Giles 1866 patent
			25001-25050	50	2	18s, 15J-17J, Giles 1866 patent
			52001-52050	50	1	18s, 15J, ¾ plate, KW, nickel
			205151-205200	50	2	18s, 17J, nickel movement
			225001-225050	50	1	18s, 15J, KW, nickel movement
John W. Lewis	16	1,550	23301-23350	50	1	18s, 15J, nickel movement, KW
			26001-27350	1350	12	18s, 15J, nickel movement, KW, Giles 1866 patent
			29001-29050	50	1	18s, 15J
			200101-200200	100	2	18s, Giles 1866 patent
Asa Fuller	53	8,050	23851-23900	50	1	18s, KW
			33401-38550	5150	31	18s, 11J-15J, KW, Giles 1866 patent, gilt movement with nickel bridge
			38951-41000	2050	10	18s, 11J, KW, Giles 1866 patent, gilt movement with nickel bridge
			45001-45050	50	1	18s, 7J, Giles 1866 patent
			64001-64200	200	4	14s, 15J, ¾ plate
			71251-71300	50	2	18s, 15J, ¾ plate
			80001-80500	500	4	18s, 15J, Giles 1866 patent
R. F. Pratt	7	325	50100-50300	200	4	10s, 15J, KW
			50476-50600	125	3	10s, 15J, KW
Chas. G. Knapp	15	625	50401-50425	25	1	10s
			60001-60600	600	14	10s, 15J, KW
Young America	3	200	60701-60900	200	3	14s, 7J, KW, ¾ plate
J. W. Deacon	36	6,575	45051-45100	50	1	14s, 15J, ¾ plate
			62151-62200	50	1	¾ plate
			73051-73100	50	1	18s, 11J-15J, KW, Giles 1866 patent
			90001-92000	2000	8	18s, 11J-13J, KW, Giles 1866 patent
			93551-94150	600	4	18s, 13J-15J, KW
			95001-96200	1200	4	18s, 15J, KW
			130251-131275	1025	3	18s, 11J, KW, Giles 1866 patent
			132751-132800	50	1	18s, 15J, KW, Giles 1866 patent
			133451-134150	700	8	18s, 11J, KW, Giles 1866 patent
			282201-282250	50	1	18s, 11J, ¾ plate
			283251-284050	800	4	18s, 15J, KW, ¾ plate
G. A. Read	93	11,755	10251-10300	50	1	

		11951-12000	50	1	18s, Giles 1866 patent
		29251-29300	50	1	7J, ¾ plate, KW
		72051-72100	50	1	18s, 7J, ¾ plate
		78001-79500	1500	7	18s, 7J, ¾ plate, Elson regulator
		79945-80000	55	2	18s, 7J, KW, ¾ plate
		100151-100300	150	2	18s, 7J, KW, Giles 1866 patent
		102151-102200	50	1	18s, 7J
		102501-102550	50	2	18s, KW, Giles 1866 patent
		103251-103350	100	3	18s, 7J, KW
		105551-105800	250	2	18s, 7J
		106501-106700	200	3	18s, KW, Giles 1866 patent
		111001-111050	50	3	18s, 7J, Giles 1866 patent
		111801-111850	50	2	18s, KW, Giles 1866 patent
		112751-112800	50	2	18s, KW, Giles 1866 patent
		113001-113750	750	4	18s, 7J, KW, Giles 1866 patent
		114351-114600	250	3	18s, KW, Giles 1866 patent
		115451-115500	50	1	
		120001-120550	550	2	18s, 7J, KW, Giles 1866 patent
		121651-122800	1150	3	18s, 7J, KW, Giles 1866 patent
		124651-124800	150	4	18s, 7J, KW, Giles 1866 patent
		125251-125900	650	3	18s, KW, Giles 1866 patent
		127051-127700	650	2	18s, 7J, KW
		128551-129000	450	3	18s, KW, Giles 1866 patent
		129801-129850	50	1	
		150351-150400	50	3	18s, KW, Giles 1866 patent
		153201-153550	350	3	
		155451-155700	250	5	18s, 7J, KW
		157401-157850	450	2	18s, 7J, KW
		158001-158900	900	1	18s, 7J, KW, Giles 1866 patent
		159801-159850	50	3	18s, 7J
		165401-165950	550	5	18s, KW, Giles 1866 patent
		166101-166950	850	6	18s, Giles 1866 patent
		167351-168000	650	1	
		169251-169300	50	2	18s, KW, ¾ plate
		282101-282200	100	2	18s, 7J, ¾ plate
		282701-282800	100		
I. H. Wright	10	7701-7800	100	3	18s, 15J
		8151-8200	50	1	18s, 11J, KW
		10151-10200	50	1	18s, 11J, KW
		70001-70050	50	1	10s, 19J, nickel movement, ¼ plate
		162401-162500	100	3	18s, 15J-19J, KW
		268051-268100	50	1	
A. J. Wood	3	241001-241050	50	1	10s, 15J, ¼ plate and bridge, special order "Fisher & Thatcher"
		263001-263100	100	2	10s, 15J, ¼ plate and bridge
North Star/Empire City	10	201-900	700	8	18s, 7J, KW, Empire City Watch Co. on dial

191

Table XI (Continued)

Model/Grade Group	Total Number Surveyed	Total Estimated Production	Estimated Allocation	Estimated Production	Number Surveyed	Notes
North Star/Empire City (continued)			79901-79944	44	1	18s, 7J, KW, Empire City Watch Co. on dial, Model marked "The Champion"
			263351-263400	50	1	10s, 15J, KW, Empire City Watch Co., Model marked "New York Belle"
Royal Gold	14	700	0901-1000	100	2	14s, 15J, Royal Gold American Watch
			7201-7250	50	1	18s, 17J
			11451-11500	50	1	18s, 11J, Royal Gold American Watch
			28101-28300	200	2	14s-18s, 15J on former
			38901-38950	50	1	18s, 15J, KW, ¾ plate, Royal Gold American Watch
			161501-161550	50	1	18s, 15J, KW, Royal Gold American Watch
			169851-169900	50	2	18s, KW, Giles 1866 patent, Royal Gold American Watch
			253051-253100	50	1	14s, 15J, KW, ¾ plate, Royal Gold American Watch
			260101-260200	100	3	14s, 15J, ¾ plate, Royal Gold American Watch
Centennial Commemoratives	6	100	Numbers assigned out of sequence — see pages 173-177 for further details.			
Railroad Special Order	10	200	Four railroads noted to date — see pages 171-172 for further details.			
Newspaper Special Order	11	200	Five newspapers noted to date — see page 171 for further details.			
Jewelers Special Order	30	2,000	See pages 169-171 for further details.			
No serial number or model name	40	2,000	Reported with incomplete information.			
Totals	680	56,524				

Notes: (1) Since the time data was accumulated in this survey and these tables were prepared, other USWC examples have surfaced (some as illustrations in this work) with serial numbers indicating the need for additional serial number blocks (production runs). This confirms the offsetting factor (correcting error) comments on page 187.

(2) The NAWCC Research Committee continues to build up data on USWC examples and will appreciate any significant information, descriptions, and serial numbers not indicated by the present survey. Please send your data to NAWCC Research Committee, c/o NAWCC, Inc., 514 Poplar St., Columbia, PA 17512.

Chapter 25
USWC Miscellaneous Data

13 Maiden Lane

A first-hand account (ca. 1871) of a visit to Giles, Wales & Co. wholesale showrooms at 13 Maiden Lane, New York City follows:

"This repository is worthy a visit, and no one that is in the business coming to this city, who contemplates purchasing, should fail to see the Aladdin-like wonders kept in store. Novelties of every description in the watch trade greet the eye, some of which are of exceeding great value. Here are watches richly embellished with various kinds of valuable stones, some whose cases are fairly encrusted with diamonds, and bearing in design and color, birds and flowers, as like as if they had been placed there by the pencil of the artist. Then we see watches, encased in plain solid gold, some of which are marvels in point of construction, possessing every device to make them useful as well as ornamental. A few points of one watch we saw were admirable, viz: the watch is wound without a key (a stem-winder, which, by the way, is a specialty of this house, and for which several patents are held; they are the most simple and the only practical ones yet produced, and eventually will be the only style of watch that will have favor with the public); the hands are set also from the stem; it strikes the hours, loud, distinct, and musically sweet; after striking the hour it strikes the quarters and the minutes, so that in the dark, its possessor can know the hour and the minute. Other watches in cases, in various forms of butterflies, etc., the works of which are not larger than a one cent piece; while in other materials, such as silver-plated ware, etc., this house stands unrivaled." (Source: September, 1871, *The Watchmaker and Jeweler*.)

The major retail outlet in New York City at this time was the C. A. Stevens Company, located at 40 East 14th Street.

Employees

It is interesting to note that H. J. Lowe was one of the original employees, initially as foreman of the Finishing Room. Another original employee, George Hart, who was foreman of the Plate Room, was made Master Mechanic at the same time that Mr. Lowe became Superintendent.

	Superintendent	*Master Mechanic*
1865-67	James H. Gerry	James H. Gerry
1867-69	William H. Learned	
1869-74	H. J. Lowe	George Hart

It is highly possible that other persons served in these two positions for short periods of time, however, they were not publicly announced as officially filling these positions. For example, a news release of December 1871 states: "Five hundred skilled workmen are constantly employed in the works, under the daily supervision of F. A. Giles, the President and founder of the Company."

Employee figures were equally difficult to obtain. Following are inputs from various sources:

Table XII
Employees and Production Capacities

	Reported Number of Employees	Reported Capacity — Number of Movements Produced per Day
October 1863	15	—
July 1867	50	25
December 1869	125	100
July 1870	315	150
September 1871	500	200
1874	600	250

Foremen and Employees

Bardeen, A. R. — Employee
Berlin, Charles — Damaskeening
Cain, H. J. — Balance Making
Carpenter, L. H. — Dials
Crowell, Gilbert — Foreman
Dodge, Charles — Employee
Elves, W. H. — Jewel Finishing
Farnsworth, Walter — Damaskeening
Gardiner, John — Pinion Finishing
Gerry, D. B. — Stem-Wind Parts
Gerry, E. S. — Escapement Making
Gerry, James H. — 1st Superintendent
Guest, William — Employee
Hart, George — Plate Room, Master Mechanic
Hull, Edgar L. — Dials
Keegan, F. W. — Employee
Learned, William H. — 2nd Superintendent
Leman, Frank — Machinist
Logan, John — Hairsprings
Lowe, H. J. — Finishing Room, 3rd Superintendent
Lowell, Frederick — Motion Work
Murray, Leonidas — Pinion Making
Parker, Thomas — Employee
Pray, John — Carpenter
Rose, Henry — Dials
Rose, William — Dials
Sanborn, Jacob — Engraver
Sheppard, William — Flat Steel Work
Smith, William — Jewel Finishing
Vanderhoff, C. — Employee
Wheeler, P. H. — Employee
Whitehouse, Charles — Employee

USWC Miscellanea

LEFT. 237. What time is it? 8:17? Look again . . . it's actually 3:43 on this watch. A mirror would help one read the time, as everything about this watch is backwards! It has a mirror image dial and the watch has been made to run backwards. The dial is a treated paper type, very similar to the type used on early Waterbury and Ingersoll watches. Originally illustrated in the William E. Drost 1966 book on "Clocks and Watches of New Jersey," it was thought to have been made up as a salesman's display piece. It could have also been a "point-of-purchase" display item used in conjunction with a mirror as a traffic stopper. Whatever its purpose or origin, it qualifies for a piece of USWC miscellanea. The movement is a standard "G. A. Read" Model 46, key wind and set, 18 size.

BELOW. 238. Apparently the USWC and Frederick Giles did not want the overseas makers to be the only ones to use "grotesque" decorations on the balance cock. This smiling face variety is on "Edwin Rollo" SN 5203, an 18s, 15j, gilt Model 40. The jewels almost appear larger on this watch, and possibly this type of finish might have been for the overseas markets distributed through the agency offices in Chaux-de-Fonds.

BELOW. 239. What happened to the gold settings on the escape wheel and pallet arbor jewels on this "Wm. Alexander" SN 22152? This grade normally has them. An example of a USWC variation. The engraving on this frosted cock is thought by some to be for a monogram or initials. Too small? Remember the USWC was a company that thought in terms of the Lord's Prayer in half the area of a seconds bit!

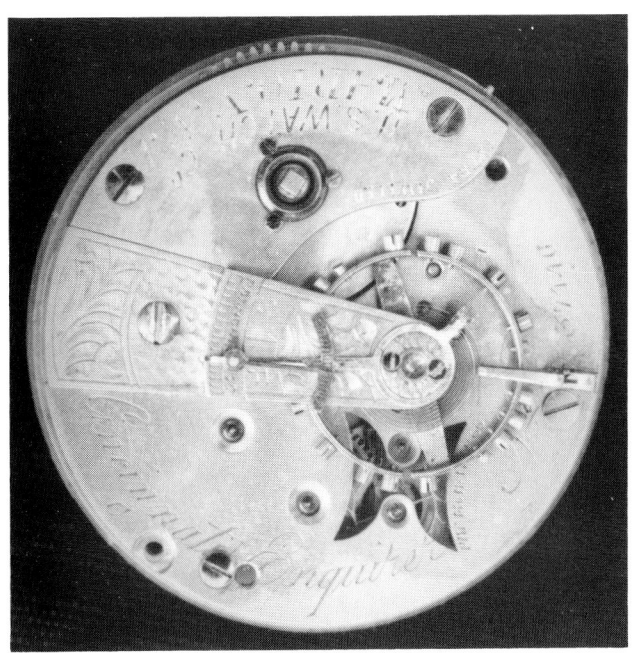 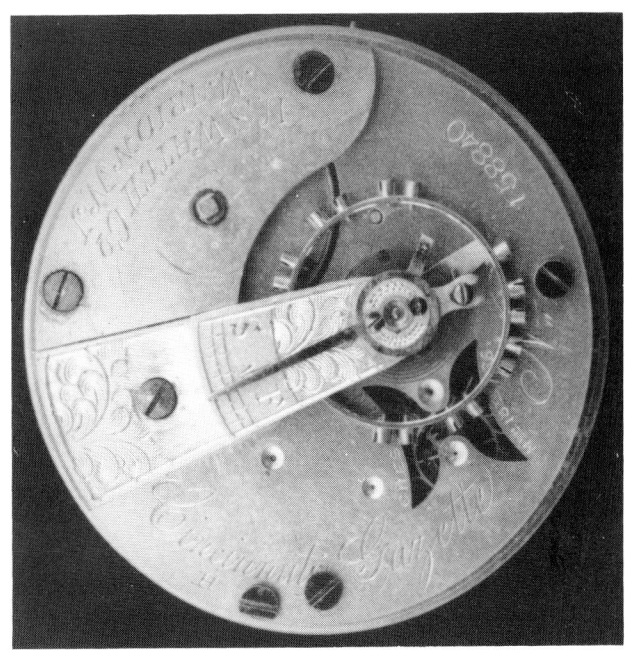

ABOVE. 240 and 241. Cincinnati 18s newspaper twosome, the "Enquirer," SN 37146 and the "Gazette," SN 158840. The "Enquirer" is a basic 11j "Asa Fuller" with nickel barrel bridge and Elson regulator (see Figures 107 and 170). The "Gazette" is a 7j "G. A. Read" and carries the letter code H, while the "Enquirer" has the letter code C. BELOW. 242 and 243. If imitation is the sincerest form of flattery, then this example speaks well for the USWC. Dial and movement views, 18s, "United States Watch Co.," SN 90799, a nice example of the so-called "Swiss fake." Other fakes have turned up signed "U.S. Watch Co., New York" and "Marvin Watch Co." in Old English script that resembles "Marion." These surviving examples explain the warning that appeared in most USWC promotional material . . . "Beware of worthless imitations with which the country is flooded." Fakes turned up to date are all key wind and key set.

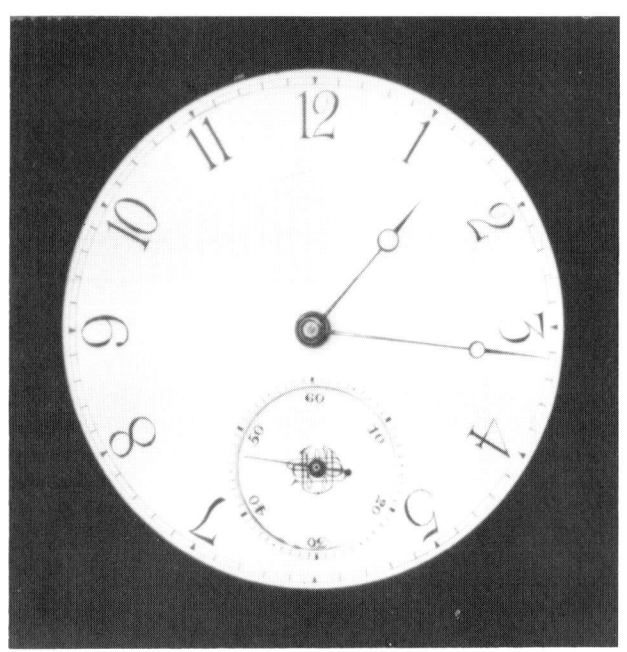

244. Another unusual USWC dial, this 10s single sunk version has Arabic numerals in a Tiffany style chapter ring and USWC monogram in seconds bit. A 16s similar dial has been recorded; both are on ¼ plate movements and represent very early Arabic numeral usage by an American watch company.

246. Interesting late "United States" grade, 18s, 15j, SN 206329, nickel damaskeened with flat, nickel plated settings. Typical flat hairspring, but unusual balance cock almost appears to be "Frederic Atherton" stock with added nickel plating and enameling. Next highest serial number, 206330, is reported to be 19j.

245. First run, first serial number "Edwin Rollo" grade, SN 4001, 18s, 15j, Model 40 in key wind, key set style. A very nice example with another probable first not quite as obvious . . . USWC usage of a simple, on the balance cock regulator style.

247. Another "Empire Combination Timer," 18s, 15j, SN 11905, gilt with nickel cock and Elson regulator. Same dial as the two other known examples, but different style movement identified as the Model 22 Empire City "Henry Harper" grade (see Figures 222-224).

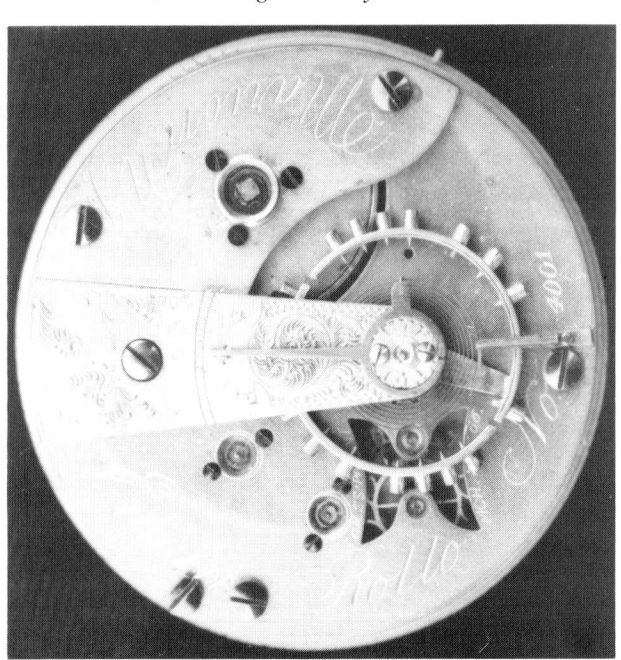

248. Late "United States" grade, 18s, 15j, SN 206292, nickel damaskeened and frosted, flat hairspring, illustrating skimpy balance cock and letter engraving style typical of many full plate USWC products in the post-Panic years. Note the contrast of this balance cock to the one illustrated in Figure 246.

250. Double sunk early 18s dial with script lettering, a variation seldom seen on USWC watches. Many American watch historians believe that nicer USWC dials for their time were without peers in the industry. USWC was among the earliest to establish a fine reputation for double sunk and special designs.

249. "A. H. Wallis" grade, 18s, 15j, SN 203474, nickel damaskeened and frosted. Above average finish for this relatively late USWC product — even the jewel settings appear to be gold, but still a poor contrast to the earlier pre-Panic 19j versions.

251. Inside rear cover of USWC marked, very heavy 18s coin silver hunting case, SN 29097, plain polish with ribbed edge. Several USWC marked cases have been recorded with case serial numbers in the 20,000 to 30,000 range probably from the same manufacturer.

ABOVE. 252 and 253. Contrast of 18s full plate, SN 0422 to 14s ¾ plate, SN 0827 "North Star" 7j grades. The 18s is key wind and set and the 14s is stem wind and key set. These "low end" products give testament to the diversity and broad scale of the USWC line. BELOW. 254 and 255. Triple variation 18s ¾ plate "A. H. Wallis," SN 54409, stem wind and lever (side) set, circa late 1872 with (1) "Empire Combination Timer" dial circa 1876-77 (see Figure 222), (2) 19j count on a Model 53½ listed as 17j, and (3) a serial number that puts it in a run of "United States" 19j and 15j models (see Figures 77 and 83). Dials are easily changed from movement to movement and jewel count variations have been recorded before. However, this example definitely establishes at least two designs for "Empire Combination Timer" dials, and that "high end" products may have both jewel count and grade variations within a single run of serial numbers.

Epilogue

In the final analysis, it was probably the desire to achieve excellence by Frederick Giles in every aspect of watch manufacture that contributed the most to the downfall of the USWC. He was not content to position USWC as an alternative to Waltham . . . he insisted on superiority in the market place as witnessed by many unique innovations such as damaskeening, frosting, and stem winding. His "United States" grades were probably the highest priced products in America and were offered in more variations than any of the competitive models. Continuous over-expenditures and production delays coupled with the financial panic of 1873 were especially damaging to Frederick Giles and his high priced product lines. The Elgin company, a contemporary organization, succeeded by being satisfied to concentrate its efforts in early years on a few good key wind models and refused to introduce stem winding until 1874. They were content to let pioneers like Frederick Giles get "the bugs" worked out of this new innovation while they observed. But, Elgin was one of the first to introduce price reductions to match the demands of a more conservative market environment. By late in 1874, the USWC was on a down-hill slide that could not be reversed, with poor finish applied to once proud products that gradually evolved to the sale of previously discarded imperfect movements in their final years.

A sad, but probably very fitting, conclusion to the USWC story has been provided by the NAWCC Museum in the form of an actual article printed in a Newark newspaper during August of 1896:

"Fortunes of the Old Watch
Factory at Marion, N.J.

EVERYBODY who has traveled between Newark and New York over the Pennsylvania Railroad has felt something of an interest in the big iron factory building on the south of the railroad, just west of the Marion station, and those who have watched it for a quarter of a century or more are prepared at any time to see a new sign upon its front as they go down to the big city in the morning and come out at night. The sign upon it now is: "For Sale — Apply to Samuel Frothingham," of some number in some downtown street in New York. It does not make any difference in this sketch just where Mr. Frothingham is located, for it is not likely that any of its readers will be moved by old associations, desire to possess white elephants or any sentimental considerations so far as to purchase the structure, especially as the impression has gone abroad that the building is hoodooed. That is what they say about it in Marion, and they seem to firmly believe it, which, under the circumstances, seems excusable, as what follows will show.

This handsome building was begun in 1866 and finished in 1867. Since then nothing has met with permanent success there, although nearly a dozen attempts have been made to establish a flourishing business within its four iron walls.

In the early days of the manufacture of watches in this country by machinery, Giles, Wales & Co., of Maiden Lane, established the United States Watch Company and built this fine structure to carry on the manufacture of just such watches as have made the work turned out at Waltham and Elgin famous throughout the world. The industry was then in an experimental stage and many costly blunders were made, but thousands of good watches were turned out and some of them are still carried with pride. The work was carried on with varying success until 1873, when the stringency in the money market, due to the Wall St. panic, coupled with a decline in the reputation of the watches, owing to the impolitic sale of a lot of movements which had been discarded for a fault in construction, brought about the winding up of the business in a manner disastrous to everybody concerned, from the humblest workman up.

They were great days for Marion when the watch company located there. The place was known merely as West End up to that time, but this title was too common-place and Marion undoubtedly sounded better.

Intimately connected with the watch company was the Marion Building Association, which planned to make the place a thriving town. The company bought up all the available land south of the railroad for nearly a quarter of a mile in breadth and half a mile or more in length, and while the factory and its numerous outbuildings were going up in the block they now occupy, other blocks were divided up into scanty building lots, 20 feet by 100 feet, and half a dozen rows of frame houses were erected ostensibly for the workmen. Then there was an enormous hotel erected nearly opposite the factory. It was in fact two hotels in one with two scales of prices. One wing was known as the Marion House and the more imposing part of the structure was called the St. James Hotel. People who stopped at the latter place were supposed to have more money than those who put up around the corner in the other wing and they were expected to spend it. Both houses flourished for a short time. Now they are tumbling to pieces from neglect, although some of the rooms are occupied as cheap tenements. There were some lively auctions of building lots in the first two years of Marion. The company bought the bare ground at a price which made each lot cost about $70, but with the aid of glib tongued auctioneers, free excursions, bands of music, and liberal luncheons, so much of a frenzy of enthusiasm was worked up that corner lots sold for $1,200, and less desirable selections went off at from $700 to $900. A large block of this property could now be purchased for from $200 to $300 a lot, and the owners would not hesitate a moment if such an offer were made. The tract is only sparsely settled now, and the tenants of the few houses are apparently all wage-earners. There are no homes of luxury to be found on the land covered by the map made at the time of the boom, and there is but one stately old mansion, which has evidently been there since the early part of the century. That part of Marion is far from being an inviting place for a home. When the west winds of Summer blow over the meadows they become heavily laden with the musty odor of Shanley's reeking manure piles, the oppressive smells from the hog slaughter-house, and the disagreeable scents from the several places upon the Hackensack meadows, where dead animals are converted into merchandise. Still people live there and raise fat, dirty and necessarily healthy children.

Of the iron factory building there is more to be said. After the United States Watch Co. succumbed, Elias Bros., of New York, essayed the manufacture of watches of a low grade to fit the oroide cases. This venture lasted nine months or so, and then the machinery was sold, much of it going for junk. After an idle spell of several years there was a sudden bustle observed in the iron railed enclosure, and it was learned that the New York Silk Mfg. Co. had secured the factory and would give employment to a thousand operatives, more or less. This company spent a great deal of money for machinery, floored the big building anew and built a big brick wing on the south side, but the enterprise failed in its second year, and there was another spell of rest, after which the Manhattan Knitting Co. tried to remove the increasing bad reputation of the building, and gave it up after a few months. Some time afterward Leo Daft secured a fresh lot of investors in his electrical schemes and the Daft Electric Light Co. secured the building, but the enterprise had no greater success than Daft's preceding ventures, and the United Electric Traction Co. took the place — machinery and all — to evolve an electric railway system. An elliptical track was laid in the enclosure, trolley poles were erected and an old horse car fitted with a motor made experimental trips around the course. Four years ago this venture was given up, and now the building stands as a monument of the millions that have been wasted in and around it, and as a temptation to small boys with a predilection for throwing stones at windows."

— Newark *Call*.

* * *

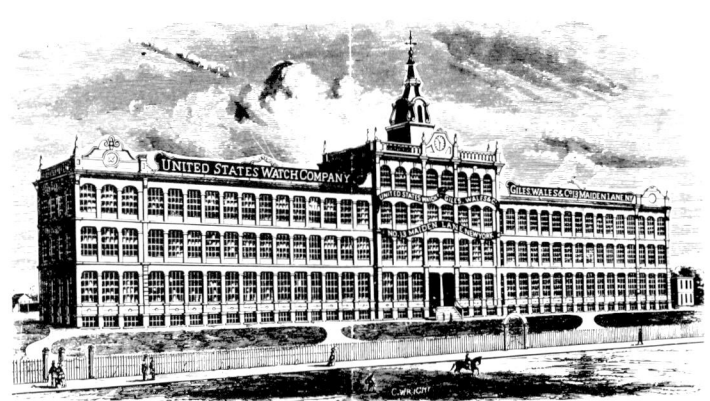

MANUFACTORY OF THE UNITED STATES WATCH COMPANY.

Illustration Credits

Color Dust Jacket

Private collection. Photo by Lloyd Koenig.

Endpapers

The National Archives, Patent Office, NAWCC Museum

Figures

Frontispiece. Burns Collection, Henry C. Wing, Jr.

Figures 2-4, 6, 17, 19, 20, 22, 29, 48, 50, 89, 90, 106, 109, 219. Burns Collection, William Muir.

Figures 5, 21, 49. Burns Collection, Henry C. Wing, Jr.

Figures 7, 18, 56, 112. The Smithsonian Institution, Carlene Stephens. Bernard Kraus.

Figures 8-11, 100, 110. The New York Historical Society, New York City. William Muir.

Figures 12, 13, 14, 71, 72. William L. Scolnik.

Figures 15, 43, 55, 57, 58, 105. NAWCC Research Committee.

Figures 16, 88, 89, 90, 99, 100, 101, 118. William Muir.

Figures 23-28, 30-42, 63, 73-75, 93-95, 122, 125, 126, 140-146, 155, 156, 158, 159, 187, 190, 192. John Wilson, photos by T. B. Jackson.

Figures 44, 45, 54, 92, 127, 196, 232, 235. Bernard Kraus.

Figure 46. The Chicago Historical Society, Bernard J. Edwards.

Figure 47. Samuel W. Jennings.

Figures 51, 61, 64, 65, 69, 82-87, 119, 120, 123, 186, 189. Private collections, photos by T. B. Jackson.

Figures 52, 168, 183. Dr. Ernest Lewis, photos by Dick Hoban.

Figures 53, 59, 97, 103, 104, 114, 133, 138, 139, 148, 151-153, 161-163, 166, 170-173, 175-182, 201-203, 208, 216, 225, 226, 234. NAWCC Museum, photos by Roy Ehrhardt.

Figures 60, 157. Irv Roth, photos by Bernard Kraus, Roy Ehrhardt.

Figures 62, 70, 121, 131, 137, 147, 154, 191, 199, 200. Dick Hoban.

Figures 66, 207. NAWCC Museum, photos by Blaine Sheffer.

Figures 67, 68, 91, 111, 149, 150, 210. Gene Fuller, photos by Dick Hoban.

Figures 76-81. Dr. W. Barclay Stephens collection, California Academy of Sciences, arrangements and photos by Ward Francillon and Dorian Clair.

Figures 96, 98, 116, 169, 209. Roy Ehrhardt.

Figures (continued)

Figures 102, 227-230. Richard J. Wagner, photos by Rick Wagner.

Figure 107. Richard Ziebell.

Figures 108, 184, 185, 204, 215, 217, 218. Frederick L. Orr / Richard Bovard.

Figures 113, 197, 198. Rockford Time Museum.

Figures 115, 132. Ward Francillon.

Figures 117, 128-130, 238, 240, 241, 244, 245, 247, 252, 253. John Fossette, photos by Vincent Angell.

Figure 124. Christie's, NY, Jonathan Snellenburg.

Figures 136, 165. Orville R. Hagans' Manuscript and Photo Library.

Figures 160, 174, 220, 221, 233, 236, 242, 243, 246, 248-250. Bill Guido.

Figures 164, 231, 239, 251. Dr. William C. Heilman, Jr.

Figures 167, 224. Private collection.

Figure 188. Private collection, photo by Lloyd Koenig.

Figure 191. Dick Hoban, photo by Bill Guido.

Figures 193, 194, 195. Henry C. Wing, Jr.

Figures 206, 223, 237. Private collections, photos by Dick Hoban.

Figure 211. *Two Hundred Years of American Clocks and Watches*, Chris H. Bailey, photo by John Garetti.

Figure 212. American Clock & Watch Museum, Chris H. Bailey, photo by Edward Goodrich.

Figures 213, 214. Irv Roth, photos by Dick Hoban.

Figure 222. Private collection, photo by Charles Foster.

Figures 254, 255. Robert M. Wingate.

Note: Photo accumulation project, 1977-1985, coordinated by NAWCC Research Committee.

Charts and Tables

Chart I, Table I. William Muir.

Chart II, Table II, Table IX. NAWCC Research Committee.

Table III, Table V, Table VI, Table XI, Table XII. Bernard Kraus.

Table IV, Table VII, Table X. Bernard Kraus and NAWCC Research Committee.

Table VIII. William Muir, Bernard Kraus, and NAWCC Research Committee.

* * *

Bibliography

Books & Major Series

Samuel Hopkins Adams, *Tenderloin*, New York, Random House, 1959. A novelized rendering of the period, which contains excellent caricatures of Alexander Williams and Anthony Comstock.

Henry Abbott, *Watches and Men*, New York, Maiden Lane Historical Society, 1933.

Henry G. Abbott (George H. A. Hazlett), *Watch Factories of America — Past and Present*, Chicago, George K. Hazlett & Co., 1888.

Henry G. Abbott, *Watch Factories of America — Past and Present*, revised edition (incomplete) published as a series in the *American Jeweler* 1903-6.

Henry G. Abbott, *The Watchmaker and Jeweler*, Chicago, George K. Hazlett & Co., 1891.

Herbert Asbury, *The Gangs of New York*, New York, Garden City Publishing Co., 1927.

John R. Asher and George H. Adams, editors, *Pictorial Album of American Industry*, New York, Asher & Adams, 1876.

Chris H. Bailey, *Two Hundred Years of American Clocks and Watches*, Prentice-Hall, Inc., Englewood Cliffs, NJ, 1975.

John W. Barber and Henry Howe, *Historical Collections of New Jersey*, New Haven, 1868. (Reprint — Spartansburg, S.C., Reprint Co., 1966.)

Edwin A. Battison, *The Auburndale Watch Co.*, Bulletin 218: Contributions from the Museum of History and Technology, Washington, Smithsonian Institution.

J. Leander Bishop, *A History of American Manufacturers, 1608 to 1860*, Philadelphia, Edward Young and Co., 1868.

Albert Sidney Bolles, *Industrial History of the United States*, New York, A. M. Kelly, 1966.

Hank Wieland Bowman, *Pioneer Railroads*, Greenwich, CT, Fawcett Publications, 1954.

Haywood Brown and Margaret Peech, *Roundsman of the Lord*, 1927.

John Carbuitt, *Biographical Sketches of the Leading Men of Chicago*, Chicago, Wilson and Sinclair, 1868.

Paul M. Chamberlain, *It's About Time*, New York, Holland Press, 1941.

Cecil Clutton and George Daniels, *Watches*, New York, The Viking Press, 1965.

Anthony Comstock, *Frauds Exposed*, 1880. (Reprint — Montclair, NJ, Paterson Smith Publishing Co., 1972.)

Augustine E. Costello, *Our Police Protectors*, 1885. (Reprint — Montclair, NJ, Paterson Smith Publishing Co., 1972.)

Charles S. Crossman, "The Complete History of Watch and Clock Making in America," published as a series (50 articles) in the *Jewelers' Circular and Horological Review*, 1885-91. The first 30 articles were reprinted in book form as *The Complete History of Watchmaking in America*, Exeter, NH, Adams Brown, n.d.

Chauncey M. Depew, editor, *One Hundred Years of American Commerce*, D. O. Haynes & Co., 1895. (Reprint — New York, Greenwood Press, 1968.)

William E. Drost, *Clocks and Watches of New Jersey*, Elizabeth, NJ, Engineering Publishers, 1966.

James Dugan, *The Great Iron Ship*, New York, Harper, 1953.

Frederick Dyer, *A Compendium of the War of the Rebellion*, New York, Thomas Yoseloff, 1959.

Roy Ehrhardt, *Pocket Watch Price Indicator Book 2 — 1977*, Kansas City, Heart of America Press, 1976.

The Essex Institute, *Vital Records of New Salem, to the End of the Year 1849*, Salem, Essex Institute, 1927.

George H. Farrier, editor, *Memorial of the Centennial Celebration of the Battle of Paulus Hook*, Jersey City, 1879.

James W. Gibbs, *The Dueber Hampden Story*, Philadelphia, 1954.

Fayette Stratton Giles, *The Industrial Army*, New York, Baker and Taylor, 1896.

Fayette Stratton Giles, *Shadows Before; or A Century Onward*, New York, Humboldt Publishing Co., 1894.

Horace Greeley & Others, *The Great Industries of the United States*, Hartford, J. B. Burr & Hyde, 1872

Thomas W. Herringshaw, *National Library of American Biography*, Chicago, 1909.

Charles T. Higginbotham, *Incidents in the American Watchmaking Industry*, published as a series in the *National Jeweler and Optician*, January-August 1912.

Charles T. Higginbotham, *Precision Time Measures*, Chicago, North American Watch Tool Supply Co., 1952.

Allen Johnson and Dumas Malone, *Dictionary of American Biography*, New York, American Council of Learned Societies, 1958.

Edward Chase Kirkland, *Industry Comes of Age*, Chicago, Quadrangle Books, 1961.

William B. Learned, *The Watchmaker's and Machinist's Handbook*, Chicago, George K. Hazlett & Co., 1897.

Walter Arndt Lucas, *From the Hills to the Hudson*, New York, Privately Printed, 1944.

James Angell MacLachlan, *Handbook of the Law of Bankruptcy*, St. Paul, MN, West Publishing Co., 1956.

James D. McCabe, Jr., *Lights and Shadows of New York Life or the Sights and Sensations of the Great City*, Philadelphia, National Publishing Co., 1872.

James D. McCabe, Jr., *History of the Centennial Exhibition*, National Publishing Co., Philadelphia, 1876.

Charles W. Moore, *Timing a Century*, Cambridge, Harvard University Press, 1945.

William A. Neiswanger, Ph.D., *Statistical Methods*, Macmillan Company, New York, Twelfth Printing, 1950.

Warren H. Niebling, *History of the American Watch Case*, Whitmore Publishing Co., Philadelphia, PA, 1971.

State of New Jersey, *The Laws of New Jersey, 1866.*

Theophilus Parsons, L.L.D., *Laws of Business*, Chapter 33, "The Law of Patents," S. S. Scranton Co., Hartford, 1878, 1909 revised edition.

Henry F. Piaget, *The Watch: Handwork Versus Machinery*, 3rd edition, New York, 1877.

Henry F. Piaget, *The Watch: Its Construction, Its Merits and Defects*, New York, C. Vintou, 1860.

Edmund L. Sanderson, *Waltham Industries*, Waltham, MA, Waltham Historical Society, 1957.

Robert Sobel, *Panic on Wall Street*, New York, MacMillan, 1960.

George E. Townsend, *Almost Everything You Wanted to Know About American Watches and Didn't Know Who to Ask*, Vienna, VA, Privately Printed, 1971.

C. G. Trumbull, *Anthony Comstock. Fighter*, 1913.

Albert Ulmann, *Maiden Lane*, New York, Maiden Lane Historical Society, 1931.

Robert Underwood & Clarence Clough Buel, editors, *Battles and Leaders of the Civil War*, New York, Thomas Yoseloff, 1956.

United States Bureau of the Census, *9th Census of the Population — 1870*, Washington, Government Printing Office, 1870.

Daniel Van Winkle, ed., *Histories of the Municipalities of Hudson County, New Jersey*, New York, Lewis Historical Publishing Co., 1924.

Harold G. Vatter, *The Drive to Industrial Maturity*, Westport, CT, Greenwood Press, 1975.

John Adams Vinton, *The Giles Memorial*, Henry Dutton & Sons, 1864.

Vroom, New Jersey Law Reports, Vol. 7.

George W. Walling, Recollections of a New York Chief of Police, New York, Caxton Book Concern, 1887. (Reprint — Montclair, NJ, Paterson Smith Publishing Co., 1972.)

Irving Werstein, *July 1863*, New York, Ace Books, 1957. (Note: While this work is somewhat novelized, its factual base is quite accurate.)

Sites Visited

Laurel Hill Cemetery, Montague, Massachusetts — Where many of the principals of this work now lie buried in a small family plot.

Maiden Lane, New York City — The old jewelry district is long gone and the past is lost.

Marion, Jersey City — Although much has changed there is still evidence of its past history.

Montague, Massachusetts — Here so little has changed that Frederick and Julia would still be at home.

New Salem, Massachusetts — New Salem was in trouble even before the Civil War. In modern times the building of the Quabbin Reservoir did little to preserve things, but the past is still found.

Shanley's Cut, Jersey City — The old railroad right-of-way and Marion junction are now the Hudson and Manhattan Tube line and a grimy industrial siding, but this ancient segment of American railway history can still be discerned.

Newspapers & Periodicals:

The American Horological Journal

The American Jeweler

The Gazette (Jersey City)

Harpers Weekly

The Jersey City Journal

The Jewelers' Circular and Horological Review

The Jewelers' Circular Weekly

Keystone

Moore's Rural New Yorker

NAWCC BULLETIN
See especially:
No. 32, February, 1950. Dr. W. Barclay Stephens, "The Newark Watch Co. and Its Career."
No. 34, June, 1950. Dr. W. Barclay Stephens, "The United States Watch Company."
No. 139, April, 1969. Frederick M. Selchow, "The Watch Company of Fitchburg, Massachusetts."
No. 190, October, 1977. Eugene T. Fuller, "U.S. Watch Company Centennial Commemoratives."
No. 14, Spring, 1984, Supplement to the BULLETIN, "American Watchmaking: A Technical History of the American Watch Industry, 1850-1930," Michael C. Harrold.

The National Jeweler and Optician

The New York Herald

The New York Sun

The New York Times

The New York Weekly News

Scientific American

Van Nostrands Engineering Magazine

The Watch Dial

The Watchmaker and Jeweler

Directories:

Boyd, *Jersey City Directory*, 1876-8.

Doggett and Rode, *New York City Directory*, 1851-3.

Dun, Barlow & Co., *Merchantile Agency Reference Book*, 1879.

Dun, Barlow & Co., *National Business Directory*, 1867.

Fellhauer-Calame, *Grand Indicateur Complet de l'Industry Horology*, Beinne, 1905.

Charles H. Folwell, *Newark City Directory*, 1863-5.

Daniel E. Gavit, *Jersey City Directory*, 1854-5.

James Gopsill, *Jersey City Directory*, 1857-79

James Gopsill, *Newark City Directory*, 1865-9.

James & William Gopsill, *Jersey City Directory*, 1855-7.

Goulding, *New York City Directory*, 1876-7.

Holbrook, *Newark City Directory*, 1872-80.

George M. Hopkins, *Atlas of the County of Hudson and the State of New Jersey*, 1873. This work contains an extensive Jersey City directory.

Rode, *New York City Directory*, 1853-4.

J. A. Ryerson, *Jersey City Directory*, 1852-4.

John F. Trow, *New York City Directory*, 1854-80.

Trade Catalogs:

American Watch Tool Co., *Precision Machinery*, Waltham, MA, ca. 1890. Reprinted by Ken Roberts Publishing Co., Fitzwilliam, NH, 1980.

A. C. Becken, *Watch Catalog*, 103 State Street, Chicago, IL, ca. 1885. Reprinted by Robert Spence, St. Louis, MO, 1964.

Cross & Beguelin, *Watch Tools and Materials*, 21 Maiden Lane, New York City, ca. 1885.

Elgin Watch Co., *Elgin Watch Materials Catalog*, Elgin, IL, 1950.

Empire City Watch Co., *Watches*, Jersey City, NJ, 1876.

Giles, Wales & Co., *Watch Materials*, 13 Maiden Lane, New York City, n.d.

L. Hammel & Co., 9 Maiden Lane, New York City, 1884.

Lapp & Flershem, 77, 79, & 81 State Street, Chicago, IL, 1884.

Sloan & Chace Mfg. Co., Limited, *Precision Machinery*, 6th Avenue and 13th Street, Newark, New Jersey, 1904.

Stark Tool Co., Waltham, MA, 1920.

Swartchild & Co., *Catalog B-232*, 29 E. Madison Street, Chicago, IL.

Waltham Watch Co., *Serial Numbers With Descriptions of Waltham Watch Movements*, Waltham, MA, 1954.

Otto Young & Co., *Tool and Material Catalog*, Heyworth Building, Chicago, IL, ca. 1905.

Maps & Atlases:

Aldridge & Wood, auctioneers, *40 Superb Building Lots at Marion*, Auction map of June 15th, 1871.

Daniel E. Gavit, *Jersey City Directory*, 1854-5.

James V. Hogan, Jr., *Survey Map of City Block 1608, Jersey City*. Survey of September 11, 12, & 14, 1925.

George M. Hopkins, *Atlas of the County of Hudson and the State of New Jersey*, Philadelphia, 1873.

George M. Hopkins, *Atlas of the County of Hudson*, Philadelphia, 1909.

Jersey City, *Tax List*, 1908-9.

Marion Building Co., *Map of June 16, 1869*.

Marion Building Co., *Land Auction Held at Marion June 20, 1872*.

Marion Building Co., *Map of August 2, 1872*.

Spielman & Bush, *Sanitary and Topographical Map of Hudson County, N.J.*, New York, The National Board of Health, 1880.

Tallis, *Views of Maiden Lane*, New York, 1872.

Manuscript and Archival Materials:

The Burns Collection — Frederick A. Giles' great grandson G. Robert Burns and his wife, Carolyn, have carefully gathered and preserved many items pertaining to the Marion venture. This material includes letters, newspaper clippings, photographs, and other memorabilia. Their willingness to share every piece of this information has earned the authors' deepest respect.

Charles S. Crossman, *Scrapbook*, New York Public Library.

The Empire City Watch Co., *Certificate of Incorporation, January 10, 1876*, New Jersey Secretary of State.

The Empire City Watch Co., *Statement by a Corporation Transacting Business in the State of New Jersey*, June 30, 1877, New Jersey Secretary of State.

William B. Fowle, *Cash Book*, Edwin A. Battison.

Frederick A. Giles, *Record of Death*, June 18, 1879, Secretary of the Commonwealth of MA.

Giles, Wales & Co., *Copartnership Agreement*, March 1, 1863, New York Secretary of State.

Manhattan District Attorney, *Register of Felonies, December 1876-July 1877*, New York Municipal Archives.

Marion Watch Co., *Certificate of Incorporation*, July 30, 1874, New Jersey Secretary of State.

New York Watch Co., *Inventory*, Richard Ziebell.

United States District Court for the District of New Jersey, Case No. 1124, *In the Matter of Frederick A. Giles, William A. Wales and George C. F. Wright*, National Archives, record group 21.

United States District Court for the District of New Jersey, Case No. 1197, *In the Matter of Frederick A. Giles and George C. F. Wright*, National Archives, record group 21.

United States Patent Codes, Revised Statute of 1870, *Requirements for Patent*, Sec. 4886 and *False Marking*, Sec. 4901; 35 U.S. Code (1985), *False Marking*, Sec. 292.

United States Patent Office, *Various Patents*, National Archives, record group 241.

The United States Watch Co., *Certificate of Incorporation*, February 27, 1865, New York Secretary of State.

George Channing Fuller Wright, *Certificate of Death*, February 7, 1884, New York Municipal Archives.

L. Zabriske, *Cash Book*, New Jersey Room, Jersey City Public Library.

Oral Sources:

Dana J. Blackwell
Carolyn G. Burns
Christabel G. Burns
G. Robert Burns
Dorian Clair
Roy Ehrhardt
John Fossette
Warren Franz
Eugene T. Fuller
Del Gantz
James W. Gibbs
Bill Guido
William C. Heilman
James Henderson
Dick Hoban
Freeman H. McMillan
Barry Ted Moskowitz
Frederick L. Orr
Maylene Rabeneck
Robert L. Ravel
Glen A. Smith
George Townsend
John Wilson
Henry C. Wing
Richard Ziebell

Index

Page numbers in **boldface** type indicate illustrations. Watch companies are indexed (1) by the first letter, or initial, e.g. California Watch Co. under C, and (2) in alphabetical sequence under Watch Companies. Watch grades are indexed (1) by the first letter, or initial in the grade name, e.g. "Asa Fuller" and "A. H. Wallis" fall under A, and (2) in alphabetical sequence under Watch Grades. Names on watches such as retail jewelers or watch grades are shown in quotation marks, e.g. "Empire Combination Timer." With the exception of the grade name, United States Watch Company is abbreviated as USWC. The Chronology on pages 17-20 may also assist reference if the approximate date is known for a certain USWC event.

"A. H. Wallis" 77, **79**, 127, **129**, 130, 131, 134, 136; description, 148-150; **149**, **150**, 179, 186; serial number allocation, 189; **197**, **198**

"A. J. Wood" 127, 130, 131, 135, 136; description, 160, **160**; 179; serial number allocation, 191

Abbott, Henry (George Henry Abbott Hazelitt) 16, 52, 72, 115, 139, 142

Academy of Music 106

Adams, John C. 40

"Adams Street" boys' watch 157

Advertising **57**, **68**, 69-71, **70**, **71**, 128, **167**

Aetna Insurance Company 52, 53

Aiken, Lambert & Co. 15

Albany, New York 25

Albert, E. 69

Allen, Henry 98

Alphabet code, USWC watches 171

Alexander, James A. 53, 121, 124, 150

American Clock & Watch Museum (Vogel Collection) 177

American Horological Journal 70, 71, 73

American Institute of New York, 1870 Fair 71; 1873 Fair, USWC silver medal award **165**

"American Watch Co." **83**, **84**, **86**, 142-143

American Watch Co. (Waltham) 29, 35, **36**, 37, 65, 83, 84, 85, 95, 130, 142, 147, 157, 199

"American Working People" 171

Amherst, Massachusetts 24

Angelica, New York 23, 25

Anti-magnetic watch protector (shield) 123, 128

"Appleton, Tracy & Co." 138

"Asa Fuller" 127, 130, 131, 135, 136; description, 155; **155**, 174, 179, **186**; serial number allocation, 190

Astor Hotel 52

Athol, Massachusetts 23, 59

Auburndale Watch Co. 101, 115, 121, 123

Audemars, Louis 40, 53

"Augustin Perrenoud, Ponts" **156**

Austin, Henry W. 61

Austin, Illinois 61

Avery, Thomas M. 95, 96

Award medals, USWC **165**, 177

Balance cock engraving **78**, **79**, **129**, **143**, **145**, **149**, **151**, **152**, **194**, **196**, **197**

Balance wheel 40, 49-50, **51**, **54**, 85

Baldwin, Oliver J. 39, 49, 53

Bankrupt sale 106

Bankruptcy, 19th Century laws 97; bankruptcy, of Giles, Wales & Co., 97-99; of Giles, Wright & Co., 103; of USWC companies, 117; of Giles Bros & Co., 123-124

Barber, John W. 47, 69

Barrel, mainspring **51**, **54**, **144**, 185

Barrel, reversible 86, **88**, **89**, 128, 130, 155, 185

Bartlett, Patton S. 40

Battison, Edwin 115

Beard & Cummings 151

Beard, S. M. & Sons, Co. 52, 97
Beard, Sylvestor M. 52, 96
"Benjamin & Ford" **170**
Bergen Hill 23, 43
Bergen Town, New Jersey 43-44, **93**
Bergen (Marion) Junction 23, 44
Berkshires 25, 177
Bienne, Switzerland 66
Berlin, Charles 65, 193
Biographical Sketches of the Leading Men in Chicago 61
Bishop, J. Leander 52
"Black Diamond" 179
Blake, Ira 40
Bliss Chronometer 71
Blooming Grove Park Association 81, 123
Boiler problems 47, 50
Boston, Massachusetts 24, 123
Boston, Concord & Montreal Railroad 171-172
Boston Watch Co. 35
Bottom, James M. 15, 26, 38, 96
Bourquin & Co. 66-67, 156
Bourquin, Edwin 49
Bowman, Ezra F. 122
Brick Church, New Jersey 160
Brown, Thomas G. 43, 52, 96, 97
Buckwalter, Benjamin K. 113
Buffalo Industrial Exposition, 1871, **68**, 165
Burbank, Samuel D. 31
Burns, Bob and Carolyn 61
Burns, Christabel G. 177
Burnside, General 26
Butterfly Opening 39, 49-50, 67, 127, **127**, 184

California Watch Company 184
Canton, Ohio 82
Cases, Watch 31, 58, 60; Magic **84**; 131-133, **131**, **132**, **133**, **197**
Castle Garden 53

Century Onward, A 123
Centennial Exhibition watches 165, 173-177, **173**, **174**, **175**, **177**; serial number allocation, 192
"Centennial, Philadelphia" 174, **175**, 176, **177**
Charts:
    Giles Family Enterprises 14
    Marion and Surrounding Area 44
Chase, Salmon P. 29
"Chas. G. Knapp" 66, **67**, 127, 130, 131, 135, 136; description, 156, **156**; serial number allocation, 190
Chaux-de-Fonds, Switzerland 32, 34, 63, 64
Chicago fire 94
Chicago, Illinois 59, 60, 94, 95, 123
Chicago Northwestern Railroad 61
Child Labor 82
Christmas, season of 1872, 94, 101; season of 1874, 97
Chronograph, jump quarter **33**
Church, Metcalf & Co. 31
"Cincinnati Enquirer" **114**, 171, **195**
"Cincinnati Gazette" 171, **195**
Cincinnati Industrial Exposition, Ohio Mechanics' Institute, 1870, 71, 165
Cincinnati, Ohio 25, 105
Civil War 25, 26, 29, 32, 33, 36
Clark, Benjamin G. 46
Clark, Samuel 103
Clay, R. J. 124
"Clubber" Williams (See Williams, Alexander)
Cole, James F. 49
Commemorative watches 173-177
Complete History of Watch and Clock Making in America, A 15
Compton, Charles W. 115
Comstock, Anthony **107**, 117
Comstock Laws 113
Conscription Act 33
Continental Life Insurance Co. 53
Cook (jeweler) 59
Cooke, Jay 29, 95
Cornell Watch Co. 183-184
Crane, Charles A. 97, 103

Crans, Lewis D. 115

"Crescent Street" 85

Crooker, Jacob 25

Cross & Beguelin 31

Crossman, Charles S. 15-16, 39, 47, 49, 52, 105, 115, 139, 142

Curtis, W. G. 124

"Cyrus Loutrel" 179

"D. C. Wilcox" 179, **181**

"D. R. Brown" **171**

Damaskeening 64-65, **65**, 86, **87**, **88**, 129, **129**, 184

Daft Electric Light Co. 200

Daniels, George W. 38

Denman, Isaac 96, 115

Dennison, Aaron L. 15, 29, 35, 38

Dials **56**, **74**, **75**, **76**, **77**, **87**, **88**, **96**, **109**, **137**, **144**, **159**, **160**, **161**, **162**, 179, **180**, 182, **194**, **196**, **197**

Dials, Lord's Prayer 75, **75**, **76**, **162**

Dials, Masonic **74**, 75

Dials, Symbol of Peace **74**, **162**

Dials, USWC, types and prices 161, 163

Dickinson & Rowden, Newark, NJ 38-39, 48

Dickinson, Charles W. 38

Dodd, Hedges & Co. 96, 97

Dodd, Jesse A. 179

Dodge, Charles T. 115

Dollar Store Business, Elias Bros. **104**, 106

Donna Emilia Huranez Swindle 106

Draft riots, New York City 33-34

Drew, Daniel 95

Dubuque, Iowa 63, 123

Dueber-Hampden Company 82, 183

Dueber, John C. 82, 108

"E. C. Hine" 179

"E. D. Howard" 184

"E. F. C. Young" 179, **179**

"Edwin Rollo" 55, 127, 130, 131, 135, 136; description, 141-142, **141**; 179; serial number allocation, 188; 194, **196**

Elgin (See National Watch Company)

Elias, Ellis H. 83-84, 104-113, 115; sawdust swindle, 106; General Average Scheme, 107-108; 109, 116-117; death, 122

Elias, Henry P. 83

Elias, Richard H. 105, 106

Elias, William M. 104, 105, 107, 108, 122

Elson, Julius 85; double index regulator, 85, **112**, 113, 128, 186

Empire City Watch Co. 105, 110, **111**, **112**, 116-118, 124, 179-182; price list, **178**; watch grades, 179

"Empire Combination Timer" 180, **180**, **181**, **196**, **198**

Employees, USWC (Partial List) 193

Enameling, USWC watches 86, 134-136; washout, 152

Enfield, Connecticut 59

England; English watches, 31-32; watch industry organization, 36

Engraving of USWC watches 55; "United States Match Co.," 55

Enos Richardson & Co. 52, 96, 97

Escapement shop 73; wheel machine, **73**

Europe 24, 32

Exhibitions, industrial and trade fairs **68**, **71**, 165, 173-177

Experimental department products **74**, 75

Fahys, Joseph 31

Faigaux, Bertha (See Giles, Bertha Faigaux)

Falkman, Pollack & Co. 31

Favre 32

"Fayette Stratton" 53, 63, **65**, **81**, 127, **129**, 130, 131, 135, 136; description, 138-139, **139**; serial number allocation, 188

Fayette Street 81

Feature, USWC watches, relative scarcity 186

"Federal Shooting" Celebration 32

Feliz & Schwitter 97

Fellows, Louis S. & Schell 31

Female employees, USWC 82

First National Bank of Jersey City 53, 102, 122, 179

Fiske, A. H. 52

Fiske & Hatch 52

Fitchburg, Massachusetts 52, 102

Fitch, Ezra C. 86

Finish, USWC watches 134-136

Fowle, William 115, 121

Francis and Loutrel 53

Frankfort Street 81

Frauds Exposed, Anthony Comstock **107**, 122

"Frederic Atherton & Co." 49, 50, **50**, 53, **53**, **57**, 127, 130, 131, **131**, 135, 136, **137**, **138**; description, 137-138; 185, 186; serial number allocation, 188

Fredonia Watch Co. 121, 183-184

Freeport Watch Co. 47

Frothingham, Samuel 199

Full plate 39-41, 49-57, **50** (first illustration), **126**, 127, 130, 134-136

Frosting, USWC watches 86, **87**, 129, **143**

Fuller, Eugene T. 173, 177

"G. A. Read" 101, 127, 130, 131, 135, 136; description, 158-159 **158**; 179; serial number allocation, 190-191, 194-195

Gazette, The 69, 71

Gendar, William & Thomas (jewelers) 24, 35, 43

"General Andrew Jackson Exhibition Watch" 177

General Average Scheme 107-108, 117

"General George Washington Exhibition Watch" 177

Geneva, Switzerland 123

"George C. F. Wright" **122**, 171-172

George C. F. Wright & Co. 122

"George Channing" 55, 127, 130, 131, 135, 136; description, 139-140, **140**; 174, 179; serial number allocation, 188

"George Washington Exhibition Watch" **174**, 176-177

Gerry, David 39, 48, 193

Gerry, Emery J. 39, 193

Gerry, James H. 40, 41, 48-49, 193

Gilbert Manufacturing Company 62

Giles, Anna 23, 59, 63

Giles, Bertha Faigaux 63, **64**

Giles, Brothers & Co. 53, **58**, 59, 62, 69, 102, 105, 123-124, **133**

Giles, Charles K. 59-60, 94, 123, 128

Giles, Daniel 23, 59

Giles, E. A. & Co. 53, 63, **63**, 102, 105, 123

Giles, Edwin 26, 63, 105, 123, 128

Giles, Emma Smalley 123

Giles Family; genealogy, 23-24; relationships, 59, 63-64, 100, 117, 119

Giles, Fayette Stratton 53, 56, 63, **64**, 66, 81-82, 94, 118-119, 123, 124, 127-128

Giles, Frederick A. **Frontispiece**, 22-24, 26-27, **26**, 32-33, 46, 49, 50, 56, 59, 63, 69, 94, 101, 102, 103, 109, 113, 117, **118**, 118-119, 127-128

Giles, Grace 103

Giles, Hanna Learned 23

Giles, Hattie C. 52, **52**

Giles, Julia M. Wright 24, **24**, 26, 50, 94, 103, 117, 118, 119, 123

Giles, Lafayette (See Giles, Fayette Stratton)

Giles, Lemira 23, 25, 59

Giles, Mary Harper 59

Giles, Prescott 23, 25, 59

Giles, Sue 23, 63

Giles, Wales & Co. 25, **28**, **30**, 29-31, **33**, 34, 47, 56, 63, 66, 69, 91, 94, 96-99, 131, **132**, 193, 199

Giles, William Alexander 23-24, 26, 59, **61**, 61-62, 102-103, 123-124

Giles, Wright & Co. 15, 101, 102, 103, 105, 110, 117, 119

Gilt damaskeening 65, **65**, 129, **129**

Gould, Jay 95

Grafulla's 7th Regiment Band 91

Grand Central Hotel, New York City **104**, 106

Grades, Watch (See Watch Grades)

Grant, Ulysses S. 71, 95

Great Eastern 32

Great Geneva Watch Company 83-84, 91, 105, 106

Greeley, Horace 34

"Greenback," origin of term 29

Greenfield, Massachusetts 25

"H. G. Cook" **169**

Hackensack River, New Jersey 43, **44**; salt marsh, **90**; **92**, 93, 101; meadows, 200

Hairspring; Breguet **55**, **78**; Flat, **79**; USWC models, 134-136

Hampden Watch Co. 82, 183

Hanover Bank 53

Harner, S. M. 34

Harper Brothers publishers 52, 97, 179

Harper, Henry 179

Harper, J. Abner 52

Harper, Mary (See Giles, Mary Harper)

Harper's Weekly **68**, 69, 71, 73

Harrison, Henry 31

Harrison, W. B. 91

Hart, George E. 48, 66, 73, 82, 96, 101, 115, 121-122, 193
   Hart & Denman 96
   Hart, George & Co. 115
   Hart, Logan & Co. 96
   Hart, Sloan & Co. 121-122

Hatch, A. S. 52, 95

Hayes, Annie C. 117

Hayes, Helen Curll (See Wright, Helen Curll Hayes)

"Henry Harper" 179, 196

"Henry Randel" 127, 130, 131, 134, 136; description, 152-153, **152**, **153**; 179; serial number allocation, 187-190

Higginbotham, Charles T. 16, 55, 71, 82

Historical Collections of New Jersey 69

Hoboken, New Jersey 24

Hogan, James V. 46

Holden, C. N. 62

"Home Watch Co." 159

Hoosac Range, Berkshires 25

Howard & Co. 84

Howard Brothers, Fredonia 121, 183-184, **183**, **184**

Howard, Edward 15, 35, 38

Howard, E. & Co. 35, **63**, 85, 130, 133, 142

Howard Hotel 25, 29, 34

Howard regulator 71

Howe, Henry 69

Hudson County **93**

"I. H. Wright" 127, 130, 131, 135, 136; description, 159-160; **160**; serial number allocation 187, 191

"I. W. Frost" 179

Illinois Watch Co. **63**, 183

Incidents in the American Watchmaking Industry 16, 55, 71

Increase in Capitalization 90

"Independent" **183**, **184**

Independent Watch Co. 121, 184

"Independence, America" **173**, 173, 174, 176, 177

Industrial Army, The 123

Ingold, Pierre Frederic 35, 38

Iowa State Fair, 71, 165

Iron Dike & Land Reclamation Co. 44

"J. L. Ogden" 179

"J. W. Deacon" 127, 130-131, 135, 136; description, 157-158; **157**; 174, 175, 179; serial number allocation, 187, 190

"James Russell & Co." 84

James, Thomas M. 103

Japy, Frederic 35

Jennings, Samuel W. 61-62

Jerome, Chauncey 61-62

Jersey City Evening News 82

Jersey City Journal 69

Jersey City, New Jersey 43, 82, 93, 102, 103, 110, 121, 124, 148

Jersey Meadows 44, 46

"Jesse A. Dodd" 179, **181**

Jeweler's contract, USWC watches 170

Jewels, USWC watches 134-136

"John Ellery" 84

"John S. Robson" **169**

"John Street" 133

"John W. Lewis" 127, 130, 131, 135, 136; description, 154-155, **154**; 179; serial number allocation, 187, 190

Jordan & Marsh Co. 37

Jurgensen Watch Co. 34, 123

Keystone Standard Watch Co. 15

Kemlo, F. 69

Kiesner 32

Klinger, John G. 98, 99

Klinger, Rupp & Held 97, 98-99

Klondike, game of 108

Lafayette 64

Lake Superior copper 61

Lapeer, Michigan 123

Lathes, watch factory 38, 39, 73, 101

Laurent, Jacques 132, 133

Laurent, Emil 133

Learned, Hanna 23

Learned, William B. 49, 52, 55, 56, 65-66, 83, 91, 193

Leman, Frank 15, 193

Linn, John 98

"Little Jewel" **183**

Litzenburg, Wayne 84

Lock, D. D. 15

Locle, Switzerland 32

Logan, John 15, 96, 101, 115, 193

Louisiana State Fair, 1870, New Orleans 71, 165

Louisville, Kentucky 25

Loutrel, Cyrus H. 53, 179

Lowe, Henry J. 52; appointed superintendent, 66; 97, 101-102, 193

"Ls. A. Bourquin" **67**

MacLachlan, James Angell 97

Machinery, USWC sale of 102, 115-116, 121-122, 184

Maiden Lane, description 30-31, **30**, **31**; 34, 43, 66, **120**

Manhattan Knitting Co. 200

Marion and its Temple of Labor 69-70, **70**

Marion Building Co. 44, **45**, 47, 90-91, **91-93**, 97, 101, 103, 109, 116, 199

Marion, Francis 23

Marion House 47, **48**, 50, 199

Marion, New Jersey 44, **44**, **45**, 46; mosquitoes 91, 93; 94, 199

Marion Watch Co. 96-97, 103, 116-117, 124

"Marion Watch Co." **96**, 127, 130, 131, 135, 136; description, 147-148, **147**, **148**; serial number allocation, 187, 188

Markham, F. P. 124

Marsh, E. A. 72, 73

Marsh, Coe & Wallis 97

Marvin Watch Co. 195

"Mary Anns" (watches) 85

"Mason Ware & Co." **169**

Material (parts) prices, USWC **163-164**

McCabe, James D. Jr. 33

McGregor, Iowa 59

Measurement procedures, USWC 73

Merchants Hotel 66

Merrill, Samuel, Governor of Iowa testimonial **71**

Milton Gold Swindle 106

Milwaukee, Wisconsin 25

Mississippi, or Maine, Central Railroad 77

Model of 1865, possible 39-40

Model of 1866, description 49

Models, Watch, USWC 134-136

Mohawk and Hudson Railroad 25

Montague, Massachusetts 24-26, 32, 50, 52, 118-119, 123, 177, 179

Montreal, Canada 133

Moore's Rural New Yorker 69, 71, 171, **171**

Moseley, Charles S. 40

Movements, USWC, sizes and styles **126**, 127, 130, 134-136

Mozart Watch 79

Muirheid, William 98, 103, 110, 115, 117

Mulligan, Hugh 177

NAWCC Museum 128, 199

N. E. Company 32

Nashua Department, Waltham 85, 86

Nashua Watch Co. 16, 36

National Watch Co. (Elgin) 36, 37, 40, **63**, 65, 85, 95, 130, 199

Negus chronometer 71

New Berne, NC, battle of 26

New England Watch Co. 115, 122

New Haven, Connecticut 62, 117

New Jersey District Court 103

New Jersey Legislature 44

New Jersey Midland Railroad 44

New Jersey Railroad 23, 43-44, 46, 171-172

New Jersey State Fair, 1872, 71, 165

New Jersey Supreme Court 83

New Orleans, Louisiana 25, 71, 165

New Salem Academy 59

New Salem, Massachusetts 23, 24, 25

"New York Belle" 178-179, 192

New York and Erie Railroad 23

New York, NY 24, 25, **30**, **31**, 72, **120**, 133, 148, 151, 152, 180, 199

New York Silk Manufacturing Co. 200

New York Standard Watch Co. 49, 124

New York Supreme Court 97

New York Times 82, 105, 106

New York Watch Co., Providence 36, 37

New York Watch Co., Springfield 39, 47, 48, 79, 82, 85

Newark, New Jersey 31, 37, 38, 39, 43

Newark Watch Works 37, 128, 184

Newell, Edward S. 84

Newspapers, USWC watches 169, 171

Nickel watch movements, 64, 65

Niebling, Warren H. 133

Nixon, John T., U.S. Judge 99, 103, 119

Non-Magnetic Watch Co. 121, 184

"North Star" 101, 127, 130, 131; description, 159, **159**; 179; serial number allocation, 187, 191; **198**

Northampton, Massachusetts 59

Oak *Leaves* 62

"Oliver Arnzen" **170**

Owners, Original Some, USWC watches 166-167

"**P**. S. Bartlett" 141

Page, Frank H. 35, 52

Paillard, Marius J. 31

Panic of 1857, 25

Panic of 1873, 95-103, 143, 149, 199

Parsons Horological Institute 121

Parts storage 79; parts, USWC prices (See Material)

Passaic River, 43, 44

Patek Philippe watches 65; Adrien Philippe, 39

Patents, Giles Brothers 127-128; Julius Elson, 85, 128, 186; Arthur Wadsworth, 128 (Also see U.S. Design and U.S. Letters Patents)

Patent Office 128

Paterson & Hudson River Railroad 23

Paterson Depot 44 (Also see Marion Junction)

"Pennsylvania Railroad" **126**, **168**, 171, 172, **172**

Peoria Watch Company 82, 121, 184

Pequignot, C. & A. 133

Perrenoud, Augustin 156, **156**

Perret, David 32

Personnel, USWC factory 49, 81-84, 193

Philadelphia, Pennsylvania 133, 173, 177

"The Philadelphia Exhibition Watch" **174**, 176

Philadelphia Watch Co. **83**

Phoenix, Arizona 124

Pictorial Album of American Industry, 1876, 70

Pike County, Pennsylvania 81

Pike, Samuel N. 44

Pillar plate **55**, **56**, 85, **87**, **88**, **89**, **98**, **129**, **143**, **145**, **149**, **150**, **182**

Pinion cutting 78

Pitkin, Henry and James 35

Plate screws 50

Platt and Brother 24

Platt, George W. & Co. 39, 66

Ponts, Switzerland 156

Post Office, U.S. Department 113

Potter, J. M. 113

Praire du Chien, Wisconsin 59

Pratt, Albert J. 31

Pratt, George F. 46

Prestige Items, USWC **78**, 131, 142, 148, 161

Pray, John 39

Prices: USWC movements, 131-136; cases, 131-133; parts, **163-164**; Empire City Watch Co. movements, **178**

Price cutting 95-96, 143, 147, 199

Printing House Square, New York City 72

Private label (See Special Order)

Production capacities, USWC 193

Production, cumulative, USWC 69, 187

Production delays 48, 53, 55, 81, 94, 95, 142, 147, 186, 199

Production estimates, USWC model/grades 187

Product line changes 83, 85, 130

Punches, part numbering 185

Putnam Machine Co. 47, 72

Quarter plate and bridge 85-95, **95** (first illustration), **126**, 127, 130, 134-136

Quarter plate, cock and bridge 66, **67** (first illustration), **126**, 127, 130, 134-136

"**R**. F. Pratt" 66, 67, 127, 130, 131, 135, 136; description, 156, **156**; serial number allocation, 187, 190

Railroad watches 85, 168, **168**, 171, 172; serial number allocation, 187, 192

Railroads 23, 25, 169-172; personnel, 180

Randel, Baremore & Billings 52, 152

Randel, Henry 52, 152

Read, G. A. 52, 158

Regulator, double index (See Elson)

Repeater, minute **33**

Richardson, Frank H. 52

Rielly, Detective NYPD 106

Roanoke Island, Battle of 26

Robbins and Appleton 29

Robbins, Royal E. 29

Rockford Watch Co. 49

Rocking bar (yoke) 53 (Also see Stem winding and setting mechanisms)

Roger, Smith & Co. 26

Roulets watches 32

"Royal Gold American Watch" 109-110, **109**, **110**, 182, **182**, **183**, 187; serial number allocation, 192

Rowden, George 38

"Rural New Yorker" 171, **171**, 192

"**S**. M. Beard" 127, 130, 131, 135, 136; description, 151-152, **151**; 179; serial number allocation 187, 189

St. Cloud Hotel, New York City 122

St. Imier, Switzerland 65

St. James Hotel 47, **48**, 50, 52, 71, 91, 94, 117, 118, 199

"St. Paul Press" 171, **171**, 192

Salt Meadow 44; salt marsh **90**

Sample, random 185; bias, 185; stratified (strata) 187

Sanborn, Jacob 55-56, 193

Sawdust swindle 106

Sawyer, Sylvanus 52, 101-102

Schell, Robert & Co. 37, 128

Schenectady, New York 25

Schroeder, Frederick 119

Screw manufacture 73

Scrip, USWC 102, **102**

Serial number blocks 187

Serial numbers, allocation by grade 188-192

Serial numbers, matching case and movement 133

Serial numbers, sample 186-187

Shanley's Cut 23, 43, 200

Sizes and styles, USWC watches **126**, 127, 134-136

Sloan and Chace 122

Smalley, Emma (See Giles, Emma Smalley)

Society for Suppression of Vice 113

South Royalton, Massachusetts 59

Spaulding, H. D. 123

Special Order Watches 169-172, **169**, **170**, **171**, **172**; serial number allocation, 192; **195**

Spring collet 38

Squire, Seth P. 24

Standard time 180

Standards, declining, USWC watches **100**, 101

Stark, John 38

Steere & Crooker 31

Stem winding and setting mechanisms 39-41, 49, **51**, 53, **53**, **54**, **55**, **56**, 85, **87**, **88**, **89**, **98**, **129**, **143**, **149**, **182**, 199

Stephens, Dr. W. Barclay 88, 142

Stock certificate, USWC 39-40, **40**

Stopwork **144**, 145

Street, John 133

Sun times 180

Surplus movements, USWC 83, 109-113, 182, 184, 185, 199

Sussfeld Lorsch and Co. **30**, 31

Sutton, William M. 103

Swiss fakes **83**, 84, **195**

Switzerland 32, 36, 53, 63, 64, 65, 66, 67

Tables:
Sales & Profits 37
USWC Movements 127
USWC Grades & Movements, 1st and 2nd Generation 130
Summary of USWC Models & Prices 134-136
Original Owners of USWC Watches 166-167
Jeweler's Contract USWC Products 170
USWC 1876 Centennial Product Contrast 176
Empire City & Equivalent USWC Grades 179
Relative USWC Feature Scarcity 186
Model/Grade Groups Ranked by Relative Scarcity 187
Model/Grade Groups with Serial Number Allocation 188-192
Employees & Production Capacities 193

Tariffs 31

Ten size USWC watch development 66-67, 85

Tenderloin District, New York City 111

Texas State Fair, 1871, Houston 71, 165

Textile industry 124

Theft, employees 82-83

Thetford Academy 59

Three-quarter plate; first design, 39-40; second design, 85-94, **87** (first illustration), **126**, 127, 130, 134-136

Tiffany & Co. 65

Trade-marks 177

Trading down 143

Train; fast and slow 50, 85; gold 16, 86, 88, 142, 164

Tremont Watch Company 36-37

Tribune 34

Twain, Mark 95

Tweed, William Marcy 95

27th Massachusetts Infantry Regiment 26

U.S. Clock & Brass Co. 61-62, **61**, 103

U.S. Design Patents:
D 2,055   39, **39**, **40**, 128
D 2,266   128
D 2,281   39, **127**, 128
D 2,291   128
D 2,525-6   128
D 3,885   128
D 3,886   128

U.S. Letters Patents:
21,864   38
43,490   127
47,412   39, 127
47,997   38
49,397   41, 128
57,495   63, 128
65,208   53, 124, 128
69,561   128
74,457   128
100,511   85, **112**, 128
145,939   86, 128
289,642   128

U.S. Trade-marks:
2,535   177
2,820   177

"U.S. Watch Co., New York" 195

USWC: founding, 35-37; construction of machinery, 37-38; stockholders, 52-53; employee relations, 56, 82-83; 96-97; demise, 116-117; movement styles & sizes, 127; models & prices, 134-136; grade descriptions, 137-160; price lists, 163-164; railroad watches, 168, **168**, 171, 172; special order watches, 169-172, **169**, **170**, **171**, **172**; commemorative watches, 173-177, **173**, **174**, **175**, **177**; production estimates, 185-187; employees, 193; miscellanea, **194-198**; an 1896 perspective, 199-200.

USWC Factory Building: **42**, **46**; construction 46-47; description, 72-79; machinery maintenance and problems, 101, 110; sale of machinery, 102, 115-116, 121-122, 184; closing, 116-117; last years, 124, 200

"Union Pacific" 171, **172**

Union Watch Co. 102

United States economy 25, 29, 37, 69 (Also see Panics)

United Electric Traction Co. 200

"United States Match Co." (engraving) 55

United States Watch Co. Act 44

"United States Watch Co." 16, 55, **55**, **57**, **68**, 77, **78**, 86, **87**, **88**, **89**, 95, **98**, **100**, 127, 130, 131, 134, 136; description, 142-148, **143**, **144**, **145**, **146**, **147**, 185,